Potentiometric Water Analysis

Potentiometric Water Analysis

Derek Midgley and **Kenneth Torrance**
*Central Electricity Research Laboratories,
Leatherhead, Surrey, England*

A Wiley–Interscience Publication

JOHN WILEY & SONS
Chichester · New York · Brisbane · Toronto

Copyright © 1978 by John Wiley & Sons, Ltd.

All rights reserved.

No part of this book may be reproduced by any means, nor transmitted, nor translated into a machine language without the written permission of the publisher.

Library of Congress Cataloging in Publication Data:

Midgley, Derek.
    Potentiometric water analysis.

    'A Wiley–Interscience publication.'
    Includes bibliographies.
    1. Water – Analysis. 2. Electrochemical analysis. I. Torrance, Kenneth, joint author.
II. Title.
QD142.M52   546'.22   77–7213

ISBN 0 471 99532 0

Typeset in IBM Press Roman by
Preface Ltd, Salisbury, Wiltshire.
Printed in Great Britain by
Unwin Brothers, The Gresham Press, Old Woking, Surrey

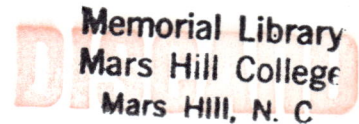

Preface

This book draws on work carried out at the Central Electricity Research Laboratories and the authors are grateful to the Central Electricity Generating Board for permission to publish it. The authors alone, however, are responsible for opinions expressed and for the selection of factual material.

Contents

PART I THEORETICAL AND PRACTICAL BACKGROUND

1 Introduction 3

2 Electrochemical Principles 6
    Concentration and Activity 6
        Single-ion activity coefficients 8
        Ion association 8
        Activity as an analytical parameter 9
    Cell Potentials and Electrode Potentials 9
        Standard potentials 10
    Reference Electrodes and Liquid Junction Potentials . . 10
    Reversible Reactions 13
    Measurement of the Cell Potential 13

3 Electrodes 16
    Metal-based Electrodes 16
        Cation-sensitive electrodes 16
        Anion-sensitive electrodes 17
            Interferences 18
            Limit of response 18
        Metal–metal oxide pH electrodes 19
        Inert metal–redox system electrodes 20
            The hydrogen electrode 21
            The quinhydrone electrode 22
    Membrane Electrodes 23
        Solid-state membrane electrodes 24
        Glass electrodes 27
        Liquid ion-exchange electrodes 29
    Indirectly Selective Electrodes 34
        Gas-sensing membrane electrodes 34
        Construction of gas-sensing membrane electrodes . . 37
    Reference Electrodes 38

The calomel electrode	39
Silver–silver chloride electrodes	40
The Thalamid electrode	41
The mercury–mercurous sulphate electrode	41
Unusual reference electrodes	41
Design and construction of reference electrodes	42
Form of liquid junction	43
Concentration of reference solution	44
Sealed reference electrodes	44
Double-junction reference electrodes	45
Temperature effects	45
Colloidal and suspension effects	46

4 Equipment — 49

Types of Meters	49
pH and pIon meters	49
Digital and analogue display	50
Portable pH meters	50
Microprocessor analysers	51
Recorder and printer outputs	51
Matching meter output and recorder input	52
Temperature compensation	52
Additional features	53
Voltmeters and electrometers	54
Industrial Monitoring Equipment	54
pH measurements	54
Measurements with ion-selective electrodes	55
Flow cells	56
Automatic Titration Apparatus	58

5 Analytical Principles — 61

Combination of Errors	63
Significance Testing	64
Comparison of means	65
One-sided and two-sided tests	66
Use of the t-test	66
Comparison of standard deviations – the F-test	67
Criterion and limit of detection	68
Within-batch and between-batch errors	69
Testing Analytical Methods	71
Precision	71
Accuracy	71
Recovery tests	71
Interference tests	74
Blanks, baselines and reference solutions	75
Control charts	76

	Calibration Graphs	77
	Linear regression	77
	Correlation coefficients	78
	Testing for linearity	78
6	**Potentiometric Titrations and Related Methods**	**81**
	Titrimetric Procedures	81
	Acid–base titrations	81
	Precipitation titrations	83
	Compleximetric titrations	83
	Redox titrations	83
	Addition titrations	83
	Competitive titrations	86
	Finding the Equivalent Volume	86
	The titration curve	86
	Titration to a fixed potential or pH	87
	Derivative titrations	89
	Gran titrations	89
	Linear titration plots	93
	Experimental limitations	95
	Single-point Titrations	98
	Known addition	98
	Known subtraction	102
7	**Potentiometric Analytical Practice**	**105**
	Ion-selective Electrodes	105
	Apparatus	105
	Sample Collection	108
	Conditioning and Storage of Electrodes	110
	Concentration Range and Units	112
	Analytical Procedures	112
	Temperature	113
	Stirring	114
	Light	115
	Flow-cell analysis	115
	Calibration	117
	Direct potentiometry	117
	Preparation of a permanent calibration graph	117
	Nernstian calibration with two standard solutions	119
	Nernstian calibration using the direct activity scale of a pIon meter	120
	Non-Nernstian calibration	121
	Known-addition and known-subtraction potentiometry	122
	Titrimetry	123
	Sources of Error	124
	Ionic strength effects	124
	Masking	124

  Interferences 125
 Precision 125
 Accuracy 126
 Response Time 126
 Tracing Faults 127
  Off-scale readings 127
  Invariant readings 128
  Loss of sensitivity 128
  Slow response 129
  Drift 129
  Noise 129

PART II ANALYTICAL METHODS

The Measurement of pH 135
Determination of Acidity and Alkalinity 147
Determination of Lithium 155
Determination of Sodium with a Sodium-responsive Glass Electrode 160
Determination of Potassium 171
Determination of Calcium Using Calcium-responsive Liquid Ion-exchange Electrodes 186
Determination of Water Hardness with Liquid Ion-exchange Electrodes . . . 194
Determination of Silver 205
Determination of Copper 214
Compleximetric Titrations 226
Determination of Cadmium 239
Determination of Lead 243
Determination of Aluminium 248
Determination of Total and Free Carbon Dioxide with a Gas-sensing Membrane Electrode 254
Determination of Nitrite and Nitrogen Oxides with a Gas-sensing Membrane Electrode 263
Determination of Sulphur Dioxide with a Gas-sensing Membrane Electrode . . 271
Determination of Ammonia with a Gas-sensing Membrane Electrode . . . 279
Determination of Total Nitrogen Using a Gas-sensing Ammonia Electrode after Kjeldahl Digestion 290
Determination of Free and Total Cyanide Using a Cyanide-selective Electrode . 298
Determination of Low Levels of Total Cyanide Using a Silver Sulphide Electrode 306
Determination of Fluoride 313
Determination of Chloride Using Electrodes Based on Silver Chloride . . . 323
Determination of Low Levels of Chloride 333
Determination of Bromide and Iodide 343
Determination of Thiocyanate 352
Determination of Sulphide 355
Titrimetric Determination of Sulphate 366

Determination of Nitrate using a Liquid Ion-exchange Electrode 374
Determination of Boron as Fluoroborate using a Fluoroborate-selective Liquid Ion-exchange Electrode. 383
Determination of Perchlorate 390
Appendix 1 Theoretical Values of the Nernstian Slope 394
Appendix 2 Debye–Hückel A and B Coefficients for Electrolytes in Water. . 395
Appendix 3 Equipment Manufacturers 396
Appendix 4 Tables of the Function, $\dfrac{1}{\left(\text{antilog}\dfrac{E_2 - E_1}{k}\right) - 1}$ 399
Index. 403

Part I

THEORETICAL AND PRACTICAL BACKGROUND

Chapter 1

Introduction

With increasing demands on water for industrial, agricultural and domestic consumption, resources are being used and re-used more intensively than ever before. With each cycle of re-use, it becomes more important to check that the water is sufficiently pure for the purpose for which it is intended and that any waste water returned to the common stock does not contain unacceptable levels of pollutants. In the past, non-specific parameters such as conductivity, pH and biological oxygen demand have largely sufficed as measures of water quality, but the need in many circumstances for the determination of specific chemical substances is becoming more pressing.

As a technique, potentiometry has the advantage of being easily adaptable to circumstances. One reason for this is the very wide range of concentrations that can be determined by the same equipment with only a minimum of procedural variations and another is the comparative ease with which measurements can be made either with battery-powered instruments in the field or with automatic apparatus for continuous monitoring inside industrial plant or in remote locations. For both applications, the robustness of many electrodes is an advantage compared with photometric techniques. It is only within the last ten years that a wide variety of ion-selective electrodes has become available, but a number of species are now commonly determined by means of such electrodes: sodium in highly pure water; fluoride and nitrate in potable waters; sulphide and cyanide in industrial effluents; potassium, calcium, and carbon dioxide in biological fluids; and chloride and ammonia in a great variety of media. The fluoride electrode has made its way into standard analytical texts together with the long-established glass electrode for pH determinations, e.g. APHA (1971), DoE (1972), and more recently the nitrate and cyanide electrodes have been tentatively recognized (APHA, 1975). Although a number of books have usefully described the theory behind ion-selective electrodes, e.g. Durst (1969) and Koryta (1975), there has been little attempt to draw together the existing knowledge specifically for the benefit of analysts. Moody and Thomas (1971) provided a valuable summary of electrode properties and in a later review (Moody and Thomas, 1973) gave an extensive list of references, but very often such papers provide too little information from which the analyst can work and fail to indicate such basic attributes of an analytical method as its precision and accuracy. The same is true of many of the methods published by manufacturers of electrodes.

Our aim has been to bring our experience to bear on published results and to stress particularly those points which are of practical interest to the analyst. We have not discussed experimental electrodes, but only those commercially available. In the analytical methods in Part II we have tried to give as much information as possible on the problems specific to the use of each electrode, but to some extent each type of sample must be considered individually. Although the methods are directed primarily to the analysis of fairly pure waters such as boiler water and feed water for steam-raising plant and of natural waters such as river waters, it is hoped that they will be of interest to analysts occupied either with other types of aqueous samples such as sea water, blood and soil extracts or with the analysis of solid samples after they have been dissolved. In the latter cases, the procedure for the analysis of pure water may need to be modified to take account of a fairly concentrated background of species other than the determinand, but the treatment of pure water samples does indicate the most desirable conditions of pH, ionic strength, etc., for potentiometric measurements.

The methods are described with manual analysis in the laboratory in mind, but the majority should be directly applicable to analysis in the field and many to continuous monitoring of a sample stream. For both of these applications, direct potentiometry often has advantages over other methods, even in cases where the potentiometric method is inferior in precision or accuracy in the laboratory. Our aim has been to make the procedure clear to the non-specialist and to point out the advantages and pitfalls of potentiometry so that the analyst can reasonably judge how useful an electrode can be in his laboratory from the details of the method in Part II taken together with Chapter 7 of Part I. The methods are grouped according to their operational similarities rather than by the nature of the determinand or the type of electrode, although the three types of classification often coincide, e.g. the alkali metals and most of the halides.

Part I contains the background common to the methods in Part II. The theory in Chapters 2, 3, and 6 is aimed at promoting an appreciation of the practical aspects of potentiometry and potentiometric titrations under the analyst's control, so that he can decide the relevance to his particular samples of a method written for a general case and how the method may need to be modified. Electrode mechanisms are not considered except in general terms to bring out the differences between the various types of electrodes. Similarly, Chapter 4 is concerned with what instruments can do and not with how they do it. Consideration should be given to the understanding of analytical results as well as to the means of obtaining them and Chapter 5 describes the basis on which analytical results are assessed before decisions are taken.

Nomenclature

Certain terms are used throughout the text and it is convenient to define them here:

The *determinand* is the substance whose concentration is being measured or determined.

The *calibration slope* of an electrode is the constant relating the observed e.m.f. to the logarithm of the determinand concentration. It is also called the *slope factor* and is said to be *Nernstian* when it agrees with the theoretical value calculated as in Appendix 1.

In titrations, the *titrant* is the solution of known concentration added to the sample and the *indicator electrode* is that used to follow the course of the titration, whether it responds directly to the determinand or not. The *equivalence point* occurs when an exactly stoichiometric amount of titrant has been added to the sample and it should be distinguished from the *end–point*, which is the operational approximation to it.

A general indication of the purity of water is given by its *specific conductivity*, which is reported in units of microsiemens per centimetre. These units are related to older ones as follows.

$$1 \ \mu S \ cm^{-1} \equiv 1 \ \mu mho \ cm^{-1} \equiv 1 \ M\Omega^{-1} \ cm^{-1}$$

Pure water has a specific conductivity of about $0.04 \ \mu S \ cm^{-1}$ at 18 °C. Otherwise pure water in equilibrium with the carbon dioxide in the air has a specific conductivity of about $0.75 \ \mu S \ cm^{-1}$. Mixed-bed deionization units are capable of producing water with a specific conductivity of less than $0.2 \ \mu S \ cm^{-1}$.

Bibliography

APHA, 1971, *Standard Methods for the Examination of Water and Wastewater*, 13th ed., American Public Health Association, Washington, D.C.

APHA, 1975, *Standard Methods for the Examination of Water and Wastewater*, 14th ed., American Public Health Association, Washington, D.C.

DoE, 1972, *Analysis of Raw, Potable and Waste Waters*, Department of the Environment, HMSO, London.

Durst, R. A. (ed.), 1969, *Ion Selective Electrodes*, National Bureau of Standards Special Publication 314, US Department of Commerce, Washington, D.C.

Koryta, J., 1975, *Ion-selective Electrodes*, Cambridge University Press, Cambridge.

Moody, G. J., and J. D. R. Thomas, 1971 *Selective Ion-Sensitive Electrodes*, Merrow, Watford.

Moody, G. J., and J. D. R. Thomas, 1973, *Selective Ion-sensitive Electrodes*, Selected Annual Review of the Analytical Sciences, Vol. 3, The Society for Analytical Chemistry, London, p. 59.

Chapter 2

Electrochemical Principles

Potentiometric analysis depends on the relationship between the concentration of the determinand, i.e., the species to be determined, and the e.m.f. of an electrochemical cell in which the determinand is one of the components of an equilibrium system. The ideal relationship is known as the Nernst equation,

$$E = E° + k \log (c) \tag{2.1}$$

where E = the measured cell potential, $E°$ = a constant for a given temperature, c = the concentration of determinand and $k = RT \log (10)/nF$ where R is the gas constant, T the absolute temperature, F is Faraday's constant and n is the number of electrons discharged or taken up by one molecule of determinand. Usually, but not necessarily, n equals the charge (with sign) on the ionic form of the determinand. Values of k are tabulated in Appendix 1.

In practice, it may be difficult to achieve the ideal relationship without defining the conditions of measurement very closely. In order to arrange a potentiometric system for the greatest convenience and accuracy, it is necessary to consider the components that make up equation 2.1.

CONCENTRATION AND ACTIVITY

The ideal relationship expressed in equation 2.1 is only approached as the solution approaches infinite dilution. For real solutions we must write

$$E = E° + k\log (a) \tag{2.2}$$

where a = the activity of the species in question, related to the concentration by the equation

$$a = c.f \tag{2.3}$$

The activity coefficient, f, is a pure number and the activity therefore has the same units as the concentration.

If equation 2.2 is re-written as follows

$$E = (E° + k \log (f)) + k \log (c)$$

it can be seen, that if f can be kept constant, the form of equation 2.1 is regained. The activity coefficient is a measure of the interaction of an ion with all the other ions present and its magnitude depends on the total ionic composition of the solution, in fact on a property known as the ionic strength, usually symbolized as either μ or I

$$I = 0.5 \, \Sigma \, c_i z_i^2 \tag{2.4}$$

where c_i and z_i = the molar concentration and charge, respectively, of the ith species and the summation is carried out over all the ionic species present. In most potentiometric analytical methods, the ionic strength is kept constant by adding to the sample an excess of 'indifferent' electrolyte, i.e., one which has no effect on the equilibria involved in the cell reaction except through the agency of the ionic strength, so that variations in the concentration of the determinand do not affect the activity coefficient significantly and the desired relationship between e.m.f. and concentration is obtained.

It is nevertheless often useful to be able to calculate activity coefficients and to know the limitations of the various ways of doing so. The simplest formula is the Debye–Hückel limiting law equation, in which A is a function of the temperature

$$-\log(f_i) = A \cdot z_i^2 \cdot I^{1/2} \tag{2.5}$$

and the dielectric constant of the solvent. The equation is fairly accurate at ionic strengths up to $10^{-4} - 10^{-3}$ mol l^{-1}, depending on the nature of the ions present, but at higher ionic strengths (up to 10^{-2} mol l^{-1}) the extended Debye–Hückel equation is needed

$$-\log(f_i) = A \cdot z_i^2 \cdot I^{1/2} / (1 + B \cdot a \cdot I^{1/2}) \tag{2.6}$$

where B is also a function of the temperature and dielectric constant and a, the 'ion-size parameter', is the distance of closest approach in solution of ions of opposite charge. When more than one electrolyte is present in the same solution, the meaning of the ion-size parameter is ambiguous and it is often adjusted empirically to give good agreement between observed and calculated values. A useful simplification of equation 2.6 is known as the Guntelberg equation

$$-\log(f_i) = A \cdot z_i^2 \cdot I^{1/2} / (1 + I^{1/2}) \tag{2.7}$$

At still higher concentrations, equation 2.6 can be extended by an extra term

$$-\log(f_i) = A \cdot z_i^2 \cdot I^{1/2} / (1 + B \cdot a \cdot I^{1/2}) - b \cdot I \tag{2.8}$$

where b is an empirical coefficient. A particularly useful form of equation 2.8 is the Davies equation, which is reasonably accurate up to $I = 0.1$ mol l^{-1} (Davies, 1962)

$$-\log(f_i) = A \cdot z_i^2 \, [I^{1/2}/(1 + I^{1/2}) - 0.3 I] \tag{2.9}$$

In mixtures of two or more electrolytes, further terms (Guggenheim and Turgeon, 1955) can be added to equation 2.8, but this is of little analytical use. It is possible to calculate activity coefficients at ionic strengths above 1×10^{-1} mol l^{-1}

(Robinson and Stokes, 1965), but it is not expected that the formulae would be relevant to the purposes of this book.

Values of the Debye–Hückel coefficients A and B are tabulated in Appendix 2 as a function of temperature for water as solvent. Experimentally determined activity coefficients for many electrolytes have been tabulated by Conway (1952), Harned and Owen (1958) and Robinson and Stokes (1965).

Single-ion Activity Coefficients

So far, we have assumed that the activity coefficient of a single ionic species has a definite meaning, as implied by equation 2.3, but there is no way of measuring such *single-ion activity coefficients* and, thermodynamically, they are neither meaningful nor necessary. The only measurable quantity of this sort is the *mean* ionic activity coefficient, $f_\pm$, and single-ion activity coefficients must be based on some extra-thermodynamic assumption. Whatever convention is adopted, equation 2.10 should be obeyed, i.e., for an electrolyte of general formula $A_m B_n$

$$(f_\pm)^{m+n} = (f_A)^m (f_B)^n \tag{2.10}$$

Mean ionic activity coefficients can be calculated from equations analogous to 2.5–2.9 by replacing z_i^2 with $z_A \cdot z_B$. If $z_A = z_B$, the calculated values of f_A, f_B and $f_\pm$ are the same. The mean ionic activity coefficients calculated from equations 2.5, 2.7, and 2.9 are compared in Table 2.1 with the observed values for aqueous solutions of calcium chloride at 25 °C.

Table 2.1 Mean ionic activity coefficients of calcium chloride in water at 25°C

Concentration (molal)	Observed	Calculated		
		Equation 2.9	Equation 2.7	Equation 2.5
0.0001	0.962	0.961	0.961	0.960
0.0005	0.918	0.917	0.916	0.913
0.0010	0.887	0.887	0.885	0.879
0.0050	0.783	0.782	0.774	0.750
0.0100	0.724	0.722	0.707	0.665
0.0500	0.574	0.576	0.518	0.402
0.1000	0.518	0.537	0.435	0.276

Ion Association

It is important to note that electrodes respond only to the concentration of the 'free' ion in solution, e.g., the fluoride electrode (p. 313) is sensitive to F^- ions but not to the coexisting species HF and HF_2^-. The concentration, c, in equations 2.1 and 2.3 refers only to the free ions and does not include any associated or complex species. Similarly the c_i terms in equation 2.4 refer to each separate ionic form in the solution. When an ion can form complexes and its total concentration is required, it is necessary to adjust the conditions in the sample so that the free ion is present as a constant proportion that can easily be related to the total (see, for example, *Determination of fluoride*, p. 313).

Activity as an Analytical Parameter

Potentiometry is the only technique that can easily indicate ionic activities and it should be decided before analysis if advantage can be made of this. In chemical equilibria, the activity is a more fundamental parameter than the concentration and may be the more useful measurement. The analyst, however, often has a sample which is at a quite different temperature and pressure from its source, in which case it is almost impossible to relate the results of activity measurements in the laboratory to the original situation, and total concentration is the most useful measurement that can be made.

CELL POTENTIALS AND ELECTRODE POTENTIALS

If two different electrodes, immersed in a solution containing species which can react at the electrode, are joined by the appropriate circuitry (see below), the potential generated across the electrodes as a result of the reaction can be measured and will be characteristic of the physico-chemical state of the electrode–solution–electrode system, i.e., it will depend in a definite way on the temperature, pressure and chemical composition of the system. If we consider electrodes composed of metals A and B immersed in a solution containing salts AX and BX of the two ions, respectively, the following reactions can occur

$$A \rightleftharpoons A^+ + e^- \tag{2.11}$$

$$B \rightleftharpoons B^+ + e^- \tag{2.12}$$

and the overall reaction will be given by the difference between the equations

$$A + B^+ \rightarrow A^+ + B$$

i.e., A is oxidized at the first electrode to give A^+, and B^+ is reduced at the second, depositing the metal B, resulting in a net displacement of B^+ by A^+ in the solution. The conventional representation of the cell is

$$A \mid AX(m_1), BX(m_2) \mid B \tag{2.13}$$

where m_1 and m_2 are the concentrations of the salts and the vertical line represents the electrode–solution interface.

The potential generated across the electrodes is given by the general formula

$$E = E^\circ - k \log \text{(product of the activities of the products)} + k \log \text{(product of the activities of the reactants)} \tag{2.14}$$

For the cell in equation 2.13 the e.m.f. is

$$E = E° - k \log(a_{A^+} \cdot a_B) + k \log(a_{B^+} \cdot a_A) \tag{2.15}$$

Since the activities of pure elements are, by definition, unity, we can see that the magnitude of the potential depends on $-k \log(a_{A^+}/a_{B^+})$ and equation 2.15 can be rewritten as

$$E = \{E_B^° + k \log(a_{B^+})\} - \{E_A^° + k \log(a_{A^+})\} \tag{2.16}$$

in which the constant term has been split into two components each associated with a particular electrode. Measurement of the cell potential can so far only relate to the ratio a_{A^+}/a_{B^+}. If a_{A^+} is kept constant, equation 2.16 will reduce to the form of equation 2.2 and we have an analytically feasible system. There are several different types of electrode (described in Chapter 3), but the cell potential can always be reduced to the form of equation 2.16, although the $E°$ term may comprise several diverse factors, depending on the mechanism of the electrode.

Standard Potentials

In equation 2.2 the constant term is equal to the e.m.f. when $k \log(a) = 0$, i.e., when $a = 1$, and is called the *standard e.m.f.*, $E°$. The standard e.m.f. is a constant for a given cell reaction at a fixed temperature and pressure and is highly characteristic of that reaction. In equation 2.16 the constant term has been split into two *standard electrode potentials*. There is no way in which the absolute values of these electrode potentials can be determined from measurements of cell potentials, and the many tabulated values (Latimer, 1952; Conway, 1952), are based on the convention that the standard potential of the hydrogen electrode (p. 21) is zero. The usefulness of the concept is that all standard e.m.f. values can be expressed as the difference of two standard electrode potentials, provided that they are defined on the same scale and regardless of their absolute values. Correspondingly, the cell potential can be regarded as the difference of two single electrode potentials or half-cell potentials. In equation 2.16 each term in square brackets is a single electrode potential and we can write

$$E = E_B - E_A$$

where $E_B = E_B^° + k \log(a_{B^+})$ and $E_A = E_A^° + k \log(a_{A^+})$

In principle, we can predict the e.m.f. of any cell from tabulated values of standard electrode potentials, the concentrations of the species in solution and one of the activity coefficient equations.

REFERENCE ELECTRODES AND LIQUID JUNCTION POTENTIALS

Assuming that activity coefficients can be kept constant, equation 2.16 may be rewritten as

$$E = \{E_B^{°\prime} + k \log(c_{B^+})\} - \{E_A^{°\prime} + k \log(c_{A^+})\} \tag{2.17}$$

If equation 2.17 is to be useful for determining B^+, the concentration c_{A^+} must be known, either explicitly or implicitly (by calibrating in the presence of a constant concentration of A^+). When both electrodes are immersed in the same solution this means that the concentration of A^+ must either be determined by a separate analysis — which is generally unacceptable as it increases the time and reduces the precision of the analysis for B^+ — or kept at an effectively constant level by adding an excess of A^+ to the sample. In the latter case, no B^+ ions or interfering ions must be introduced as impurities with the added A^+ ions. By far the most common arrangement, however, is to have the A-electrode immersed in a separate solution that contains a constant high concentration of A^+ ions and is prevented from mixing with the sample solution except by diffusion. The arrangement electrode + solution + diffusive barrier is a reference half-cell, often termed a *reference electrode*, and is usually constructed so that it and the sensing B-electrode can be dipped together into the sample solution. The interface between the reference half-cell and the sample solution is called a *liquid junction*. The most common forms of reference half-cells consist of an electrode responsive to chloride ions immersed in a potassium chloride solution, for which the half-cell potentials have been well characterized (Ives and Janz, 1961). Particular examples are discussed in Chapter 3.

The conventional representation of such a cell is

$A \mid AX(m_1) \parallel BX(m_2) \mid B$

where the double vertical lines indicate the liquid junction. Equation 2.17 can now be rewritten

$$E = E_B^{o'} - E_{ref} + k \log (c_{B^+}) \tag{2.18}$$

where $E_{ref} = E_A^{o'} + k \log (c_{A^+}) =$ a constant. When the potential of a cell with a liquid junction is measured, it is not generally equal to the potential calculated from the standard potentials and concentrations in the two half-cells, due to the existence of a liquid junction potential, E_j. In all cells with a liquid junction, we must allow for the E_j term, thus even in otherwise ideal circumstances we would have instead of equation 2.1,

$$E = E^\circ + k \log (c) + E_j \tag{2.19}$$

The liquid junction potential depends on the mobilities and concentrations of all the ions in the solutions on either side of it and also on how the two solutions are brought together physically to form the junction. There have been many attempts to find a suitable expression for the liquid junction potential (Bates, 1973), often with reasonable accuracy, but since the overall composition of the sample solution must be known, these are of little interest to the analyst. Procedures have been devised for 'eliminating' liquid junction potentials, usually by extrapolation of results obtained in different conditions, but again these would rarely be practical for analytical purposes. The first consideration for the analyst is not that the liquid junction potential is precisely calculable, or small, but that it is effectively constant so that a reliable calibration can be made.

A convenient means of estimating liquid junction potentials and of assessing the factors that go to make them up is Henderson's equation

$$E_j = -k \cdot \frac{\Sigma \lambda_i (c_i^R - c_i^L)/z_i}{\Sigma \lambda_i (c_i^R - c_i^L)} \cdot \log (\Sigma \lambda_i c_i^R / \Sigma \lambda_i c_i^L) \qquad (2.20)$$

where c_i, λ_i and z_i are the concentration, ionic mobility, and charge (with sign), respectively, of the ith ion. The superscripts R and L refer, respectively, to the right- and left-hand sides of the cell A | AX(m_1) | | BX(m_2) | B described above. Apart from the trivial case when the two solutions are identical, the liquid junction potential will be zero only if

$$\Sigma \lambda_i (c_i^R - c_i^L)/z_i = 0$$

which occurs when the current across the junction is carried equally by the anions and the cations. Such a junction is said to be *equitransferent*.

Since the right-hand (sample) side of the cell is variable in composition, E_j can only approach constancy if the left-hand terms of equation 2.20 are much larger than the right-hand terms. The reason for the use of concentrated potassium chloride solution in reference electrodes now becomes apparent. As $\lambda_{K^+}/\lambda_{Cl^-} = 0.96$ at 25 °C, the solution is almost equitransferent. At concentrations of 3.0 mol l^{-1} or greater the denominator of the pre-logarithmic term of equation 2.20 is effectively constant (and large) at $-(\lambda_{K^+} + \lambda_{Cl^-})c_{KCl}^L$ for dilute samples, reducing E_j still further. By adding a proportion of potassium nitrate, as advocated by Grove-Rasmussen (1949; 1951), a more nearly equitransferent solution is obtained, since $\lambda_{K^+}/\lambda_{NO_3^-} = 1.04$ at 25 °C. Proprietary equitransferent mixtures are available from Orion Research Inc., but potassium chloride solutions are adequate for most purposes.

Consideration of equation 2.20 shows that the presence of a relatively high (about 0.1 mol l^{-1}) concentration of background electrolyte in the sample, such as may be added in an analytical procedure to control the ionic strength, will tend to increase the liquid junction potential, unless this electrolyte is itself equitransferent, but the constancy will probably improve. Because of the high mobility of hydrogen ions and, to lesser extent, hydroxide ions, samples with a very high or low pH are the most likely to cause problems. Colloidal and particulate matter can have a large destabilizing effect on liquid junction potentials and before measurements are made in solutions containing such materials it is necessary to check that a valid calibration procedure is being used.

It should be emphasized that equation 2.20 is not the definitive expression for the liquid junction potential, but a simplified form of the general equation 2.21 below, assuming that ionic mobilities and activity coefficients are constant across the junction, which is not generally true. It does, however, enable the principles underlying the liquid junction potential to be demonstrated and provides a good approximation when the ionic strengths of the two solutions forming the junction are equal.

$$E_j = -k \int_L^R \frac{t_i}{z_i} \, d\log(a_i) \tag{2.21}$$

where $t_i = c_i\lambda_i/\Sigma\lambda_i c_i$ is the *transference number* and a_i is the activity of the ith ion.

REVERSIBLE REACTIONS

Potentiometric measurements are only analytically useful if the cell reaction occurs reversibly. In thermodynamic terms this means that the energy released by the reaction is converted to the maximum extent possible into useful work, rather than being dissipated as heat. This has implications as to whether an electrode is actually useful and how the measurement of potential must be made. The cell reaction is reversible if, when the electrodes are connected through an external circuit to a source of e.m.f. such that the cell potential is exactly balanced and no current flows, there is no chemical or other change in the cell. If the external e.m.f. is increased infinitesimally, current will flow and a change proportional to the quantity of electricity passing will occur in the cell. If the external e.m.f. is decreased correspondingly, the current should pass in the opposite direction and whatever process occurred should be exactly reversed.

In practice, most metal electrodes do not behave reversibly in aqueous solution at room temperature, either because the reaction is too violent, e.g.,

$$Na \longrightarrow Na^+_{(aq)} + e^-$$

or because the activation energy of the reaction is so large that the reaction will not occur spontaneously at all. The tabulated values of standard electrode potentials (Latimer, 1952) are often obtained indirectly by calculation from other thermodynamic data and do not necessarily mean that a practical electrode exists. Most electrodes used in potentiometry depend on either silver or mercury electrodes for the final step of transferring charge from the solution to the external circuit. Even when an electrode can work reversibly, it will do so only if the current drawn through it is very small ($10^{-6}-10^{-12}$ A), which imposes limits on the circuitry which it is possible to use.

MEASUREMENT OF THE CELL POTENTIAL

Ideally, the potential observed when the electrodes of a cell are connected externally through a measuring circuit should exactly equal that predicted theoretically. In practice, this is rarely achieved, but the differences that do occur in a carefully designed circuit are so small that they can be ignored except for the most rigorous theoretical studies. It is important, however, to understand how such differences arise, so that the necessary precautions can be taken when an analytical procedure is being developed.

The most important aspect of the measurement of e.m.f. is that it is done without disturbing the chemical equilibrium within the cell. This implies that no current is allowed to pass or, as is normal, that the current that flows through the cell is so small that no change in chemical conditions can be detected within the normal period required to make a measurement. The maximum permissible current for a given accuracy in e.m.f. limits the use of a millivoltmeter for analytical potentiometry and has led to the special circuitry incorporated in pH and pIon meters (Chapter 4). The exact value of the current that can be drawn from the cell before the e.m.f. shifts perceptibly from its zero-current value is not usually precisely known and it depends on a number of factors, including the nature of the electrodes and their susceptibility to polarization, the concentrations of the reactants, the volume of the cell solution and total resistance of the cell. In cells containing membrane electrodes of high resistance the last factor is most important, as the potential drop across the cell, ΔV, due to the passage of even a minute current can lead to considerable errors. For example, if ΔV is required to be less than 0.1 mV and the cell resistance is 100 MΩ, by Ohm's law the current must be less than $10^{-4}/10^8 = 10^{-12}$ A. The input resistance of a voltage-measuring instrument designed to measure a maximum of 1 V from the above cell to an accuracy of 0.1 mV must be so high as not to cause a current in excess of 10^{-12} A to be drawn, i.e., the input resistance must be at least $1/10^{-12} = 10^{12} \Omega$. Most commercial millivoltmeters designed for potentiometry have high input resistances of the order of $10^{12}-10^{13} \Omega$ and a precision of 0.1–1 mV and are adequate for cell resistances of 100 MΩ or more. Although the problems of obtaining the correct instrument for measuring the potentials of cells containing electrodes of high resistance have largely been overcome by recent advances in electronics, there is still considerable difficulty in achieving noise-free signals from these cells. The weak electrical signal being transmitted from the cell to the millivoltmeter requires a signal cable with an insulation resistance several orders of magnitude greater than the resistance of the membrane electrode that is usually the highest source of resistance in the cell. Most electrodes of this nature are fitted with a screened co-axial cable, but great care may be necessary to avoid electrical or magnetic interference from ancillary apparatus such as stirrers and the arrangement of cables may need some consideration. Cells through which the sample is flowing are particularly prone to electrical noise and may also suffer from the presence of troublesome earth loops, especially if the electrodes are used in-line or in conjunction with water-circulating thermostats.

Bibliography

Bates, R. G. 1973, *The Determination of pH,* Wiley, New York and London.
Conway, B. E., 1952, *Electrochemical Data,* Elsevier, Amsterdam.
Davies, C. W., 1962, *Ion Association,* Butterworth, London.
Grove-Rasmussen, K. V., 1949, 'The diffusion potential between dilute solutions and concentrated solutions of potassium chloride plus potassium nitrate', *Acta Chem. Scand.,* 3, 445.
Grove-Rasmussen, K. V., 1951. 'Equitransferent salt bridge for elimination of the diffusion potential compared with saturated potassium chloride salt bridge', *Acta Chem. Scand.,* 5, 422.

Guggenheim, E. A., and J. C. Turgeon, 1955, 'Specific interaction of ions', *Trans. Faraday Soc.*, **51**, 747.
Harned, H. S., and B. B. Owen, 1958, *The Physical Chemistry of Electrolytic Solutions*, 3rd ed., Reinhold, New York.
Ives, G. J., and D. J. G. Janz, 1961, *Reference Electrodes, Theory and Practice*, Wiley–Interscience, New York and London.
Latimer, W. M., 1952, *Oxidation Potentials*, 2nd ed., Prentice-Hall, Englewood Cliffs, N.J.
Robinson, R. A., and R. H. Stokes, 1965, *Electrolyte Solutions*, 2nd ed., revised, Butterworth, London.

Chapter 3

Electrodes

Modern ion-selective electrodes are membrane electrodes in which the current is carried by ions, but the classical metallic electrodes with electronic rather than ionic transport are still useful for many purposes, especially as components of reference electrodes, which deserve considerable attention in themselves. No attempt is made here to describe every type of ion-selective electrode, as many have had almost no practical applications and are not commercially available. Some of these electrodes have been briefly described by Koryta (1972; 1975) and Moody and Thomas (1971). Commercially-available electrodes for particular ions are listed in the appropriate analytical methods in Part II.

Whatever the nature of the electrode, the electrode potential should reduce to the form of the Nernst equation, as discussed in Chapter 2

$$E = E^\circ + \frac{k}{z} \log c$$

where k/z, the calibration slope, is theoretically equal to $(RT/zF) \ln (10)$. (R = the gas constant; T = the absolute temperature; F = the Faraday constant; z = the charge on the ion.)

METAL-BASED ELECTRODES

Cation-sensitive Electrodes

Only a small number of metals can be used to make electrodes suitable for potentiometric analysis. Metals with negative standard electrode potentials will tend to react with water, displacing hydrogen, and a stable potential cannot be obtained, e.g. the alkali metals and alkaline earth metals. Some metals do not react with water because they are kinetically, if not thermodynamically, passive and these could be used as selective electrodes (e.g. zinc and lead electrodes) for their own ions but they have found virtually no practical applications because of their susceptibility to oxidation. By amalgamating metals with mercury, their sensitivity to oxidation and reactivity towards water may be reduced sufficiently to enable potentiometric measurements to be made in carefully controlled conditions (Bennetto and

Willmott, 1971) but analysis with such electrodes would not normally be convenient and other methods would be preferred.

Of the metals with positive standard electrode potentials, many are inert and cannot be used for the determination of their own ions, although they may be used to monitor other types of oxidation–reduction equilibria, e.g. the platinum electrode. Copper metal is sensitive to trace concentrations of oxygen in the solution and this leaves only silver and mercury as practical ion-selective electrodes. Both have been used extensively as indicator electrodes in potentiometric titrations and as the bases of reference electrodes, but they have been little used for the direct potentiometric determination of silver and mercury.

The electrode potential corresponds to the reaction

$$M \longrightarrow M^{m+} + me^-$$

and is given by

$$E = E° + \frac{k}{m} \log a_m$$

where a_m is the activity of the ion of charge m.

Anion-sensitive Electrodes

If an excess of a sparing soluble salt, M_pX_q, is added to a solution containing a concentration c mol l^{-1} of the anion X^{x-}, so that the solution is in equilibrium with the solid salt, the concentration of metal ions, M^{m+}, in the solution is given by the solubility product equilibrium

$$K_s = (a_M)^p (a_X)^q$$

where a_M and a_X are the activities of the M^{m+} and X^{x-} ions at equilibrium. If an electrode of the metal M is now immersed in the solution, its potential will be determined by a_M, which is related to a_X by the solubility product

$$E = E_M° + \frac{k}{m} \log a_M = E_M° + \frac{k}{p \cdot m} \log K_s - \frac{q \cdot k}{p \cdot m} \log a_X \tag{3.1}$$

As long as there is solid M_pX_q present, the electrode potential will depend on activity of the anion X^{x-} and the electrode can be considered as an X-selective electrode. Equation 3.1 can now be rewritten

$$E = E_X° - \frac{k}{x} \log a_X \tag{3.2}$$

where

$$E_X° = E_M° + \frac{k}{p \cdot m} \log K_s$$

In practice, it is unnecessary for the entire solution to be saturated with salt

M_pX_q, provided the solution in the immediate vicinity of the electrode is so saturated. Coating the electrode with a layer of the salt, either electrolytically or by chemical reaction, provides a source of the salt and ensures that the solution near the electrode is saturated with it.

Electrodes of this sort are known as *electrodes of the second kind* and include the silver–silver chloride, silver–silver bromide, silver–silver iodide, silver–silver sulphide and mercury–mercurous chloride electrodes. The silver–based electrodes named above have all been used as ion-selective electrodes and several others have been proposed, e.g. silver–silver carbonate, silver–silver oxalate. The mercury-based electrodes are not generally used as ion-selective electrodes, partly because they are inconvenient to handle, but mainly because they are slow to equilibrate; they have, however, been widely used in reference electrodes. A variation of this type of electrode in the Thalamid reference electrode (p. 41), which is a thallium amalgam–thallous chloride electrode.

Interferences

As these electrodes are anion-selective only by virtue of the control of the metal ion activity through the solubility product, any other anion that forms an insoluble salt with the metal is a possible interferent. If an anion Y^{y-} forms a sparingly soluble salt M_iY_j with a solubility product $K_s' = (a_M)^i(a_Y)^j$, it will interfere when

$$\frac{(a_Y)^{j/i}}{(a_X)^{q/p}} > \frac{(K_s')^{1/i}}{(K_s)^{1/p}} \tag{3.3}$$

i.e. when the activity of Y^{y-} rather than that of X^{x-} controls the activity of the metal ion. In practice, however, the transfer of control from one ion to the other does not show the sharp transition implied in the above relationship and the interferent may have an effect at lower concentrations than predicted, while the electrode may still respond to the determinand at higher interferent concentrations than predicted (Harzdorf and Keim, 1976).

Strong oxidizing or reducing agents will also interfere, as the electrodes are sensitive to redox equilibria in the same way as inert metal electrodes.

Limit of response

The limit of detection of the electrodes is set ultimately by the solubility of the sparingly soluble salt. On immersion of the electrode in the sample solution some material will always dissolve off the electrode so that the solubility product equilibrium is satisfied. As long as the concentration, c, of determinand in the sample is much greater than that gained by dissolution of the membrane the relationship between the e.m.f. and c is Nernstian, but as c decreases and becomes of the same order as $^{p+q}\sqrt{K_s}$, a plot of e.m.f. against log (c) becomes curved. This plot may still be used for analysis, but the precision and accuracy of the method will be lower in this region than in the range of Nernstian response. Although electrodes are said to be 'non-Nernstian' at low concentrations, it should be remembered that a plot of e.m.f. against log $(c + c_s)$ where c_s is the concentration arising from dissolution of the membrane, will still be linear. For an isovalent insoluble salt $(p = q)$, if $c_s = c/n$ then $c = n\sqrt{2K_s}/\sqrt{2+n}$. An error of 0.25 mV is equivalent to a 1% error in concentration for a univalent ion and, for measurements made to that accuracy, dissolution of the membrane will cause a significant deviation from the ideal calibration graph when

$$c_s = c/100$$

i.e. when

$$c = 100\sqrt{2K_s}/\sqrt{102} \simeq 14\sqrt{K_s}$$

In the absence of interferences, the e.m.f. in the curved part of the calibration grapth may be calculated if the solubility product is known. Equation 3.2 may be rewritten

$$E = E_X^\circ - \frac{k}{x}\log f_X - \frac{k}{x}\log [X] \qquad (3.4)$$

where f_X is the activity coefficient of X and [X] is the total concentration of X, which is equal to the original concentration in the solution, c, plus the concentration arising from dissolution of the sparingly soluble salt, c_s.

Considering the case where $p = q$, the concentration of the metal ion, which also comes from the sparingly soluble salt, is also equal to c_s. The solubility product equilibrium can be written therefore

$$(c + c_s)c_s f_X f_M = K_s,$$

which can be solved for $c_s = 0.5(-c + c\sqrt{1 + 4K_s/c^2 f_X f_M})$. Substituting for [X] in equation 3.4 gives

$$E = E_X^\circ - \frac{k}{x}\log f_X - \frac{k}{x}\log\left[\frac{c}{2} + \frac{c}{2}\sqrt{1 + \frac{4K_s}{f_X f_M c^2}}\right] \qquad (3.5)$$

If $c^2 \gg K_s$, $[X] \simeq c$ and the electrode has a Nernstian response to the concentration, c, of anion in the sample, since c_s is negligible. As the sample solutions become more dilute, c_s can no longer be neglected and a calibration of E against log c will be curved. When this happens, the Nernst equation can no longer be used to calculate the concentration of determinand in the sample, but the concentration can still be read off the calibration graph. Even if $c < c_s$, a useful calibration can still be obtained, provided great care is taken to optimize the experimental conditions (see *Determination of Low Levels of Chloride* in Part II).

Metal–Metal Oxide pH Electrodes

These electrodes are similar in principle to the anion-sensitive electrodes in the previous section and may indeed be regarded as hydroxide-selective electrodes. If a solution is saturated with a sparingly soluble metal oxide, M_pO_q, the activity of metal ions, M^{m+}, in the solution is determined by the equilibrium

$$M_pO_q + qH_2O \rightleftharpoons pM^{m+} + 2q\,OH^-; \qquad K_s' = (a_{m+})^p(a_{OH^-})^{2q}$$

The e.m.f. of a metal electrode immersed in the solution is given by equation 3.6.

$$E = E_M^\circ + \frac{k}{m}\log a_{m+}$$

$$= E_M^\circ + \frac{k}{p\cdot m}\log K_s' - \frac{2q}{p\cdot m}\cdot k\log(a_{OH^-}) \tag{3.6}$$

As $2q = p\cdot m$ and $(a_{H^+})(a_{OH^-}) = K_w$, the autoprotolysis constant for water, equation 3.6 can be re-written as

$$E = E_M^\circ + \frac{k}{p\cdot m}\log K_s' - k\log K_w + k\log(a_{H^+})$$

$$= E_{MO}^\circ - k\cdot \text{pH} \tag{3.7}$$

in which E_{MO}^0 contains all the constant terms.

Many metal–metal oxide systems have been studied, but few are of practical use. Metals that form soluble oxides (e.g. the alkali metals) or are inert (e.g. gold) do not make useful electrodes, nor do metals such as aluminium, which form highly coherent *passive* oxide films. Ives (1961) has considered several reasons for the poor performance given by almost all electrodes of this type, including the formation of non-stoichiometric oxides and the coexistence of higher and lower oxides. The mercury–mercuric oxide electrode is reproducible and has been used as a reference electrode, but it is doubtful if the silver–silver oxide electrode can be similarly applied. The electrode most widely used as a pH sensor is the antimony–antimony oxide electrode, which is available in commercial form (Electronic Instruments Ltd. 1224; Balsbaugh 2210–1; Beckman 39027; Radiometer P301). This electrode is rugged and does not require a pH meter with a high input impedance to determine the e.m.f., but it has been superseded for most purposes by the glass electrode. It has been favoured for measurements in situations in which electrodes are easily fouled, as its performance can be restored by exposing a new metal surface on which an oxide film is formed by immersion of the electrode in water. The disadvantages of the electrode are its sensitivity to oxygen, the dependence of its potential on the rate of stirring at its surface, the reaction of the metal oxide with hydroxy-acids, its sensitivity to the presence of neutral salts and its poor precision (0.1 pH units when properly calibrated). Its uses and performance have been reviewed by Ives (1961), Stock, Purdy and Garcia (1958) and Bates (1973).

Inert Metal–Redox System Electrodes

When both the oxidized and reduced forms of a substance are soluble in water, an equilibrium may be set up between the two forms and the electrons in an inert metal electrode (usually platinum) immersed in the solution. The e.m.f. of the inert electrode will be determined by the ratio of the activities of the reduced and oxidized forms, a_{red} and a_{ox}, respectively

$$E = E^\circ - \frac{k}{n} \log \left[\frac{a_{red}}{a_{ox}} \right]$$

where n is the number of electrons discharged or taken up by one molecule of determinand. If either a_{red} or a_{ox} is known, the other can be calculated from the e.m.f. If the activity of one form is kept constant, a calibration can be obtained between the e.m.f. and the activity of the other and the electrode may, therefore, be considered as a selective electrode for the species of variable activity. For an electrode of this sort to be useful, the oxidation or reduction must occur reversibly and should not simultaneously proceed by a competing mechanism, e.g. by reaction with oxygen in the atmosphere.

Several electrodes of this type have been made, but only two, the hydrogen electrode and the quinhydrone electrode, have been extensively used as ion-selective electrodes, although the platinum electrode is widely used as a sensor in oxidation−reduction titrations.

The Hydrogen Electrode

When hydrogen gas is bubbled through a solution of hydrogen ions in which a platinum electrode is immersed, the e.m.f. is given by

$$E = E^\circ + \frac{k}{2} \log \frac{(a_{H^+})^2}{P_{H_2}} \qquad (3.8)$$

where P_{H_2} is the partial pressure of hydrogen, which is fixed by maintaining equilibrium with a known partial pressure of hydrogen in the gas phase. Equation 3.8 can be re-written

$$E = E^\circ - \frac{k}{2} \log P_{H_2} - k\, \text{pH}$$

and thus, analytically, the hydrogen electrode can be considered as a pH electrode.

The hydrogen electrode is most important thermodynamically, as it is the primary standard against which all other electrodes are compared and its standard potential is defined, by convention, as zero at all temperatures. Analytically, however, it has certain disadvantages compared with the glass electrode which now dominates the field of pH measurement. Oxygen rapidly poisons the electrode and it must be purged from the system before measurements are made; if the sample contains volatile species such as ammonia or carbon dioxide its pH is liable to be changed in the process, as these species may be lost as well as the oxygen, and the purging process increases the time taken for an analysis. Other substances that may be reduced at the electrode will interfere, e.g. silver ions can be reduced to silver metal and organic compounds, such as aromatic carboxylic acids, containing double bonds may be hydrogenated. The need for a source of purified hydrogen gas also makes the electrode far less convenient for field work than the glass electrode.

Hills and Ives (1961) have described the hydrogen electrode very thoroughly and Whitfield (1971) has summarized some recent developments, including the use of palladium electrodes in which the metal is charged with hydrogen by electrolysis, giving a more compact apparatus (Bucur and Stoicovici, 1966; Vasile and Enke, 1965; Fleischmann and Hiddleston, 1968).

The Quinhydrone Electrode

Quinhydrone is a 1:1 molecular complex of *p*-benzoquinone (quinone) and its hydroquinone (quinol). When dissolved in water, these two components form an electrochemically reversible oxidation–reduction system in which hydrogen ions also participate

$$C_6H_4O_2 + 2H^+ + 2e^- \rightleftharpoons C_6H_4(OH)_2$$

If a platinum or gold electrode is immersed in the solution, its potential is given by

$$E = E^\circ + \frac{k}{2} \log \frac{a_Q(a_H)^2}{a_{QH_2}}$$

$$= E^\circ + \frac{k}{2} \log \frac{c_Q}{c_{QH_2}} + \frac{k}{2} \log \frac{f_Q}{f_{QH_2}} - k \cdot pH \tag{3.9}$$

where the subscripts Q and QH_2 refer to quinone and hydroquinone, respectively, and *a*, *c* and *f* to activities, concentrations and activity coefficients. As the two components are added as the complex quinhydrone, $c_Q = c_{QH_2}$ and, as they are non-electrolytes, $f_Q = f_{QH_2} = 1$ (however see below) so the second and third terms on the right-hand side of equation 3.9 are equal to zero. It can now be seen that the quinhydrone electrode acts as a pH electrode. The advantages of the quinhydrone electrode are that it is easily prepared and does not require a supply of purified gas (cf. the hydrogen electrode), nor do acidic solutions need to be purged of oxygen. Unlike the glass electrode, the quinhydrone electrode has a low impedance and the e.m.f. can be measured on simple equipment. A useful advantage over the glass electrode is that the quinhydrone electrode can be used to measure the pH of hydrofluoric acid solutions (Warren, 1971; Entwistle, Weedon and Hayes, 1973). On the other hand, the useful pH range is smaller than with the hydrogen or glass electrodes and the glass electrode has fewer interferences.

Errors of the quinhydrone electrode Hydroquinone is a weak acid and its dissociation may cause errors which become appreciable at pH values above 6 to 9, depending on the composition of the sample. If the sample solution has little buffer capacity, the dissociation may actually change the pH of the solution and there would be a further effect on the e.m.f. because c_Q and c_{QH_2} will no longer be equal. In addition, the e.m.f. may be partly controlled by other reactions, i.e.

$$C_6H_4O_2 + H^+ + 2e^- \rightleftharpoons C_6H_4O(OH)^-$$

and

$$C_6H_4O_2 + 2e^- \rightleftharpoons C_6H_4O_2^{2-}$$

Hydroquinone is a reducing agent and reacts irreversibly with oxygen at a rate that becomes significant at about pH 8, thus in alkaline solutions it is necessary to purge the solution of oxygen. Because of these reactions of hydroquinone, results above pH 8 should be treated with great circumspection and even at pH 6–8 errors may be present.

Quinone is a weak base and its protonation to form $C_6H_4O_2H^+$ will cause errors in strongly acidic solutions. As an oxidizing agent, its reactions with reducing agents in the sample solution can cause errors, but such reactions are often kinetically unfavourable and of minimal importance. Quinone reacts with ammonia, amines and amino acids, resulting in readings that drift in the direction of decreasing pH; the effect is negligible below about pH 6.

Both quinone and hydroquinone show deviations from ideality in the presence of electrolytes, i.e. their activity coefficients are not equal to unity. Up to an ionic strength of 0.1 mol l^{-1}, however, the ratio of the two activity coefficients hardly deviates from unity and, therefore, the effect on the e.m.f. in equation 3.9 is negligible.

Analogous electrodes can be prepared from substituted quinones and their corresponding hydroquinones, e.g. thymoquinone, toluquinone and tetrachloroquinone (chloranil). The subject of the quinhydrone series of electrodes has been treated at length by Janz and Ives (1961).

MEMBRANE ELECTRODES

In all types of electrodes discussed previously, an essential feature of the mechanism was a transfer of electrons from species dissolved in the sample solution directly to the metal electrode. The dissolved species involved in the electron transfer did not have to be the determinand, but, if not, its concentration had to be controlled through some equilibrium by the concentration of determinand. With membrane electrodes, the sample solution is not involved in the electron transfer; instead, current is carried by ionic transport across a membrane from the sample solution to a reference solution in which the electron transfer takes place in constant conditions. The selectivity of a membrane electrode for a particular ion is determined by the mechanism of ionic transport across the membrane and not by the electron transfer process. Ideally, only one species of ion should be capable of carrying current across the membrane; this need not be the determinand provided its concentration is governed by an equilibrium involving the determinand itself. There are basically two types of membrane electrode: solid-state and ion-exchange electrodes, although the latter are usually sub-divided into glass electrodes and liquid ion-exchange electrodes.

Solid-state membranes are composed of crystalline solids of which the determinand is a component. The determinand either carries the current or controls the concentration of the current-carrying ion through a solubility product equilibrium at the membrane surface. In ion-exchange electrodes, selectivity is conferred by the ion-exchange equilibrium between solution and membrane and, except for some glass electrodes, the determinand carries the current.

Solid-state Membrane Electrodes

Practical solid-state membranes contain no more than two crystalline compounds, although membranes with more are theoretically possible and a few have been produced experimentally. If the membrane comprises only one crystalline compound it may take the form of a single crystal of that compound, or be made from many small crystals formed into a disc by compression or sintering. A third type of structure is also possible: the crystals are dispersed in an inert resin (usually silicone rubber) which is then allowed to set, forming what is conventionally known as a *heterogeneous* membrane. Membranes of this sort have been extensively studied and reviewed (Pungor and Tóth, 1970; Covington, 1969), but most commercially made electrodes have membranes made from single crystals or by compression. Membranes composed of two different crystalline compounds are made by compression or sintering of an intimate mixture of the two sorts of crystals, only one of which need be ionically conducting. Both types of crystal must be in equilibrium with the solution and they invariably have an ion in common. Membranes are sometimes made of a mixture of crystals even though only one is electrochemically necessary, e.g. chloride electrodes may incorporate silver sulphide as well as silver chloride as the mixture makes a mechanically stronger membrane. The compositions of the commoner types of solid-state membranes are given in Table 3.1.

Table 3.1 Solid-state membrane electrodes

Determinand	Membrane material
Fluoride	LaF_3*
Chloride	$AgCl$ or $AgCl + Ag_2S$
Bromide	$AgBr$ or $AgBr + Ag_2S$
Iodide	AgI** or $AgI + Ag_2S$
Thiocyanate	$AgSCN + Ag_2S$
Sulphide	Ag_2S
Silver	Ag_2S***
Copper	$Cu_{1.8}Se$* or $CuS + Ag_2S$
Lead	$PbS + Ag_2S$
Cadmium	$CdS + Ag_2S$

*Single crystals only.
**Not suitable for compression.
***Any of the chloride, bromide or iodide membranes will also respond to silver.

The membrane is set at the end of a plastic tube which is then filled with a solution containing the ion to which the electrode responds. The electrode is completed by immersing a reference electrode in the filling solution, which must, therefore, also contain the ion to which the reference electrode responds. Usually a silver–silver chloride reference electrode is used and the standard potential of the M-selective electrode is given by

$$E = E^\circ_{Ag, AgCl} - k \log a_{Cl^-} - \frac{k}{z} \log a_M$$

where $k = RT \ln(10)/F$, z = the charge, with sign, on the ion M and a_M = the activity of ion M in the filling solution. Membranes containing silver salts can be attached directly to a metallic conductor without an internal reference electrode and filling solution and in this case the standard potential is characteristic of the membrane material itself. (Note that this is not possible with heterogeneous membranes containing silver salts.)

A feature of all the membranes is that the current is carried exclusively by one ionic species. In the simplest case, the current-carrying species is also the one to which the electrode responds, e.g. silver, fluoride and some copper electrodes; this is thus the *primary response* which all electrodes have. The electrode may also have a *secondary response* to the current-carrying ion's counter-ion, i.e. the oppositely charged constituent of the crystal. Electrodes acquire their secondary response through a solubility product mechanism. The solubility product of an insoluble crystal $M_p X_q$ is a constant defined as

$$K_s = (a_M)^p (a_X)^q \tag{3.10}$$

where a_M and a_X are the activities of M and X in a solution in equilibrium with solid $M_p X_q$. For the example of a sulphide electrode immersed in a solution of sulphide ions with activity a_S, the activity of silver ion at the surface of the silver sulphide membrane is given by $a_{Ag} = (K_{Ag_2S}/a_S)^{1/2}$ and the e.m.f. is therefore

$$E = E^\circ + k \log(a_{Ag})$$
$$= E^\circ + \tfrac{1}{2} k \log(K_{Ag_2S}) - \tfrac{1}{2} k \log(a_S) \tag{3.11}$$

Thus the same electrode is used for determining silver in silver solutions and sulphide in sulphide solutions. Similar relationships can be shown for the chloride, bromide, iodide and thiocyanate electrodes, all of which have a primary silver response. The solubility product mechanism also explains how the principal interferences arise. For instance, in a solution containing both chloride and bromide ions, if the ratio $a_{Cl}/a_{Br} > K_{AgCl}/K_{AgBr}$ the silver bromide electrode immersed in the solution will tend to be converted to silver chloride, the silver activity will be governed by the silver chloride solubility product and the bromide response of the electrode will be lost. It is possible for any anion which forms an insoluble silver salt $Ag_p X$ to interfere with an electrode composed of $Ag_q R$ if

$$a_X/a_R > (K_{Ag_pX})/(K_{Ag_qR}) \tag{3.12}$$

This relationship implies that the response switches sharply from one ion to the other, but in the case of chloride, bromide and iodide electrodes the transition is more gradual, although eventually the sensitivity to the original ion is lost completely (Rechnitz and Kresz, 1966). Comparison with electrodes of the second kind shows that equation 3.11 has the same form as equation 3.1 and that the two types of electrode use many of the same compounds. The electrodes have different mechanisms, however, and the standard potentials will not be the same even if the same insoluble salt is used in both types. The interference effects are similar (cf, relationships 3.3 and 3.12), but membrane electrodes are less subject to redox effects. The limits of response are determined by the same consideration in each case.

Electrodes with a tertiary response may be made by incorporating two sorts of crystals in the membrane. Lead, cadmium and some copper electrodes are of this type, consisting of silver sulphide together with the appropriate metal sulphide. The primary and secondary responses are to silver and sulphide, respectively. The metal ion activity in solution governs the sulphide activity at the membrane surface through the appropriate solubility product — $a_S = K_{MS}/a_M$ — which controls the silver activity as before. We obtain for the e.m.f.

$$E = E^\circ + \tfrac{1}{2} k \log (K_{Ag_2S}) - \tfrac{1}{2} k \log (K_{MS}) + \tfrac{1}{2} k \log (a_M) \tag{3.13}$$

In principle, many electrodes could be made in this way but only three are sufficiently practical for commercial production. If an ion is present at such a concentration that it can displace the metal from its sulphide salt it will interfere in a manner similar to that of the anions discussed above.

Halide-selective electrodes sometimes incorporate silver sulphide in the membrane as well as the silver halide in order to reduce the electrode's sensitivity to light and increase the resistance of the membrane to scratches. As the primary silver response is still determined by the silver halide solubility, the e.m.f. of such electrodes takes the form of equation 3.11 rather than equation 3.13

$$E = E^\circ + k \log (K_{AgX}) - k \log (a_X)$$
$$= E^\circ + \tfrac{1}{2} k \log (K_{Ag_2S}) - \tfrac{1}{2} k \log (a_S)$$

The secondary response of an electrode is not necessarily useful, e.g. the fluoride electrode is not suitable for the direct (as opposed to titrimetric) determination of lanthanum, and copper electrodes with Ag_2S–CuS membranes cannot be used for determining sulphide, because of the solubility of copper sulphide in sulphide solutions.

The solubility products of the materials composing the membranes determine the limits of detection in the same way as for electrodes of the second kind, i.e. according to equation 3.5 or an exactly analogous equation for a cation-responsive electrode. Many of the compounds used in electrodes have very low solubility products, but their limits of detection are much higher than would be predicted by the above argument. One cause is the presence in the sample of a substance that reacts with the counter-ion of the determinand in the membrane (e.g. by complexation, oxidation or reduction), thus displacing the solubility equilibrium and causing more determinand to dissolve.

Construction of solid-state electrodes

The form of construction in which the membrane is set at the end of a plastic tube containing a reference electrode and a reference electrolyte solution is suitable for all types of solid-state membrane electrodes. Membranes containing silver salts can be directly contacted by a metallic conductor, thereby doing away with the reference electrode and reference electrolyte solution and making an electrode that works equally well in any attitude. An advantage of the former type of construction is that the membranes can be set in interchangeable caps, so that if the membrane is damaged only the cap needs to be replaced and not the entire electrode. Alternatively, the same body could be used with a series of different caps (and their corresponding reference electrolyte solutions) for analyses of different determinands done on an occasional basis. Philips, Activion, Metrohm, Simac and Tacussel make electrodes with such interchangeable caps.

A third type of construction is that in which a substrate electrode is impregnated with the membrane materials. The Radiometer F3001–F3004 metal-selective electrodes consist of a graphite-PTFE conductor impregnated with metal sulphides (see Table 3.1). When a new membrane is required, the old one is shaved off the graphite electrode and the freshly exposed surface impregnated with more metal sulphides. There is no need for the electrode to be impregnated with the same metal sulphide each time; thus an electrode could serve in turn for the determination of silver, copper, lead or cadmium (the range available at present). A similar range of anion-selective electrodes is available (chloride, bromide, iodide, thiocyanate, Radiometer F3005–F3008) based on the corresponding silver salts. A disadvantage of this type of electrode is that it usually requires to be conditioned before use, unlike electrodes with ready-prepared membranes. The Simac E12K3/1C is another electrode that can be activated to make it sensitive to different determinands, in this case, chloride, bromide, iodide and sulphide.

Glass Electrodes

Glass electrodes were the earliest ion-selective membrane electrodes and they are still the most widely used. The electrode consists of a thin membrane of the ion-selective glass blown on the end of a glass tube having no ion-selective properties (or set at the end of a plastic tube, although this is not usual). The tube is filled with an internal reference solution in which is immersed the internal reference electrode, connected by a screened cable to the pH meter. The silver–silver chloride electrode is almost always used as the reference electrode, although some pH electrodes have a Thalamid electrode instead. The internal reference solution contains the ion to which the reference electrode responds, in practice this is chloride ion, and the ion to which the electrode as a whole responds. A cell containing a glass electrode responding to an ion M can be represented as follows

reference electrode	test solution a_M	glass membrane	internal reference solution a'_M, a'_{Cl}	AgCl	Ag

The glass electrode potential is given by

$$E_{gl} = E°_{Ag,AgCl} - k \log a'_{Cl} - k \log a'_M + k \log a_M \qquad (3.14)$$

As the composition of the internal reference solution is fixed, the constant terms in equation 3.14 can be grouped to give equation 3.15

$$E_{gl} = E°_{gl} + k \log a_M \qquad (3.15)$$

The ion to which the electrode responds depends on the composition of the glass membrane. No anion-sensitive glass electrodes have been made and only electrodes for univalent cations are practical. The earliest electrodes were pH sensitive and had compositions typified by that of Corning 015 glass (21.4 mole % Na_2O, 6.4 % CaO, 72 % SiO_2). These electrodes were inaccurate above about pH 9 and most current commercial electrodes use lithia instead of soda, which extends the range to pH 13. The early glasses also had a very high specific resistance, which meant that the membranes had to be blown very thin and were consequently very fragile. Modern glasses incorporate tantalum or niobium oxides and sometimes also uranium oxide or rare earth oxides, which lower the specific resistance of the glass and enable quite robust electrodes to be made.

The non-ideal behaviour of the early electrodes at high pH was due to interference by sodium ions or other alkali metal ions. By incorporating alumina in the glass, it is possible to increase this interference effect to the extent that electrodes can be made with useful sensitivities for sodium and other alkali metal ions. A typical sodium-selective glass has the composition 11 mole % Na_2O, 18 % Al_2O_3 and 71 % SiO_2 and a typical potassium-selective glass is 27 mole % Na_2O, 4 % Al_2O_3 and 69 % SiO_2. Electrodes of similar selectivity can be prepared by using lithia instead of soda. Although these glasses are said to be sodium or potassium selective, they are still more responsive to hydrogen ion than to any other and the pH of a sample must be adjusted to a high enough value to suppress the hydrogen ion response before the electrodes can be used to determine alkali metal ions. The interference effect of a univalent cation N on an electrode responding to another univalent cation M is given by

$$E = E° + k \log (a_M + K_{MN} a_N) \qquad (3.16)$$

K_{MN} is the selectivity coefficient for the glass in a particular solution and its value depends on the ion-exchange equilibrium constant for the ions M and N between glass and water and the diffusion coefficients of the ions in the glass. Selectivity coefficients for some commercially-produced electrodes are tabulated in the methods for the determination of sodium and potassium. These tables also show that sodium-selective electrodes have a useful response for lithium ions and potassium-selective electrodes may be used for ammonium, rubidium, caesium and thallous ions. All the values of K_{MN} were obtained empirically and are accurate only over a limited range of solution composition. With some glasses, it is found that K_{MN} changes with age.

Divalent and trivalent metal ions do not interfere with glass electrodes, even if they are included in the formulation of the glass. The mobility of such ions in the glass is so low as to be negligible and all the current passing across the glass membrane is carried by alkali metal ions. The resistance of glass electrodes is in the range 1–500 MΩ and, therefore, their potentials must be measured with an instrument having a sufficiently high input impedance to ensure that the current acrosss the membrane is in the picoampere range. Larger currents can disturb the calibration of the electrode for many hours.

Glass electrodes for pH measurements are produced in a wide variety of shapes, sizes and formulations for clinical, industrial and laboratory analyses. They are also used as auxiliary sensing electrodes in gas-sensing membrane electrodes for ammonia, carbon dioxide and sulphur dioxide. Several types of sodium-selective electrode and a few types of potassium-selective electrode are also commercially available. The theory and uses of glass electrodes have been extensively discussed by Bates (1961; 1973), Mattock (1961) and Eisenman (1967).

Liquid Ion-exchange Electrodes

A liquid ion-exchange electrode can be considered as an analogue of the glass electrode having, in place of the hydrogen ion sensitive glass membrane of a pH electrode, a membrane consisting of a water-immiscible organic phase which has selective ion-transport properties. This membrane is interposed between the aqueous sample solution and an internal reference solution in which a suitable reference electrode is immersed.

Under ideal conditions, the potential across the membrane, E_M, is

$$E_M = \frac{k}{z} \log \frac{a_s}{a_I}$$

where a_s and a_I are the activities in the sample and internal reference solutions, respectively, of the ion of charge z. If a_I is constant, the membrane potential is a function only of the activity of the ion in the sample

$$E_M = \text{constant} + \frac{k}{z} \log a_s$$

When the membrane potential is combined with that of the internal reference electrode (usually a silver–silver chloride electrode), the half-cell equation for the complete electrode is

$$E_{½} = E° + \frac{k}{z} \log a_s$$

Where $E°$ is the sum of the constant portion of the membrane potential and the internal reference potential. If the complete liquid ion-exchange electrode is part of the following cell

Ag | AgCl | KCl + a_I | organic membrane | sample || reference electrode

then changes in activity of the determinand (a_s) in the sample produce changes in the cell potential according to the Nernst equation.

The organic phase consists of a solvent in which is dissolved a second substance, the exchanger, whose properties largely, but not solely, determine the response of the electrode. Suitable exchangers are usually identified after a series of trial-and-error experiments because there is insufficient information on their chemical behaviour in organic solvents for their selection to be made on theoretical grounds, except in general terms. Two main properties required are that the distribution of exchanger between membrane and aqueous solutions should overwhelmingly favour the former and that in the membrane the exchanger should form stronger complexes with the ion to which the electrode responds than with other commonly occurring ions.

In one class of electrode the exchanger, a large organic ion, is present in the membrane as the salt of the ion to which the electrode responds. In membranes with this type of exchanger, it is considered that only ions of opposite charge sign to the large organic ion enter the organic phase. The charge is transferred across the membrane by a complex formed from the entering ion, the exchanger and, perhaps, the solvent. In another class of exchanger, the neutral carriers, large organic molecules (e.g. polyesters, cyclodepsipeptides and polyethers) form complexes with cations by ion—dipole interactions. These compounds are electrically uncharged under conditions in which complexation takes place and the 1:1 complexes formed take the charge of the cation which is held in a central position relative to the organic molecule. Both the determinand cation and its co-ion must be present in the organic phase to preserve electroneutrality and mechanisms have been proposed in which the cation exchange between the membrane and the aqueous solution in contact with the membrane is fast while anion transfer across the same solution—membrane interface is slow. The structures of some macrocyclic ligands suitable for making neutral carrier membrane electrodes are given by Koryta (1975). Transport of the determinand across the membrane takes place by diffusion of the charged complex aided perhaps by a carrier-relay mechanism involving uncomplexed molecules or solvent molecules (Morf et al., 1973; Eyal and Rechnitz, 1971).

If the membrane is to have an acceptable life-span for practical analytical purposes, the solvent must have certain properties:

(a) The partition of the exchanger between solvent and aqueous sample must strongly favour the former or the membrane will soon be depleted of exchanger and so lose its effectiveness.

(b) The solvent should neither have a high solubility in water nor be very volatile, as in either case the exchanger will in time be deposited as a solid, leading to a breakdown of the ionic conduction mechanism and loss of electrode function.

(c) The solvent, even if sufficiently involatile and immiscible with water, should have a relatively high viscosity so that the membrane as a whole is not dispersed in the aqueous test solution, as otherwise the life of a membrane would be very short. This particularly applies to electrodes having the exchanger supplied from a reservoir.

The chemical properties required of a solvent cannot be clearly summarized since there is doubt as to its role. The solvent is not merely an inert matrix, as it can have considerable influence on the selectivity of electrodes, e.g. in those based on calcium *bis*(di-*n*-decylphosphate) there is little selectivity between divalent metal ions with decanol as solvent but considerable calcium selectivity when di-*n*-octylphosphate is used (Ross, 1969). In most cases it is assumed that the solvent plays a minor part in the transport of the charged species across the membrane but it may have an important role at the aqueous—organic interface during the transfer of the ionic species from one phase to the other.

The selectivity displayed by an electrode is a measure of the affinity the organic phase has for the ion and in many systems little difference is observed between ions of similar charge. In order for an electrode to have the widest possible Nernstian dependence on the ion M^{z+} in the presence of a second ion N^{z+}, the following equilibrium must be far to the left

$$N^{z+}_{aq} + MR_{org} \rightleftharpoons NR_{org} + M^{z+}_{aq}$$

where R represents the exchanger, in this case an anion of equal and opposite charge. The theory of the response of such a system, in which the anions of the sample do not enter the organic phase, has been reported by Sandblom, Eisenman and Walker (1967) and summarized by Eisenman (1968). Additional theoretical papers have been published dealing with anion selectivity (Reinsfelder and Schultz, 1973) and electrodes based on neutral carrier exchangers (Ciani, Eisenman and Szabo, 1969; Boles and Buck, 1973). The resulting theoretical expressions are complex and involve parameters not normally available, such as mobilities and association constants in the organic phase. However, for most analytical purposes the following empirical equation gives a useful guide to the behaviour of the electrode

$$E = E^{\circ} + \frac{k}{z_M} \log (a_M + K^{Pot}_{MN} a_N^{z_M/z_N}) \qquad (3.17)$$

where K^{Pot}_{MN} is the potentiometric selectivity coefficient and the smaller its value the more selective the electrode is for the ion M of charge z_M. Equation 3.17 may be compared with the corresponding equation 3.16 for glass electrodes. General expressions of this type can be written for any number of interfering ions in the same solution

$$E = E^{\circ} + \frac{k}{z_M} \log (a_M + \Sigma K^{Pot}_{MN_i} a_{N_i}^{z_M/z_{N_i}})$$

but they have little practical use since K^{Pot} is not a rigorous constant and cannot be applied over a wide range of concentrations of interfering ions. In addition, there is considerable controversy over the method of measuring K^{Pot} and the values obtained for the same electrode system can vary by an order of magnitude (Koryta, 1972). It is advisable, therefore, to use values of K^{Pot} only as a guide to the tolerable level of an interfering ion in an analytical method and not as a correction factor for the calculation of the true concentration of a determinand in the presence of a known level of an interferent.

The process by which the membrane potential is established involves the transfer of ions between the sample and the membrane and, therefore, changes the concentration of determinand in the sample. The e.m.f. that is measured is dependent on the final activity of determinand in the sample after equilibration with the membrane. At high concentrations in the sample, the initial and final activities are not measurably different but at low initial concentrations, sufficient determinand is extracted from the organic phase to raise the final concentration significantly and so shift the e.m.f. from that expected from an ideal electrode. When the concentration of the determinand in the sample is very low, it will be insignificant compared with the amount extracted from the organic phase in establishing the equilibrium and consequently the response of the electrode becomes constant. In the absence of interfering ions, this partition equilibrium finally sets the lower limit of the working range of most liquid ion-exchange electrodes; this limit is generally of the order of 10^{-5} mol l^{-1}.

The designs of commercial liquid ion-exchange electrodes are varied and their construction often complex. The assembly of Orion series 92 electrodes required considerable expertise as the electrode body had to be fitted with a small Millipore-type membrane, then filled with both liquid exchanger and aqueous internal reference solutions, each in its own reservoir. The latest Orion electrodes (series 93) have modular tips which can be screwed onto a common body for ease of interchange. The module contains both the organic and aqueous phases in porous plastic reservoirs, together with the silver—silver chloride internal reference electrode and the organophilic porous membrane, and itself requires no assembly.

The most popular design of liquid ion-exchange electrode has both the exchanger and the solvent incorporated into a plastic membrane during the membrane-casting process (Moody and Thomas, 1971). Electrodes having this type of membrane are manufactured by Philips, Metrohm, Activion, EDT Supplies and Simac. These electrodes usually have detachable tips which allow the membranes to be replaced after a period of use which is estimated to be 4—6 months. Many of the manufacturers market a number of different sensing membranes which fit the detachable tip of their particular electrode, thus making a versatile sensing system when used in conjunction with the appropriate internal aqueous reference solutions. Polymer or plastic membranes are relatively robust and are easily made but, unfortunately, the technique cannot be universally applied since the solvent/exchanger combination may be incompatible with the chemicals used in the preparation of the polymer.

The chemicals used in many electrode systems are not disclosed by the manufacturers but from the performance details it is probable that essentially the same systems are common to a number of electrode types. Sufficient information is available to give an indication of the types of solvents and exchangers that have been found to be useful for the most popular ions

Calcium A typical electrode uses di-*n*-octylphenylphosphonate/calcium-*bis*(di-*n*-decylphosphate) as the solvent/exchanger system. The properties of this exhanger are altered by changing the solvent to decanol and, as the resulting electrode is equally sensitive to most divalent cations, it has found an application as a combined calcium and magnesium or 'water hardness' electrode.

A PVC-membrane electrode for calcium based on the neutral carrier N,N'-di[(11-ethoxycarbonyl)undecyl]-N,N',4,5-tetramethyl-3,6-dioxaoctane amide with o-nitrophenyl n-octyl ether as solvent (Ammann et al., 1975) had selectivity coefficients similar to those given by Philips for their plastic-membrane calcium electrode.

Potassium The most common exchanger used in these electrodes is the neutral carrier valinomycin with a solvent such as diphenyl ether; combinations of this type have been used in polymer membranes. Earlier electrodes used tetra(p-chlorophenyl)borate as exchanger but the selectivity for potassium over sodium was not as great as that obtained with valinomycin.

Ammonium Ammonium-responsive neutral carrier electrodes commonly use mixtures of monactin and nonactin in a solvent such as diphenyl ether. These large cyclic molecules have the ability to bind cations selectively within their rings. Electrodes with considerable selectivity for ammonium over sodium have been made but potassium constitutes a serious interference.

Sodium The selectivity coefficients listed by Philips for their plastic membrane sodium electrode are similar to those reported by Ammann, Pretsch and Simon (1974) for a PVC-membrane electrode containing a neutral carrier with either dibenzyl ether or o-nitrophenyl n-octyl ether as solvent.

Lithium A PVC-membrane electrode for lithium, based on the neutral carrier N,N'-diheptyl-N,N',5,5-tetramethyl-3,7-dioxanone diamide and *tris*-(2-ethylhexyl)-phosphate as solvent, was reported by Güggi et al. (1975) and had selectivity properties similar to those of the Philips plastic membrane lithium electrode.

Chloride Orion manufacture a chloride electrode which has the advantage over the solid-state silver halide type of being relatively unaffected by sulphide. This electrode uses a long-chain amine hydrochloride, dimethyldistearylammonium chloride, as the ion-exchanger in an unspecified solvent.

Nitrate, Perchlorate, Tetrafluoroborate The Orion nitrate electrode has a substituted o-phenanthroline complex as the ion-exchanger, [Ni-(o-phen)$_3$]$^{2+}$ and 2-nitro-p-cymene as the solvent. Similar systems are used for perchlorate and tetrafluoroborate electrodes, the complexes being [Fe-(o-phen)$_3$]$^{2+}$ and [Ni-(o-phen)$_3$]$^{2+}$, respectively.

A number of other neutral carrier electrodes for ions such as barium and lead are available but no detailed information is available on the solvent exchanger systems which have been used.

INDIRECTLY SELECTIVE ELECTRODES

An electrode with a selectivity for a particular ion can be used to determine other substances that are not themselves electroactive, provided that these substances take part in reactions that produce or remove the electroactive ion, the concentration of which is measured by the electrode. An example of this type of electrode is the so-called cyanide electrode, in which the silver iodide in the membrane is slowly dissolved by the cyanide in the sample. This electrode, which is discussed more fully in the section on determining cyanide, depends on the establishment of a steady-state reaction at each concentration of cyanide. A second class of indirectly selective electrodes consists of the gas-sensing membrane electrodes, which are described below. These electrodes show how one type of ion-selective eclectrode (the pH glass electrode) can be made to respond indirectly to several different substances which selectively change particular sets of controlled chemical conditions. In a third class, the enzyme electrodes, the ion-selective electrode is coated with a gel containing an enzyme. The enzyme catalyses the degradation of a substance in the sample solution and the electrode responds to one of the degradation products, the concentration of which is proportional to the concentration of substance originally in the sample. Substances such as urea, glutamine and asparagine can be determined by an ammonium-selective glass or neutral carrier electrode coated with the appropriate enzyme — urease, glutaminase and asparaginase, respectively. The enzyme reaction itself can be very specfic, but the electrode will still suffer from the same interferences as when operating in a directly ammonium-selective role, e.g. sodium and potassium ions. Llenado and Rechnitz (1971) coated a cyanide electrode with β-glucosidase and so obtained a doubly-indirect response to amygdalin, which is decomposed with the production of cyanide. Enzyme electrodes have been reviewed by Guilbault (1971; 1975).

Gas-sensing Membrane Electrodes

Gas-sensing membrane electrodes are not true membrane electrodes, as no current passes across the membrane. They are complete electrochemical cells whose internal chemistry is monitored by an ion-selective electrode as it is changed by the determinand passing from the sample across the membrane to the inside of the cell. Since, in operation, these devices are handled exactly like electrodes, using the same equipment and having the same sort of Nernstian response, the misnomer is acceptable and will be used throughout this discussion. It should be noted, however, that one manufacturer prefers the term 'gas-sensing membrane probe' (Bailey and Riley, 1975).

The construction of an ammonia electrode is shown schematically in Figure 3.1. The sensing surface of a flat-ended glass pH electrode is pressed tightly against a hydrophobic polymer membrane that is acting as a seal for the end of a tube containing ammonium chloride solution. Only a thin film of solution remains between the membrane and the electrode and this is virtually sealed off from the bulk of the solution, although not so as to prevent the passage of current. A

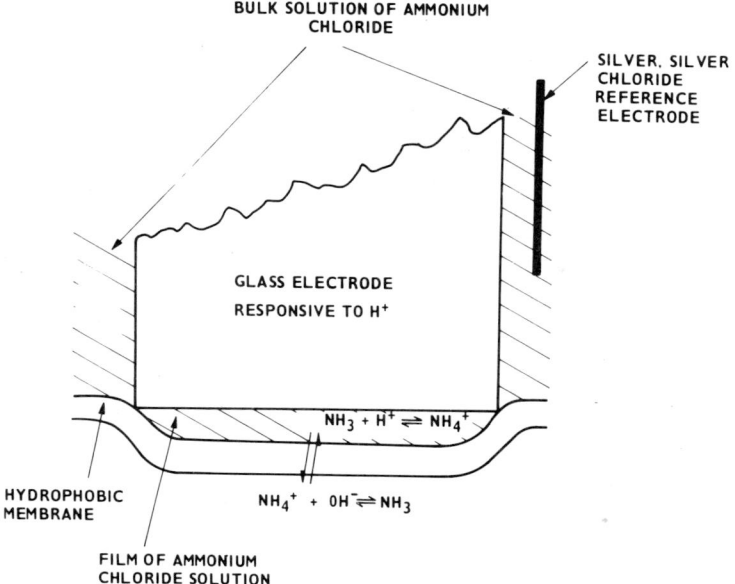

Figure 3.1 Gas-sensing ammonia electrode

silver–silver chloride electrode is immersed in the bulk solution. The membrane permits the diffusion of free ammonia (NH_3), but not of ions, between the sample and the film of solution. At equilibrium, the partial pressures of ammonia on either side of the membrane are equal, and so changes in the ammonia concentration in the sample are reflected by concentration changes in the film of solution, and thus indirectly by pH changes arising from the ammonia–ammonium equilibrium

$$NH_4^+ \rightleftharpoons NH_3 + H^+$$

The relationship between the e.m.f. and the ammonia concentration can be established by considering the cell

Ag | AgCl | mM NH_4Cl ‖ mM NH_4Cl, νM NH_3 | glass electrode

where mM is the molar concentration of the ammonium chloride solution and ν is the number of moles of ammonia that have diffused through the membrane per litre of solution in the film. The potential generated by this cell can be described by

$$E = E^\circ_{\text{glass}} + k \log(a_{H^+}) - E^\circ_{\text{Ag-AgCl}} + K \log(a_{Cl^-}) \tag{3.18}$$

Collecting the constant terms, we obtain

$$E = E' + k \log(a_{H^+}) \tag{3.19}$$

The dissociation constant, K, of the ammonium ion is

$$K = a_{H^+} \cdot a_{NH_3}/a_{NH_4^+}$$

and we can substitute for a_{H^+} in equation 3.19 and obtain

$$E = E' + k \log (K \cdot a_{NH_4^+}/a_{NH_3}) \tag{3.20}$$

As the system is effectively at constant ionic strength, the activity coefficients can be assumed to be constant. Collecting the constant terms, we have

$$E = E'' + k \log (c_{NH_4^+}) - k \log (c_{NH_3}) \tag{3.21}$$

The dissociation constant is so small that changes in the ammonium ion concentration in the film of solution due to changes in the ammonia equilibrium across the membrane also become very small and for practical purposes the ammonium ion concentration can be considered to be constant. For example, for an internal filling solution of 0.06 mol l^{-1} ammonium chloride, the ammonium ion concentration is constant to within 1 % for ammonia concentrations in the sample varying from 10^{-3} to 3×10^4 ppm (Midgley and Torrance, 1972). Within this range, equation 3.21 can be re-written as

$$E = E^{\circ}_{gas} - k \log (c_{NH_3}) \tag{3.22}$$

where E°_{gas} contains all the constant terms originating from equation 3.16. At equilibrium, the ammonia concentrations in the film of solution and in the sample are equal and equation 3.22 therefore relates the e.m.f. to the composition of the sample. In order to determine the ammonia content of the sample, it is necessary to obtain a calibration in which a constant proportion of the dissolved ammonia is present as free ammonia. This is achieved by adjusting the pH of the sample to a constant value; usually sodium hydroxide is added to give a pH of about 12 and, therefore, 99.9 % conversion to free ammonia. Lower pH values give lower conversions and the limit of detection will be increased. The option of determining the free ammonia rather than the total is also open to the analyst.

Other substances that can cross the membrane and change the pH in the film of solution will interfere with the electrode, e.g. volatile amines such as methylamine. The more basic a substance, the greater its potential interfering effect, e.g. cyclohexylamine (pK = 10.66) is a more serious interferent than morpholine (pK = 8.33), but a relatively involatile substance may not be able to interfere to the full extent predicted by its basicity, e.g. cyclohexylamine does not interfere as much as the more volatile, but equally basic, methylamine. Acidic gases such as carbon dioxide could interfere, producing a negative bias since they reduce the pH on dissolution in water, but treating the sample with base to convert all the ammonia to the free form ensures that acidic gases are retained in the sample solution in their ionic forms.

Analogous electrodes can be made for acidic gases (Severinghaus, 1968; Ross, Riseman, and Krueger, 1973). Consideration of the appropriate electrochemical cell and working through equations similar to equations 3.18–3.22 produces a final equation

$$E = E^{\circ}_{gas} + k \log (c_{gas})$$

which differs from equation 3.20 only in the sign of the logarithmic term (since the

Table 3.2 Gas-sensing membrane electrodes

Determinand	Reaction	Internal filling solution[a]	Internal electrode
NH_3	$NH_3 + H_2O = NH_4^+ + OH^-$	NH_4Cl	pH
CO_2	$CO_2 + H_2O = HCO_3^- + H^+$	$NaHCO_3$	pH
SO_2	$SO_2 + H_2O = HSO_3^- + H^+$	$NaHSO_3$	pH
NO_2	$2NO_2 + H_2O = NO_3^- + NO_2^- + 2H^+$	$NaNO_2$	pH
H_2S	$H_2S + H_2O = HS^- + H^+$	Citrate buffer	S^{2-}
HF	$HF + H_2O = F^- + H^+$	1 M H^+	F^-

[a]Other species may need to be present, depending on the internal reference electrode.

pH is changed in the opposite direction). In this case, a strong mineral acid is added to the sample to liberate the determinand and only other volatile acidic gases interfere. The chemistry of electrodes for acidic gases is summarized in Table 3.2.

So far only protonic equilibria have been considered, but if the glass electrode in Figure 3.1 is replaced by an ion-selective electrode of a different type, the range of gas-sensing electrodes can be increased. With solid-state sulphide and fluoride electrodes, gas-sensing electrodes for hydrogen sulphide and hydrogen fluoride can be made. In both these cases, an increase in the gas concentration produces an increase in the anion concentration which is determined in the film and the e.m.f. follows the equation

$$E = E^\circ_{gas} - \frac{k}{n} \log (c_{gas})$$

and since the sulphide electrode is divalent, $n = 2$ for the H_2S electrode. Both HF and H_2S could be determined by gas-sensing electrodes with internal pH electrodes, but the number of possible interferents is much smaller if the ion-selective electrodes are used.

Construction of Gas-sensing Membrane Electrodes

Many more gases could, in principle, be determined than are shown in Table 3.2, but gases that diffuse only slowly across the membrane will make electrodes with such long response times as to be impracticable. In order to minimize the response time, the volume of solution inside the electrode that has to equilibrate with the sample should be kept as small as possible. For this reason the sensing electrode has a flat end that is pressed against the polymer membrane to squeeze out all but a film of the filling solution. This film must neither dry out nor mix readily with the remaining bulk of the filling solution and in order to stabilize it, a spacer of tissue paper or cellophane is sometimes interposed between the membrane and the electrode. This spacer must not carry acidic or basic groups or else it would interfere with the measurements.

The membrane consists of material such as PTFE, polypropylene or silicone rubber about 25–100 μm thick. Two types of membrane are used — homogeneous and microporous. In homogeneous membranes the gas dissolves in the material of which the membrane is made and diffuses through the organic matrix itself, whereas in microporous membranes the gas diffuses through very small pores in the matrix. Diffusion coefficients in the air–gas mixtures in the pores are several orders of magitude greater than in the organic polymer matrix and microporous membranes will have faster responses. Homogeneous membranes are impermeable to water, but a difference in osmotic pressure across a microporous membrane will cause water transport, resulting in a concentration or dilution of the electrolyte in the film of solution inside the electrode and, therefore, a change in e.m.f. It is possible to balance the osmotic pressures by adding an inert electrolyte to either the sample or internal filling solution, but if samples with widely varying osmotic pressures are expected, it may be more convenient to use a homogeneous membrane and accept the slower response time. Detergent in a sample will also enable water to cross microporous membranes and homogeneous ones are better for such solutions.

REFERENCE ELECTRODES

All potentiometric cells contain two electrodes — one, the sensing electrode, changes its potential in response to the concentration of the species being determined, while the other, the reference electrode, makes a constant contribution to the overall potential. This constancy can only be achieved if the concentration of the species to which the reference electrode responds is itself kept constant. When the sensing and reference electrodes are both immersed in the same solution they form a *cell without liquid junction* (see p. 10); in such a case it would be always necessary to add to the sample a solution containing the ion to which the reference electrode responds. The following considerations apply to this arrangement:

(a) The reference electrolyte added to the sample must not introduce with it any of the determinand or an interferent.
(b) If the samples contain the ion to which the reference electrode responds, the concentration of reference electrolyte added must be high enough to swamp any variation in the final concentration of this reference ion.
(c) The sample must not contain substances that will interfere with the reference electrode.
(d) The sample must not contain substances that will react with the reference electrolyte.
(e) Neither the sensing nor reference electrode must release into the solution substances likely to interfere with the other.

Even if the above conditions can be fulfilled, this procedure complicates an analytical method and is rarely adopted in practice; instead, the reference electrode is immersed in a separate solution of its own reference electrolyte. Contact between the sample and reference solutions is arranged so that they can mix only by diffusion. Usually the solutions are separated by a piece of porous ceramic or plastic material, or by a tightly fitting ground-glass sleeve. This arrangement constitutes a *cell with liquid junction* (p. 10) and the assembly of the reference electrode in its reference electrolyte, in a container with an outlet to the sample solution through a suitable diffusive barrier, is a reference half-cell. In principle, any electrode could serve as a reference electrode, but for most purposes only two are normally considered — the silver—silver chloride and calomel electrodes. The advantages of these electrodes are that they are inexpensive, their properties are very well defined and their impedance is low. Most ion-selective electrodes have high impedances and conventional pH or pIon meter circuits could not cope with a high-impedance reference electrode as well. The use of half-cells containing calomel or silver—silver chloride electrodes is so widespread that these half-cells are now what is conventionally meant by 'reference electrodes', although the Thalamid and mercury—mercurous sulphate electrodes are also included in the term.

The fullest account of the preparation and properties of reference electrodes is given by Ives and Janz (1961) and much useful information is contained in the books by Mattock (1961) and Bates (1973). Covington (1969) has considered reference electrodes in relation to ion-selective electrodes and Midgley and Torrance (1976) in respect of continuous analysis.

The Calomel Electrode

The calomel electrode consists of a pool of mercury with its surface completely covered with mercurous chloride. The electrode is immersed in a solution of potassium chloride in a reservoir connected to the test solution through one of the junctions described below. Calomel electrodes are available from probably every electrode manufacturer, often in a variety of forms. The concentration of potassium chloride used is largely a matter of convention. American practice tends to favour saturated potassium chloride, giving the 'saturated calomel electrode' or 'SCE'. In Europe, concentrations of 3.8, 3.5, and 3.0 mol l^{-1}, giving the '3.8M calomel electrode' etc., are commonly used, especially for on-line industrial monitoring. For special purposes 1.0 and 0.1 mol l^{-1} potassium chloride may be used. Potassium chloride is normally used as the electrolyte because it is almost equitransferent and, therefore, minimizes the liquid junction potential (see p. 12), for which purpose a high concentration is also desirable. The saturated calomel electrode would, therefore, appear to be the best, but the tendency of the calomel electrode to take a long time (hours) to recover from a change in temperature is particularly marked in the case of the saturated version. For laboratory use, when changes in the ambient temperature are likely to be small, or when the temperature is controlled, the saturated calomel electrode is quite satisfactory, but in situations subject to large fluctuations in temperature a potassium chloride concentration of 3.0–3.8 mol l^{-1} gives a more reliable performance. The calomel electrode has only a limited life at high temperatures and it is inadvisable to use it above 70 °C for long periods. The half-cell potentials of the common forms of calomel electrode at several temperatures are given in Table 3.3.

Table 3.3   Standard and half-cell potentials of calomel electrodes

Temperature (°C)	$E°$ (mV)	Half-cell potential (mV) at KCl concentrations[a] of:				
		0.1	3.0	3.5	4.0	Saturated
5	272.9	–	–	–	–	–
10	271.9	336.2	260.2	255.6	–	254.3
15	270.8	336.2	–	–	–	251.1
20	269.5	335.9	256.9	252.0	–	247.9
25	268.1	335.6	254.9	250.1	245.9	244.4
30	266.5	335.1	253.0	248.1	243.8	241.1
35	264.7	334.4	–	–	–	237.6
40	262.9	333.6	248.7	243.9	239.3	234.0

[a] Concentrations expressed in mol l^{-1} at 25 °C.

Silver–Silver Chloride Electrodes

These electrodes are formed by electrolytic deposition of silver chloride on a silver wire or on a silver-plated platinum wire. The electrodes are immersed in potassium chloride solutions in the same sort of reservoirs with the same sort of liquid junctions as the calomel electrode. Since concentrated chloride solutions dissolve silver chloride through the formation of chlorosilver complexes, the potassium chloride solutions are usually pre-saturated with silver chloride before the electrode is immersed. The electrode recovers from changes in temperature more quickly than the calomel electrode and can be used at high temperatures for long periods. The compactness of the silver–silver chloride electrode has led to its almost exclusive use as the internal reference electrode of ion-selective electrodes and in the reference half of combination electrodes. The electrodes are widely available. The standard potential and some common half-cell potentials are listed in Table 3.4.

Table 3.4   Standard and half-cell potentials of silver–silver chloride electrodes

Temperature (°C)	$E°$ (mV)	Half-cell potentials (mV) in KCl solutions:	
		3.5 mol l^{-1}	Saturated
5	234.1	–	–
10	231.4	215.2	213.8
15	228.6	211.7	208.9
20	225.6	208.2	204.0
25	222.3	204.6	198.9
30	219.0	200.9	193.9
35	215.7	197.1	188.7
40	212.1	193.3	183.5

The Thalamid Electrode

Like the calomel and silver–silver chloride electrodes, the Thalamid electrode is essentially a metal in contact with a saturated solution of its chloride salt. In this case the metal, thallium, is present as a 40% amalgam and the surface is covered with solid thallous chloride. The electrode is immersed in saturated potassium chloride solution. As the amalgam electrode is attacked by oxygen, access of the atmosphere is restricted by placing the electrode and its saturated potassium chloride solution in a glass tube with a porous plug at one end and inserting the tube in the same sort of reservoir containing potassium chloride solution as is used for the other reference electrodes. The potassium chloride solution in the reservoir need not be saturated also, but can be 3.8 mol l^{-1} etc., as before. The advantage of the Thalamid electrode is that it resumes its equilibrium potential after temperature changes with scarcely any time lag. It is suitable for measurements in the range 0–135 °C. The potential of a Thalamid half-cell is approximately 800 mV more negative than those of the corresponding calomel or silver–silver chloride half-cells and with some pH or pIon meters there may be difficulty in getting a reading on scale if the Thalamid electrode is used with an ordinary sensing electrode. Sensing electrodes with matching Thalamid internal reference electrodes should be used. Ranges of Thalamid reference electrodes and pH electrodes with internal Thalamid reference electrodes are available from Schott & Gen. and Electrofact N.V. for both laboratory and industrial use. The electrode is rarely used, compared with the calomel and silver–silver chloride electrodes.

The Mercury–Mercurous Sulphate Electrode

This electrode looks exactly like a calomel electrode – a pool of mercury is covered with solid mercurous sulphate and is immersed in 1.0 mol l^{-1} sodium sulphate solution or in saturated potassium sulphate solution. This electrode is principally used when chloride contamination from the above reference electrodes would be harmful. The standard potential is about 350 mV more negative than that of the calomel electrode. Electrodes are available in several forms from Electronic Instruments Ltd., Chemetric, Beckman, Radiometer, and Leeds & Northrup.

Unusual Reference Electrodes

When a conventional type of reference electrode would introduce an unacceptable level of contaminant into the sample solution, an unusual reference electrode may be used. In the determination of ammonia with an ammonium-responsive glass electrode, Goodfellow and Webber (1972) used a calomel electrode with a 0.1 mol l^{-1} hydrochloric acid solution as the reference electrolyte, since alkali metal solutions would have produced an interference effect. Normally hydrogen ions interfere with the glass electrode even more than alkali-metal ions, but in this case the addition of triethanolamine buffer to the sample suppressed the hydrogen ion interference. To avoid interference with the potassium ion-selective electrode,

Orion Research use a reference electrode with a filling solution of 2.0 mol l^{-1} lithium chloroacetate plus 0.2 mol l^{-1} lithium chloride. In the Orion industrial monitoring equipment, fluoride electrodes are sometimes used as the reference electrode. Before resorting to the use of unconventional electrodes, the analyst should consider double-junction reference electrodes (below), but if these are unacceptable the electrode adopted should have as stable a potential as possible, in terms of both long-term drift and short-term noise. For these reasons, liquid ion-exchange electrodes should be avoided if possible. Solid-state silver halide electrodes are unlikely to be useful in a situation where the conventional reference electrodes are unsuitable, and the reactivity of the sulphide ion will generally disqualify the sulphide electrode. The divalent metal electrodes have not been extensively studied, but they might be adequate. The fluoride electrode is probably the most likely to be used as a reference electrode. Glass electrodes may be useful in cells without liquid junction, especially if low reactivity of the electrode material is a requirement, but the pH of the solution would have to be controlled very carefully if a constant potential is to be obtained. None of the ion-selective electrodes can be expected to match the conventional reference electrodes in the stability of their potentials.

Design and Construction of Reference Electrodes

The performance of a reference electrode depends not only on its chemical properties, but on the physical arrangement of the liquid junction, where the reference solution meets the sample solution. The main requirements for a good junction are: (i) The junction can be made reproducibly. (ii) There is no effect from stirring or from streaming of the sample solution. (iii) The junction is not clogged by particulate matter in the sample, or at least can easily be cleaned. (iv) Solution from one sample is not retained in the junction and carried over to the next.

An important factor is that there should be a steady flow of solution from the reference side to the sample side of the junction. The design of the electrode must either allow for a sufficient hydrostatic head of solution in the reference electrode or provide some other means of maintaining the flow, e.g. by using gas or springs to pressurize the reservoir. Problems are most likely to arise when the electrodes are used in-line or when the rate of flow of sample to the electrodes is regulated only by a simple throttle valve. Pressure surges in the main system can then cause disturbances at the reference electrode unless this is itself sufficiently pressurized. This is a problem mainly of process pH measurements, for which a hydrostatic head of 10–50 cm of solution is common, while even more elaborate mechanical pressurization may be needed. As the flow of sample to on-line ion-selective electrodes is generally closely regulated because of the need for the metered addition of reagent solutions, pressure fluctuations are small and a small head of electrolyte solution (1 cm) is adequate. The electrode should have an air breather at the top so that the flow of solution can be maintained. However, see the section on *Sealed Reference Electrodes*.

Form of Liquid Junction

The electrolyte solution from the reference electrode should flow into the sample at a constant rate and in a regular manner, since turbulence produces fluctuations in the potential. The hydrostatic pressure of the reference solution is one factor in achieving the right conditions, but this must be considered in relation to the flow-resistance and the geometry of the liquid junction itself. There are many different ways of establishing a liquid junction; those of most importance for analytical applications are as follows:

(a) The tube containing the reference electrode ends in a ground-glass cone fitted with a matching ground-glass collar or sleeve. A hole in the middle of the ground face of the cone leads to the reservoir of electrolyte solution which diffuses through the ground-glass joint to the sample solution. The rate of flow of electrolyte is relatively high. This 'sleeve-type' junction can be easily cleaned and re-formed and is, therefore, useful for measurements in samples that are liable to cause fouling, e.g. emulsions, solutions of proteins or soaps and samples with a high suspended solids content. Unless the fit between cone and sleeve is close, the flow of solution will be irregular and the potential will be unstable. Mattock (1961) found that sleeve-type reference electrodes gave signals four to five times noisier than those obtained with ceramic plug types (b) in both acidic and basic solutions, although the difference was less at intermediate pH values. The best electrodes of this sort are extremely satisfactory, but the quality of commercially produced ones is very variable. Virtually every manufacturer makes an electrode with this style of junction. Orion reference electrodes have plastic bodies, but may be included in this category.

(b) The wall or end of the tube containing the reference electrolyte has a plug of porous ceramic set into it. This arrangement gives a much lower flow rate than the sleeve type and the variation in rates between individual electrodes from a given manufacturer is fairly wide. The robustness of this design suits it for industrial applications where an electrode must work without close attention from the analyst and it is probably the best general-purpose design. Its main disadvantage is that, once fouled, it is very difficult to clean. Electrodes with this type of junction are made by almost every manufacturer.

(c) The end of the tube containing the reference solution is drawn out into a capillary and usually bent into a J-shape. The reservoir has to be fitted with a tap which is closed in normal operation. The rate of flow is moderate to high and interdiffusion of sample and reference solution in the tube may occur. The advantages of this arrangement are that opening the tap on the reservoir flushes out the junction completely with the electrode still in place and that its electrical resistance is very low. This type of junction is not often used in analysis.

(d) An asbestos fibre is set into the wall of the tube containing the reference electrolyte. Beckman and Corning make electrodes with this junction, which is convenient to use provided the sample does not contain material likely to clog the small orifice.

(e) A palladium annulus set in the glass at the end of the junction tube allows the reference solution to seep into the sample solution at a controlled rate. Beckman make electrodes with this type of junction.

(f) Leeds & Northrup make an electrode in which the base of the glass tube is cracked in a controlled way, leaving a path through which the reference solution can diffuse to the outside.

Concentration of Reference Solution

In calomel, silver–silver chloride, and Thalamid reference electrodes the reference solution must contain chloride ions. In order to minimize the liquid junction potential the cation should have ideally the same mobility as the chloride ion and the concentration of reference electrolyte should be as high as possible. The almost universal choice as the cation is potassium (see p. 12). The highest concentration of potassium chloride available is a saturated solution, and this is normal for laboratory work, but less suitable for application to on-stream continuous monitoring, where the equipment may be subjected to large variations in temperature and be left unattended for a considerable time. A drop in temperature causes potassium chloride to crystallize out of solution and form a plug of solid material which prevents or impedes the flow of reference solution and may thus lead to a noisy signal. Precipitation reduces the concentration of chloride in solution and the reference e.m.f. will change as a result, in addition to the inescapable effects of the changes in E° and the Nernst slope with temperature. Saturated solutions, therefore, give electrodes with larger temperature coefficients than are necessary and it is also found that the time an electrode takes to regain its original potential once the temperature has returned to normal is much longer when the solution is saturated. The concentration of potassium chloride in industrial reference electrodes for use in pH cells is commonly 3.0 mol l^{-1} (Pye), 3.5 mol l^{-1} (Schott) or 3.8 mol l^{-1} (EIL and Beckman). Ion-selective electrodes are usually kept at a constant temperature in industrial monitors and saturated reference solutions are suitable in this situation.

Sealed Reference Electrodes

The time spent preparing reference solutions and replenishing the reservoirs of reference electrodes in on-stream instrumentation can result in considerable maintenance costs. Sealed reference electrodes last for six months or more and are then thrown away as they cannot be refilled; the saving in maintenance costs should pay for the electrode. As the electrodes are sealed the reference solution cannot easily flow through the junction into the sample and a number of desirable characteristics for a reference electrode are lost. Although such electrodes would not be expected to be better than the conventional types, for many purposes they are perfectly adequate and far more convenient. Because in the sealed system, there is a tendency for the reference and sample solutions to interdiffuse, saturated potassium chloride filling solutions are used, with a large excess of solid potassium chloride to prevent dilution of the reference solution by the interdiffusion. The junctions are of the ceramic-plug type with a low leakage rate. These electrodes have the disadvantages of saturated potassium chloride types (see above) and the additional one that material can be carried over from one sample to another in the ceramic plug because there is no flow of solution to wash it away. This is of particular importance in pH measurements, since buffer solutions are normally of greatly different composition from the samples being analysed. The electrodes may take several hours to recover from the effect of immersion in a buffer solution, during which they are effectively unavailable for measurement, although an e.m.f. can always be obtained. With ion-selective electrodes, when measurements are made in a constant background of added electrolyte, the sample and standard solutions are of the same nature and similar problems are far less likely to arise. Sealed reference electrodes include the Electronic Instruments RK 28, the Philips R15, the Leeds & Northrup 117300 and 117304, the Beckman 39406 and 39407 and the Schott 9801 series.

Double-junction Reference Electrodes

When conventional reference electrodes are unsuitable because they release ions that will either interfere directly with the sensing electrode or react with one of the species in the sample solution, either a special reference electrode must be devised (see above) or the conventional one must be separated from the sample by a solution containing ions that are compatible with both the reference and sensing halves of the cell. This arrangement is a double-junction reference electrode. The conventional electrode is in the form of a glass or plastic tube with the usual potassium chloride filling solution connected via a porous plug to an outer tube which is filled with the inert bridge solution. The inner part is sometimes a sealed unit that cannot be refilled. The outer liquid junction is most often of the ground-glass sleeve type, permitting easy wash-out of the bridge solution. The inner electrolyte will diffuse into the bridge solution and may eventually build up in concentration sufficiently to resume the interference; it is, therefore, advisable to replace the bridge solution completely at regular intervals, e.g. weekly. It should be noted that in a properly designed flow cell, interference can usually be avoided by siting the reference electrode downstream of the sensing electrode. Commercially-available double-junction electrodes are the Orion 90–02, the Philips R44 and R44D, the Corning 47606700, the Electrofact R116/100G, the Leeds & Northrup 117414 and the Beckman 40452, 40454 and 19013.

Temperature Effects

If e.m.f. measurements are made in conditions of varying temperature, the choice of reference electrode can greatly influence the size of the errors produced. The change in standard e.m.f. with temperature for the common reference electrodes is in the order

calomel < silver–silver chloride < Thalamid

Unfortunately, the same order is found for the range of temperature in which the electrodes can be used. Calomel electrodes start to decompose at about 80 °C, while Thalamid electrodes have been used at 135, °C. When an electrode is subjected to a change in temperature and then returned to its original state, it takes a finite time for the original e.m.f. to be recovered. This time is longest for the calomel and shortest for the Thalamid electrodes. Both the magnitude of the e.m.f. change and the recovery time are increased if a saturated salt is used as the reference solution. A way of reducing the problem is to use a *remote junction* reference electrode. The electrode is immersed in a reservoir of reference solution connected through a length of flexible tubing to a glass tube with a ceramic plug at the end. Only the junction tube is immersed in the sample, the electrode and reservoir remain at ambient temperature and are not influenced by the temperature of the sample.

It is instructive to consider that the ion-selective electrode nearly always contains an internal reference electrode and that the $E°$ shift of the complete potentiometric cell will be minimized if the internal and external reference electrodes are matched; most ion-selective electrodes have silver–silver chloride internal references. Ideally, the reference solutions should also be matched, but the internal filling of a sensing electrode will usually be too dilute to make a good liquid junction, besides containing other species that would detract from the stability of the reference electrode.

Colloidal and Suspension Effects

The potential of electrodes immersed in a suspension can differ from that observed when they are in the supernatant solution. Most, if not all, the difference can be attributed to an effect on the liquid junction potential between the sample and reference solutions. Differences of up to 240 mV between identical reference electrodes have been reported (Jenny *et al.*, 1950) when one was immersed in an ion-exchange resin sediment and the other in the supernatant solution. This phenomenon is confined to situations where the particles are highly charged, as with ion-exchange resins and certain soils. When it occurs, it is difficult to obtain a valid calibration and measurements have to be made on the supernatant solution or a filtrate. A corresponding effect occurs in highly-charged colloidal solutions. Bower (1961) showed that the suspension effect in soil pastes could be reduced by using a relatively low concentration of potassium chloride in the reference solution. Above 2.0 mol l^{-1} KCl the effect was approximately constant for a given suspension but it could be largely eliminated by using a 0.1 mol l^{-1} potassium chloride reference solution.

Bibliography

Ammann, D., M. Güggi, E. Pretsch, and W. Simon, 1975, 'Improved calcium ion-selective electrode based on a neutral carrier,' *Anal. Lett.*, **8**, 709.

Ammann, D., E. Pretsch, and W. Simon, 1974, 'A sodium ion-selective electrode based on a neutral carrier', *Anal. Lett.*, **7**, 23.
Bailey, P. L., and M. Riley, 1975, 'Performance characteristics of gas-sensing membrane probes', *Analyst* **100**, 145.
Bates, R. G., 1961, 'The glass electrode', In *Reference Electrodes, Theory and Practice*, G. J. Ives and D. J. G. Janz (eds.), Academic Press, New York and London, Ch. 5.
Bates, R. G., 1973, *Determination of pH, Theory and Practice*, 2 nd edn., Wiley, New York and London.
Bennetto, H. P., and A. R. Willmott, 1971, 'Electrochemical measurements with amalgam electrodes', *Quart. Rev.*, **25**, 501.
Boles, J. H., and R. P. Buck, 1973, 'Anion responses and potential functions for neutral carrier membrane electrodes', *Anal. Chem.*, **45**, 2057.
Bower, C. A., 1961, 'Studies on the suspension effect with a sodium electrode', *Soil Sci. Soc. Amer. Proc.*, **25**, 18.
Bucur, R. V., and L. Stoicovici, 1966, 'Palladium–hydrogen reference electrode with its own gas reserve', *J. Electroanal. Chem.*, **11**, 152.
Ciani, S., G. Eisenman, and G. Szabo, 1969, 'Theory for the effects of neutral carriers such as the macrotetralide actin antibiotics on the electric properties of bilayer membranes', *J. Membrane Biol.*, **1**, 1.
Covington, A. K., 1969, 'Reference electrodes', in *Ion Selective Electrodes*, R. A. Durst (ed.), National Bureau of Standards Special Publication 314, US Department of Commerce, Washington, D.C., Ch. 4.
Eisenman, G., 1967, *Glass Electrodes for Hydrogen and Other Cations*, Edward Arnold, London/Marcel Dekker, New York.
Eisenman, G., 1968, 'Similarities and differences between liquid and solid ion exchangers and their usefulness as ion specific electrodes', *Anal. Chem.*, **40**, 310.
Entwistle, J. R., C. J. Weedon, and T. J. Hayes, 1973, 'The determination of hydrofluoric acid and nitric acid contents of pickling bath liquors using ion-selective electrodes', *Chem. Ind. (London)*, 433.
Eyal, E., and G. A. Rechnitz, 1971, 'Mechanistic studies on the valinomycin-based potassium electrode', *Anal. Chem.*, **43**, 1090.
Fleischmann, M., and J. N. Hiddleston, 1968, 'A palladium–hydrogen probe for use as a micro-reference electrode', *J. Sci. Instrum.*, 667.
Goodfellow, G. I., and H. M. Webber, 1972, 'The determination of ammonia in boiler feed-water with an ammonium-selective glass electrode', *Analyst*, **97**, 95.
Güggi, M., U. Fiedler, E. Pretsch, and W. Simon, 1975, 'A lithium ion-selective electrode based on a neutral carrier', *Anal. Lett.*, **8**, 857.
Guilbault, G. G., 1971, 'Enzyme electrode probes', *Pure Appl. Chem.*, **25**, 727.
Guilbault, G. G., 1975, 'Ion selective and enzyme electrodes', *Bull. Soc. Chim. Belg.*, **84**, 679.
Harzdorf, C., and H. Keim, 1976, 'Über Selektivitätskonstanten von Halogenidelektroden zweiter Art', *Z. Anal. Chem.*, **279**, 263.
Hills, G. J., and D. J. G. Ives, 1961, 'The hydrogen electrode', in *Reference Electrodes, Theory and Practice*, G. J. Ives and D. J. G. Janz (eds.), Academic Press, New York and London, Ch. 2.
Ives, D. J. G., 1961, 'Oxide, oxygen and sulphide electrodes', in *Reference Electrodes, Theory and Practice*, G. J. Ives and D. J. G. Janz (eds.), Academic Press, New York and London, Ch. 7.
Ives, D. J. G., and G. J. Janz (eds.), 1961, *Reference Electrodes, Theory and Practice*, Academic Press, New York and London.
Janz, G. J., and D. J. G. Ives, 1961, 'The quinhydrone electrode', in *Reference Electrodes, Theory and Practice*, G. J. Ives and D. J. G. Janz (eds.), Academic Press, New York and London, Ch. 6.

Jenny, H., T. R. Nielsen, N. T. Coleman, and D. E. Williams, 1950, 'The measurement of pH, ion activities and membrane potentials in colloidal systems', *Science*, **112**, 164.

Koryta, J., 1972, 'Theory and applications of ion-selective electrodes', *Anal. Chim. Acta*, **61**, 329.

Koryta, J., 1975, *Ion-selective Electrodes*, Cambridge University Press.

Llenado, R. A., and G. A. Rechnitz, 1971, 'Improved enzyme electrode for amygdalin', *Anal. Chem.*, **43**, 1457.

Mattock, G., 1961, *pH Measurement and Titration*, Heywood, London.

Midgley, D., 1975, 'Investigations into the use of gas-sensing membrane electrodes for the determination of carbon dioxide in power station waters', *Analyst*, **100**, 386.

Midgley, D., and K. Torrance, 1972, 'The determination of ammonia in condensed steam and boiler feed-water with a potentiometric ammonia probe', *Analyst*, **97**, 626.

Midgley, D., and K. Torrance, 1976, 'An assessment of various types of reference electrode for use in continuous potentiometric analysis with particular application to highly pure waters', *Analyst*, **101**, 833.

Moody, G. J., and J. D. R. Thomas, 1971, *Selective Ion Sensitive Electrodes*, Merrow, Watford.

Morf, W. E., D. Ammann, E. Pretsch, and W. Simon, 1973, 'Carrier antibiotics and model compounds as components of selective ion-sensitive electrodes', *Pure Appl. Chem.*, **36**, 421.

Pungor, E., and K. Tóth, 1970, 'Ion-selective membrane electrodes', *Analyst*, **95**, 625.

Rechnitz, G. A., and M. R. Kresz, 1966, 'Potentiometric measurements with chloride-sensitive and bromide-sensitive membrane electrodes', *Anal. Chem.*, **38**, 1786.

Reinsfelder, R. E., and F. A. Schultz, 1973, 'Anion selectivity studies on liquid membrane electrodes', *Anal. Chim. Acta*, **65**, 425.

Ross, J. W., 1969, 'Solid-state and liquid membrane ion-selective electrodes', in *Ion Selective Electrodes*, R. A. Durst (ed.), National Bureau of Standards Special Publication 314, US Department of Commerce, Washington, D.C., Ch. 2.

Ross, J. W., J. H. Riseman, and J. A. Krueger, 1973, 'Potentiometric gas sensing electrodes', *Pure Appl. Chem.*, **36**, 473.

Sandblom, J., G. Eisenman, and J. L. Walker, 1976, 'Electrical phenomena associated with the transport of ions and ion pairs in liquid ion-exchange membranes, Parts I and II', *J. Phys. Chem.*, **71**, 3862 and 3871.

Severinghaus, J. W., 1968, 'Measurement of blood gases: P_{O_2} and P_{CO_2}', *Ann. N.Y. Acad. Sci.*, **148**, 115.

Stock, J. T., W.C. Purdy, and L. M. Garcia, 1958, 'The antimony–antimony oxide electrode', *Chem. Rev.*, **58**, 611.

Vasile, M. J., and C. J. Enke, 1965, 'Preparation and thermodynamic properties of a palladium–hydrogen electrode', *J. Electrochem. Soc.*, **112**, 865.

Warren, L. J., 1971, 'The measurement of pH in acid fluoride solutions and evidence for the existence of $(HF)_2$', *Anal. Chim. Acta*, **53**, 199.

Whitfield, M., 1971, *Ion Selective Electrodes for the Analysis of Natural Waters*, Australian Marine Sciences Association, Sydney, N.S.W.

Chapter 4

Equipment

TYPES OF METERS*

pH and pIon meters

The pH meter is a familiar piece of laboratory apparatus and it can be applied equally well to measurements with ion-selective electrodes. A pIon meter is a pH meter with one or two small modifications to simplify measurements with ion-selective electrodes. Both kinds of meter measure millivolts and have an appropriate display, either digital or analogue. The pH meter also converts the millivolt signal to a pH scale, i.e. one showing an increase of one unit for a decrease of approximately 60 mV in the e.m.f. The pH scale requires the use of two controls — the calibration control and the slope control. The latter may not always be identified as such on the meter, as it often acts in the same way as the temperature compensation control. The slope and temperature compensation controls adjust the number of millivolts equivalent to one pH unit. The calibration control relates the measured e.m.f. to a fixed point on the pH scale. Provided only that the buffer or calibration control has a wide enough range of adjustment, the pH scale can be used for any univalent positive ion, i.e. measurements with a sodium electrode can be read as a pNa ($-\log c_{Na}$) scale, with a potassium electrode as a pK scale, etc. Measurements with electrodes responding to divalent or negative ions cannot be related directly to the pH scale. A pIon meter is one that has either a scale, or scales, analogous to the pH scale for ions of various charges or a scale that can be calibrated to read directly in terms of concentration or activity. The EIL 7030 and 7050 pH/pIon meters have both types of pIon scale. Almost all pIon meters also have pH and millivolt scales. On some pIon meters, the pIon scales may be labelled in terms of particular ions, e.g. pF, pS, pCa, etc. but the same scale may be used for all ions of the same charge and sign; pIon meters with such specifically-assigned scales may be incapable of dealing with all possible combinations of charge and sign. The scales so far cover only charges of ±1 and ±2, as electrodes for ions of higher charge have yet to be commercially produced. Direct

---

*The addresses of manufacturers whose equipment is mentioned in this chapter are listed in Appendix 3.

activity scales are expressed in relative terms only, and have to be calibrated before use — any units may be used, not only mol l^{-1} but ppm or oz gal^{-1}.

It is impossible usefully to describe such a wide and rapidly changing market as that of pH meters. The most sensitive ones cost twenty times as much as the simplest models, but are not always the most versatile. In a working life of up to ten years, a meter may be put to many uses and it pays to consider all the features that may be required in the meter.

Digital and Analogue Display

The convenience of the digital display cannot be disputed, but its cost means that, at present, a meter with an analogue display, i.e. with a needle moving over the face of a printed scale, will probably be more versatile than a digital meter of the same price. The most sensitive digital meters read to 0.1 mV or 0.001 pH units over a range, generally, of ±1999.9 mV or 0—14 pH units. Less sensitive ones read only to 1 mV or 0.01 pH units, but on some of these the reading at any point in the scale can be expanded to give 0.1 mV or 0.001 pH units, while losing the first figure of the display, e.g. pH 11.23 would expand as 1.234. Meters without this expansion facility are at present rather poor value for money. Some portable meters can only be read to 0.1 pH units. The Corning 101 and Orion 901 are digital meters which also have direct concentration displays. Some meters also indicate the type of reading being made, i.e. mV, relative mV, pH, or whether the meter is only on standby. Most digital meters display the polarity sign automatically on the millivolt scales. The Philips PW9414 and Radiometer PHM64 have pIon ranges, analogous to pH ranges, for univalent and divalent anions and cations. The rate at which the e.m.f. is sampled to provide the digital reading may be varied on some meters. It is sometimes possible to hold a given reading on the display indefinitely.

Analogue meters are usually scaled 0—14 pH units, with the smallest division on the scale equivalent to 0.1 units giving an estimated accuracy by interpolation of 0.02 pH units. The millivolt scale runs typically from 0 to 1400 mV, with a polarity switch, or —700 to +700 mV without one. The smallest division is 10 mV, allowing readings by interpolation to 2 mV. Many meters have the facility of expanding the scale so that the precision of the reading can be increased by between four and ten times, but this is often only possible by calibrating the normal and expanded scales independently. The Pye 290, Leeds & Northrup 7415 and EIL 7050 pH meters can give an expanded reading on any part of the scale without altering the calibration in any way. Ion concentration/activity scales are a feature of a number of analogue meters (Orion 400 series, EIL 7030 and 7050, Leeds & Northrup 7410, 7413 and 7415); the extremes of the range are in the ratio of 100 or 1000 to 1. The Philips 9413 and Radiometer PHM53 meters have pIon as well as pH scales.

Portable pH meters

Battery-powered pH meters weighing only 2 kg or so are now available. Most of these meters have an analogue output, but digital models are also made. The Orion

400 series incorporates a concentration/activity scale for direct readings with ion-selective electrodes. Small portable meters are not capable of the highest precision, but they can be extremely useful for fieldwork and for making checks on industrial plant where there may be no convenient power supply. The Simac 61 SI and Radelkis OP-107 meters have pIon scales for univalent and divalent anions and cations.

Microprocessor Analysers

Instruments combining a pH meter and a microprocessor are now being developed. An example is the Orion 901, which can automatically calculate the concentration of determinand from direct potentiometry or known addition potentiometry and give a digital read-out in concentration units. Blank and volume corrections can be applied automatically. The more usual pH and millivolt modes are also available.

Recorder and Printer Outputs

Both continuous monitoring and automatic analysis demand a means of recording the results automatically. All but the simplest meters have an analogue output for transmitting the e.m.f. or pH to a chart recorder and some digital meters also have an 8-bit BCD (binary coded decimal) output that can be used to operate a printer or interface with a computer. Modern chart recorders are versatile instruments with multi-range switches enabling them to receive a wide variety of potentiometric signals, but it is advisable to check that a particular combination of pH meter and recorder will give the desired performance. A full-scale deflection (f.s.d.) of 100 mV on the recorder is probably the most commonly used. Meters may have either a potentiometric or an amperometric output. In the former, the output is equal to the actual millivolt reading times a factor usually not greater than unity. For a meter in which the output potential equals the input potential, a full-scale deflection of 100 mV will be obtained with a 100 mV recorder, an f.s.d. of 10 mV with a 10 mV recorder etc., whereas for a meter in which the output potential is only one-tenth of the input, 100 mV f.s.d. calls for a 10 mV recorder and so on. If the recorder is set to record over the input range 0–100 mV, but the electrode potential is varying between 150 and 250 mV, an on-scale recording can only be obtained if a back-off potential is applied between one of the output terminals of the meter and the corresponding input terminal of the recorder. In the example quoted an opposing potential of 150 mV is required. This adjustment may be made internally by changing the zero control on the recorder or by measuring on the relative millivolts scale of the meter, or an external source of e.m.f. may be placed between the meter and the recorder. If the electrode potential ranges over more than 100 mV and the same sensitivity of recording is desired, the back-off device must be calibrated so that an accurately known e.m.f. can be opposed to the signal output. Some chart recorders have a back-off facility built into them.

An amperometric output has advantages for range switching, as the deflection on the recorder is proportional to the deflection on the meter on any range. This is particularly useful with meters such as the Pye 290 and Philips PW9413 that have scale expansion facilities in any part of the range without the need for altering the calibration. The output is usually 1—1.5 mA at full scale on any range.

Matching Meter Output and Recorder Input

Many meters with a potentiometric output have an output span control and many recorders also have a control that enables the recorder span to be varied from a fixed value. With the meter input terminals shorted, the recorder is set to zero by means of its own zero control. The meter is then supplied with an appropriate e.m.f. and the span control on either the meter or the recorder adjusted until the deflection on the recorder corresponds to the desired scale. The meter input should be shorted again to check that adjusting the span control has not changed the zero; if it has, the recorder is zeroed again, as before, and then the span checked and adjusted and so on until consistency is achieved.

In order to use a meter with an amperometric output in conjunction with a potentiometric recorder, a suitable resistance must be connected across either the output terminals of the meter or the input terminals of the recorder. The resistance is calculated as follows: Suppose the meter has a maximum output of 1.4 mA and the recorder has a span of 20 mV, then by Ohm's law the resistance required is $20/1.4 \, \Omega = 14.28 \, \Omega$. The chances of finding such a resistor are small and a variable resistance will generally be used. The meter — recorder combination is zeroed as above, a known e.m.f. supplied to the meter and the deflection on the recorder set to the correct position by adjustment of the variable resistance. As before, the zero should be checked and adjusted, etc. If the recorder has a variable span control, a resistor of approximately the correct value may be used. In the above example, say 14 or 15 Ω. After zeroing, the deflection is set by adjusting the recorder span control. Again the zero and the deflection are checked until consistency is achieved.

Temperature Compensation

The standard potentials and calibration slopes of electrodes change with temperature and all pH meters have some means of correcting at least for the latter effect. In the simplest meters, this consists of a manual adjustment for the change in the slope factor $RT \ln(10)/F$, the control being calibrated directly in terms of temperature. More advanced meters have the option of automatic compensation, achieved by a circuit incorporating a thermocouple immersed in the sample solution. The object of such a control is to adjust the pH scale on the meter so that the electrode can be calibrated at different temperatures, rather than to permit electrodes calibrated at one temperature to be used at another. When ion-selective electrodes are being used, this control can often be considered as a means of correcting for a non-theoretical but linear response, and on pIon meters this dual function is usually made explicit.

Only a relatively few meters are able to correct for changes in standard potentials, so that electrodes calibrated at one temperature may be used for measurements over a range of temperatures; these are meters with an *isopotential* control setting. Although all combinations of ion-selective and reference electrodes have an isopotential point, applications of this concept have so far been restricted to pH measurements. When the isopotential correction is applied, simultaneous correction for a non-theoretical slope is possible only if the meter has a separate control for that purpose, e.g. the Radiometer PHM64. On some meters, e.g. the EIL 7030, it is possible to make a slope correction when working at constant temperature with ion-selective electrodes, as the isopotential control can be switched out on the pIon and direct-activity scales. As before, adjustment for changes in temperature may be performed manually or automatically.

A few pH meters, such as the EIL 7060 and Radiometer PHM64, have a variable isopotential control, so that they can be used with a number of different combinations of electrodes; the EIL 7050 has a particularly wide range in this respect. More commonly, the meter has a fixed isopotential setting which is suitable only for certain combinations of pH and reference electrodes, e.g. some Corning, Philips and Pye meters. It is recommended that both the glass and reference electrodes be obtained from the same manufacturer as the pH meter in these cases, since errors will be introduced by using electrodes with different characteristics. It is often the wrong choice of reference electrode that is the problem: most manufacturers assume the use of a calomel electrode, but the concentration of potassium chloride in the calomel electrode and whether the temperature of the reference electrode varies with that of the solution or is kept constant (because a remote junction is used) have also to be considered. It is regrettable that, when a meter has a fixed isopotential setting, its value is not always clearly marked on the body of the meter, as an inexperienced operator may easily be led into error.

Temperature compensation circuits work only on the pH, pIon and direct activity ranges of a pH meter and not on the millivolt, expanded millivolt or relative millivolt ranges. In addition, isopotential corrections may be inoperative on pIon and direct activity ranges and in some cases even on an expanded pH range.

Additional Features

Features additional to the main ones described above may increase the versatility or ease of operation of pH meters. Some digital meters have an automatic standardization control, so that with the electrodes in a standard solution the relative milivolts scale is automatically set to zero at the touch of a button rather than by turning a knob until zero is observed on the scale. On the pH scale, the same control will set the meter to read a predetermined pH value when the electrodes are immersed in the appropriate buffer solution. Meters with this feature include the Orion 801A and 901, Corning 119 and Beckman 3550. On any meter, the calibration control knob should not be so stiff as to make adjustment difficult, nor should it be placed in such a position that it it liable to be adjusted accidentally — a locking control is a great advantage.

When the input terminals are shorted the meter should read 0 mV, but electronic drift may cause deviations from this reading. All meters have a control to reset the zero when necessary and this is usually recessed and slotted, so that it can only be turned by a screwdriver. Meters with a variable potentiometric output for a chart recorder have a similar adjustment screw.

All portable meters should have a battery-level indicator.

A number of meters have terminals that enable them to be used for Karl Fischer titrations of water content.

There is no international standard fitting for the plugs on pH, ion-selective or reference electrodes and there is a corresponding variety of input sockets on pH meters. Most ion-selective electrodes have the US standard plug which fits meters such as those made by Beckman, Corning, Leeds & Northrup and Orion. The reference socket on these meters takes a 2 mm pin. The other common plug is the co-axial type, fitting meters made by Electronic Instruments, Pye and Philips; a 3 mm reference pin is usual on these meters. Radiometer and Metrohm each have their own types of plugs. Adaptors to convert US and Radiometer sockets to take co-axial plugs are available, but a co-axial socket cannot readily be made to take the other plugs. For most purposes, there is no advantage to any one sort of plug or socket. Some meters allow the reference signal from a combination electrode to be carried along the screening of the pH or ion-selective half of the combination electrode, which therefore requires only a high-impedance plug.

Voltmeters and Electrometers

Special meters are not necessary for making measurements with ion-selective electrodes, but they are so much more convenient than the alternatives that it is generally worthwhile to have one. Voltmeters have too low an input impedance for use with most ion-selective electrodes, but they are suitable for measurements with silver—silver chloride, hydrogen, antimony—antimony oxide and other low-impedance electrodes. Electrometers have a sufficiently high input impedance to cope with any electrode that a pH meter can deal with, and may be excellent instruments in themselves, but they lack some of the facilities available on a pH meter. The scale can normally only be read in millivolts and may be such that a discrimation of 0.1 mV can only be obtained over a restricted range, say 0—10 or 0—100 mV. No temperature compensation of any kind is fitted.

INDUSTRIAL MONITORING EQUIPMENT

pH Measurements

Although industrial pH instruments are more rugged than the laboratory equipment described above, they do not differ in principle and have generally the same features — scale expansion, automatic or manual temperature compensation, etc. A feature often found on industrial amplifiers is the provision, in addition to the recorder output, of a second relatively heavy duty output (up to 5 A) for such purposes as operating alarms or control valves. The amplifier unit may not have its own display and will therefore have to be used with a separate display unit or a chart recorder. Such pH 'transmitters' are used when the display is required at a place distant from the point of measurement. As pH electrodes have high impedances, the amplifier must be connected to them by fairly short leads if capacitance effects are not to interfere with the measurement and therefore only the amplified signal can be transmitted any significant distance. As the amplifier will have to be located near the electrodes, it may need protection from dust, moisture, vibration or extremes of temperature. A number of companies make specially modified transmitters for intrinsically safe operation in flammable atmospheres. Only the electrodes and the transmitter are located in the hazardous area, while the power pack, alarm units etc. are in a safe area. The electrical signal from such apparatus may be converted into a pneumatic one to operate controls valves in the danger area.

Measurements with Ion-selective Electrodes

pH systems are sufficiently stable for calibration to be performed at intervals of a week or more, but most ion-selective electrodes require more frequent attention and this has led to the development of equipment that automatically calibrates the electrodes at regular intervals. Such apparatus is more complicated and expensive than is required for pH measurements, for which manual calibrations are more economical, and therefore industrial pH equipment is distinct from that used for ion-selective electrodes, in contrast to laboratory practice.

A typical ion-selective monitor is shown schematically in Figure 4.1. A pump delivers the sample solution and any buffer or ionic strength adjustment solution to

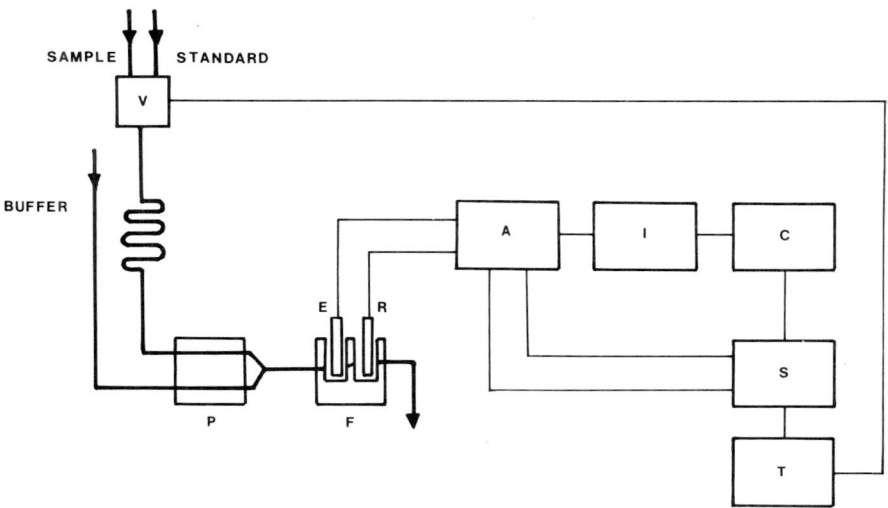

Figure 4.1 Schematic flow diagram of an ion-selective monitor. A = amplifier; C = controller; E = sensing electrode; F = flow cell; I = indicator/recorder; P = pump; R = reference electrode; S = servo-potentiometer; T = sequence timer; V = valve

the flow cell and a valve allows a standard solution to be introduced automatically under control of the sequence timer. Because of the difficulty of correcting for the effects of changes in temperature, the liquid-handling unit usually includes some means of maintaining the sample and standard solutions at a constant temperature by the time they reach the electrodes.

The electronics unit includes the amplifier, a display panel and a number of optional modules for automatically standardizing the electrodes. Before any measurements are made on the sample, the electrodes are allowed to stabilize in the standard solution under the control of the operator. The potential developed by the electrodes serves as a reference point for future calibrations. If the amplifier has a slope control, a second standard solution may be introduced so that the correct calibration slope can be established, otherwise a Nernstian response will be assumed. Thereafter, the timer will automatically cause the valve in the liquid-handling unit to switch at some preset interval from the sample to the standard solution. This solution is pumped for a sufficiently long time for the electrodes to reach a steady potential, which is compared with the reference potential found from the initial calibration. The compensating servo-potentiometer is then driven so that the amplifier output is restored to its original level. In this way, any shift in the standard potentials of the electrodes or electronic drift of the amplifier can be automatically corrected. The entire sequence requires about 30 min, during which any alarm or control circuits would be disabled. Manual initiation of the standardization sequence is also possible. There is no facility for checking or correcting the calibration slope, but experience shows that this is a far less important source of error than changes in the standard potential.

Although the above description shows the general features required of an ion-selective monitor, individual details will vary between the products of different manufacturers and because of the nature of the electrodes. Glass and many solid-state electrodes are so stable that fairly infrequent manual calibration is adequate for many purposes, but liquid ion-exchange and gas-sensing membrane electrodes are more likely to need frequent and preferably automatic standardization. Manufacturers of industrial ion-selective monitors are less numerous than those of laboratory equipment or industrial pH monitors. Among them are found Electronic Instruments Ltd., Leeds & Northrup, Foxboro, Beckman, Orion, and Serck Controls.

Flow Cells

The flow cell represents one of the most critical features of continuous monitoring. The wash-out characteristics of the flow cell must be compatible with the available flow of sample if adequate response times are to be achieved. Some pH flow cells require 500 ml of sample per minute, which can often be inconveniently large. As the sample stream generally has to be dosed with a reagent before analysis by ion-selective electrode, flow cells have been developed with low sample requirements (<10 ml min^{-1}) in order to reduce the capacity of the reagent reservoir and/or the frequency of refilling it. A common arrangement is for the sample to be supplied to a constant head reservoir, while a portion is continuously removed by a metering pump and fed to the flow cell. A second pump injects any reagents required into the metered sample stream, which then passes through a mixing coil or mixing chamber so that the dosed sample is homogeneous. This general arrangement, in which measurements are made on a small sample continuously removed from the main stream, is known as *on-line* or *on-stream* monitoring. When the electrodes are immersed directly in the main stream, measurements are *in-line* or *in-stream*. In this situation, it is impossible to add reagents without contaminating the main sample, the volume of which would normally make dosing impractical in any case, and the electrodes will therefore measure the activity of free determinand. All pH measurements are of this nature, but activity measurements are not generally accepted for other determinands. In process streams where the concentration of a minor component or an impurity is being measured, the concentrations of the major components may be sufficiently constant for the stream itself to act as a pH or ionic strength adjustment solution, thus enabling an in-line measurement of the total concentration of determinand to be made. Before this could confidently be done, the suitability of each process stream would have to be considered individually with respect to a given determinand. In general, therefore, ion-selective electrodes are only used on-line, while pH and redox measurements are commonly made in-line.

Cells for in-stream measurements are of two kinds — *pipework* cells which form an integral part of the pipe carrying the sample stream and *dip* cells, which are used for immersing electrodes in tanks, culverts, ponds and other situations in which the sample is open to the atmosphere. Pipework cells must be made of a material compatible with the rest of the pipe and the mountings of the electrodes must be designed to withstand the maximum pressure expected in the pipe. Cells for use at pressures up to 50 p.s.i.g. (3.5 kg cm^{-2}; 4.5 atm) may be made of a variety of materials — PVC, polypropylene, plastics reinforced with glass fibre or stainless steel — and are available from many manufacturers. Cells for use at high pressures (10–25 atm) are invariably made of stainless steel. Cells working at high pressure may require special glass electrodes and special reference electrodes. It is often the case that fluctuations in pressure are more troublesome than high pressures as such. In this case, it may be necessary to resort to one of the self-compensating types of reference electrodes, e.g. the Serck Bellomatic series or the Electronic Instruments 1360–500 fitted with an EIL GPR pressure regulator for the 7600 series flow cells. Special systems, e.g. the EIL 2877 and the Pye/Ingold 768-35 enable the electrodes to be removed from a sealed vessel for replacement or calibration without having to stop the process being monitored, even if this occurs at high temperatures and pressures. Dip cells are usually made of plastics material, as they are used only at atmospheric pressure. Cells of this type are available from almost all manufacturers of industrial pH equipment, but one model of particular interest is the Philips WMZ 1008 floating immersion assembly, which keeps the electrodes at a constant depth below the surface when the water level changes.

A frequent problem with in-stream measurements is fouling of the electrodes, which results in increased response times, inaccuracy or virtually complete loss of

Table 4.1 Selection of cleaning systems

Nature of fouling	Ultra-sonic	Mechanical		Chemical			Hydro-dynamic
		Brush	Scraper[a]	Acid	Base	Emulsifier	
Oils and fats		x		x		x	
Resins, wood pulp	x				x		
Latex emulsions			x				
Fibres, e.g. paper, textiles							x
Solid suspensions							x
Crystalline precipitates (carbonates)		x	x	(x)			
Amorphous precipitates (hydroxides)	x	x	x	(x)			

[a] Applicable only to metal electrodes, including antimony–antimony oxide electrode.

sensitivity as the sample may be unable to reach the sensitive surface of the electrode. As fouling can be the major cause of outage and maintenance costs and may also damage the electrodes beyond repair, a variety of different systems has been devised to overcome it. None, however, can be guaranteed to work in a particular situation and a cure may often only be effected by trial and error, if at all. The most suitable system depends on the nature of the fouling and Table 4.1 sets out general guidelines. Brush and ultrasonic cleaning systems are made for most flow cells. The chemical and hydrodynamic methods are relatively unusual and are produced by Polymetron Ltd. In the worst cases, the glass electrode is replaced by the antimony–antimony oxide electrode, which is robust enough to survive vigorous mechanical cleaning.

AUTOMATIC TITRATION APPARATUS

A simple automatic titrator can be made by using a motor-driven syringe to inject the titrant solution into the sample at a constant rate while the output from the pH meter is recorded on a chart recorder. The point of inflection of the titration can then be read in terms of chart divisions, which are proportional to time and therefore to the volume of titrant delivered. Constant-speed apparatus such as this, however, is not very convenient, since additions slow enough to enable the end-point to be precisely defined would make the overall titration very long, while rapid titrations would suffer from having the end-point smeared out. In modern titrators, the syringe is driven by a stepping motor, the frequency of which is proportionately controlled by feedback from the pH meter, so that at the start of the titration the titrant is added fairly quickly, but as the end-point is neared and the potential begins to change more rapidly, the rate of addition of titrant is reduced until it reaches a minimum at the point of inflection of the titration curve. After the end-point, the procedure works in reverse, the rate of addition of titrant gradually increasing again. Such an arrangement gives a good compromise between speed and precision and, in addition, the feedback to the proportional controller represents the first derivative of the titration curve, $\delta E/\delta V$. This function can be plotted instead of the titration curve itself, giving an even more precise end-point. Once the titrant is delivered at variable rates, the paper feed of the chart recorder must be coupled to the syringe motor if an accurate representation of the titration curve is to be obtained.

The requirements of automatic titrators have led to the development of special pH meters and chart recorders. The Metrohm Potentiographs (E536 and E576) combine both in one compact instrument, while Radiometer have adopted a modular approach that gives great flexibility and allows their pH meters to be adapted for use in the titrator. The pH meter part of the titrator has the usual features — slope and buffer calibration controls, temperature compensation, millivolt and pH ranges — as well as the additional controls needed to regulate the titration. The display available from special titrators does not make them suitable for direct potentiometry where great precision is required, e.g. the Metrohm titrigraphs have no display other than their recorder charts, which give 100 or 250 mV full scale on the most expanded ranges. The Radiometer TTT60 titration control unit can convert certain Radiometer pH meters into titrators, thus the PHM64 research pH meter, which has a digital display reading to 0.1 mV or 0.001 pH or pIon units, can be made into an extremely versatile instrument for both potentiometry and titrimetry. The simpler PHM63 and PHM62 digital pH meters can be similarly converted, as can the very simple PHM61 analogue meter. The Radiometer Servograph or Titrigraph recorders are used to draw the titration curve. Most of these systems can plot either the titration curve or its first derivative, but the Tacussel Titrimax system can also plot the second derivative curve ($d_2 E/dV^2$) for even greater sensitivity in determining end-points. The latest development in automatic titrators is the application of a built-in microcomputer to control the titration, as in the Radiometer TTT61 titrator as part of the DTS633 digital titration system. The microcomputer stores the coordinates of the simulated titration curve (no recorder is needed) and calculates the first and second derivatives. The zero points on the second-derivative curve are obtained by interpolation and printed out for each end-point found. The microcomputer also controls the addition of titrant.

The titrant is delivered from a syringe burette mounted on a motor unit which is controlled by feedback from the pH meter. Interchangeable syringes are available with capacities in the range from 0.25 to 50 ml. The volume delivered is shown on a digital counter, one digit being equal to 0.04–0.1% of capacity, depending on the make of burette. The Radiometer ABU12 and ABU13 autoburettes also have BCD outputs for connection to an external printer or computer. Other features that may be found on automatic burettes are variable speed controls, manual override and automatic refilling of the burette. The whole process can be further automated by using a carriage that holds up to fifty samples at once and automatically moves each fresh sample into place when the previous titration is finished. The electrodes are automatically rinsed between each titration. In this way a large number of titrations can be carried out without intervention or supervision from laboratory staff. The Metrohm E553 and Radiometer ATS1 are examples of completely automatic systems.

Automatic titrations can also be carried out in a simpler way, in which titrant is added until the meter records a pre-set pH or millivolt value, e.g. pH 7 in the titration of a strong acid with a strong base. No recorder is required. The Mettler automatic titrator works in this way, as can the Radiometer and Metrohm titrators, if desired. Industrial automatic titrators generally work on this principle, although they may also show the titration curve on a chart recorder. The observed equivalent volume can be compared automatically with a pre-selected value and the difference used to actuate alarms or control valves. The Metrohm E440 and Electronic Instruments Ltd. 8110 are examples of this kind of instrument. The EIL analyser incorporates daily automatic standardization (with manual override) and allows for automatic treatment of the sample with a reagent and/or a diluent before the titration starts. A timer controls the sequence of events and one cycle may take as little as four minutes. Up to six sample streams can be switched successively to the titrator.

More sophisticated industrial process titrators are now being developed, e.g. the Ionics model 1800 DigiChem titrator, which has digital computation circuits that calculate one or two end-points from the derivative curve, automatically subtract a titration blank and can give answers directly in engineering units.

Chapter 5

Analytical Principles

The two basic factors in assessing the suitability of an analytical method are its accuracy and precision, two terms which are sometimes confused. Accuracy is concerned with how close an experimentally-determined result is to the true value, while the precision describes the variation or scatter of results obtained when a sample is analysed repeatedly. In practice, if a sample is analysed n times there will be n different results — $x_1, x_2, \ldots x_n$ — and the best estimate of the true concentration, c, is the *mean* of the results, μ_n

$$\mu_n = \Sigma x_i/n \tag{5.1}$$

The *accuracy* is defined as $c - \mu_n$ or in relative terms $100(c - \mu_n)/c\%$. The reliability of μ_n increases with the number of times the analysis is repeated, but as it is impracticable to repeat every analysis several times the analyst needs some measure of the confidence he can have in the result of a few repeats or even of a single determination, i.e. he needs to know the precision of the method.

Consider a sample of concentration 1 ppm that is analysed n times by the same method. If the results are rounded to the nearest 0.05 ppm, the relative frequency with which each value of concentration occurs can be plotted against the concentration, as in Figure 5.1. It can be seen that the greatest frequency is found at the mean and that approximately the same frequency occurs at concentrations equally distant from the mean. If it were possible to perform an infinite number of analyses with an infinitesimally divided concentration scale, the frequency distribution would be the continuous plot also shown in Figure 5.1. This is an example of a *normal* or *Gaussian* distribution. Other types of distribution are possible, but the Gaussian distribution is usually adequate for results in analytical chemistry. The shape of the curve is defined by equation 5.2

$$P(y) = \frac{1}{\sqrt{2\pi}} \exp(-y^2/2) \tag{5.2}$$

where $P(y)$ = the probability of finding a result $y = (\mu - x)/\sigma$, y = the deviation of a result, x, from the mean, μ, expressed in multiples of σ, the *standard deviation*, which is characteristic of the distribution. The Gaussian distribution has the

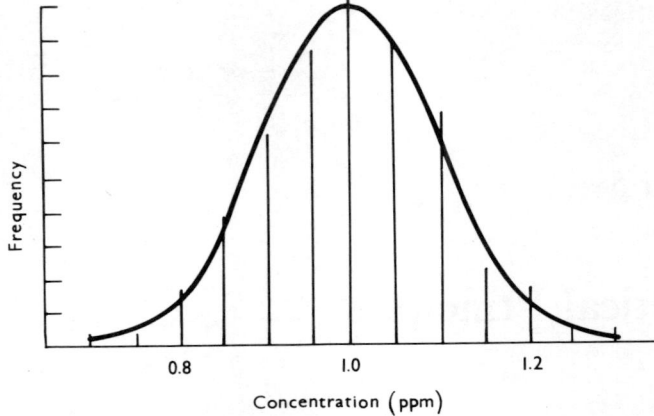

Figure 5.1 Approximation of discrete analytical results to a normal distribution curve

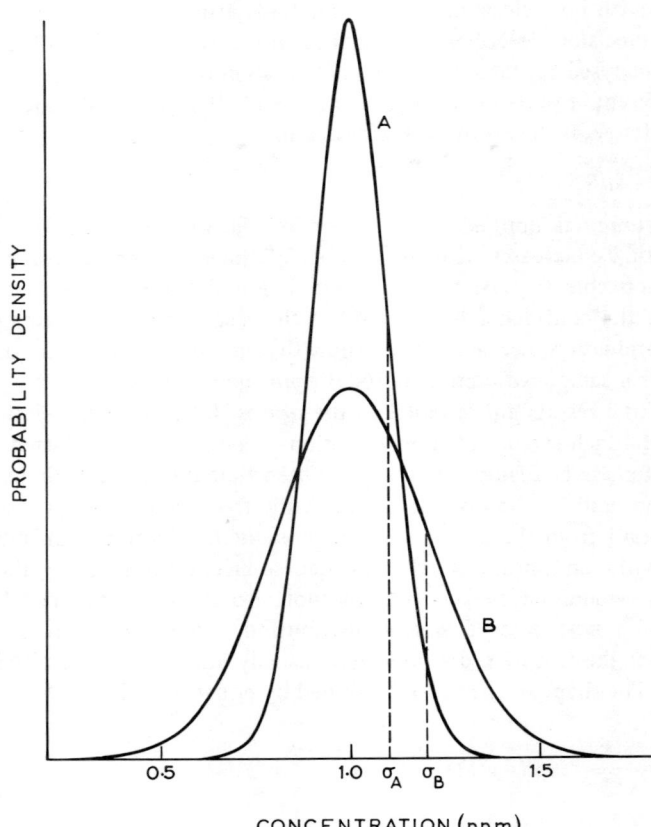

Figure 5.2 Normalized distribution curves for analyses with standard deviations of 0.1 ppm (curve A) and 0.2 ppm (curve B)

property that 68.3% of all results lie within one standard deviation of the mean and 99.7% within three standard deviations. In many of the tests, below, we shall use limits containing 95% of the results, which occur at 1.96 σ but are conveniently rounded to 2 σ. The standard deviation is sometimes called the precision and it defines the confidence we can place in an analytical result. For example, Figure 5.2 shows two normalized distribution curves which represent the results of two different analytical methods. The mean is the same for both distributions, i.e. the two methods are equally accurate. The area under both curves is the same, as the total probability is always unity, but a far higher proportion of the results represented by curve A lie close to the mean than is the case for curve B. Curve A has a standard deviation of 0.1 ppm, i.e. there is only one chance in twenty of obtaining a result more than 2 x 0.1 ppm different from the mean. An alternative statement would be that the 95% confidence limits for method A are ±0.2 ppm. The same limits for method B are ±0.4 ppm. Other things being equal, the method with the smaller standard deviation is clearly to be preferred. Other measures of precision are the *relative standard deviation* $S = \sigma/\mu$ and the *variance* $s = \sigma^2$. Statistical texts have to be read with care as there are many variations on the symbols used for the different measures of precision.

The standard deviation is rigorously a property of the continuous function only; for a real situation in which the number of measurements, n, is finite, the best estimate of the standard deviation is given by

$$\sigma_n = \sqrt{\frac{(\mu - x_i)^2}{n - 1}}$$

$$= \sqrt{\frac{n\Sigma x_i^2 - (\Sigma x_i)^2}{n(n - 1)}} \tag{5.3}$$

As only a finite number of results is available, the mean of those results, μ_n, is only an estimate of the true mean, μ, i.e. if the n tests were repeated again we would obtain a different value of μ_n and so on until we obtain a distribution curve for μ_n. If the individual results x_i had a Gaussian distribution, so will the mean results, and the standard deviation of the mean in one batch of n determinations will be related to the standard deviation for a single result in that batch, σ_n, by

$$\sigma(\mu_n) = \sigma_n/\sqrt{n} \tag{5.4}$$

COMBINATION OF ERRORS

If a function z is a combination of the variables x and y, each of which has its own standard deviation, and the constants a, b and c, which by definition have zero standard deviation, the standard deviation of z is calculated as follows

(1) Addition $\quad z = ax + by + c \quad$ or subtraction $\quad z = ax - by - c$

$$\sigma_z^2 = a^2 \sigma_x^2 + b^2 \sigma_y^2$$

(2) Multiplication $\quad z = c \cdot x^a \cdot y^b \quad$ or division $\quad z = c \cdot x^a/y^b$

$$\left(\frac{\sigma_z}{z}\right)^2 = a^2 \left(\frac{\sigma_x}{x}\right)^2 + b^2 \left(\frac{\sigma_y}{y}\right)^2$$

(3) Logarithms $\quad z = a \ln x \quad$ or $\quad z = b \log x$

$$\sigma_z^2 = a^2 \left(\frac{\sigma_x}{x}\right)^2 \quad \text{or} \quad \sigma_z^2 = \left(\frac{b}{2.3}\right)^2 \left(\frac{\sigma_x}{x}\right)^2$$

SIGNIFICANCE TESTING

The analyst is frequently required to judge whether the difference between two quantities is due to a bias or whether it has arisen purely because of the variations associated with all experimentally observed results. Significance tests calculate the frequency with which a difference of a given magnitude is likely to appear as a result of random error, e.g. when the same sample is analysed by different methods or by different analysts or on different occasions it is unlikely that the same result will be obtained each time. If there is only a small chance that the difference could have been caused by random errors, the analyst must conclude with a given *degree of confidence* or at a given *confidence* or *significance* level that there is a bias. Because such conclusions are based on probabilities, the analyst must accept that a decision based on such results has a calculable chance of being wrong.

What is an acceptable risk of making a false decision is a matter for personal judgment, but for most analytical purposes a 5% chance is accepted. Where the consequences of a false decision are more serious, 1% or even 0.1% limits may be adopted. Note that confidence limits may be expressed in two complementary ways, thus a statement may be true 'with 95% confidence' or 'at the $P = .05$ level', the percentage confidence being given by $100(1 - P)$.

In all the tests given below the standard deviation is critical, but in practice this parameter can only be estimated from a finite number of results. In making the tests, therefore, allowance must be made for the uncertainty of this estimate by taking into account the number of *degrees of freedom* associated with the estimate. The number of degrees of freedom, N, is given by the total number of results less the number of restraints on the system, e.g. in determining the mean of n results there is one restraint (the experimental mean is assumed to coincide with the true mean), thus

$$N = n - 1$$

As the number of degrees of freedom increases, the greater the confidence in the estimate of the standard deviation and the smaller the difference that can be discriminated with a given confidence. This is reflected in Tables 5.1 and 5.2 by the decrease in the value of the tabulated function with the increase in the degrees of freedom.

Comparison of Means

Student's t-test

This is used to find whether the difference between the means of two independent sets of results is statistically significant at a given probability level. This test may be applied to analyses of the same sample by different methods or by different analysts using the same method. The general formula is given by equation 5.5, where $\bar{x}$ and $\bar{y}$ are the two means and σ_m is the standard deviation of the difference between the means

$$t = \frac{|\bar{x} - \bar{y}|}{\sigma_m} \tag{5.5}$$

The calculation of σ_m depends on the circumstances of the test, as described below.

Equal standard deviations for both sets of results

The chances that two experimentally obtained estimates of the standard deviation will be exactly equal are very small, but it is enough that these values are consistent, as determined by the F-test (see below).

If there are p results $x_1, x_2, \ldots, x_p$ in one set, q results $y_1, y_2, \ldots, y_q$ in the other and σ_x and σ_y are the standard deviations of single results in each set,

$$\sigma_m^2 = \frac{\Sigma x^2 - (\Sigma x)^2/p + \Sigma y^2 - (\Sigma y)^2/q}{p+q-2} \cdot \frac{p+q}{p \cdot q}$$

$$= \frac{(p-1)\sigma_x^2 + (q-1)\sigma_y^2}{p+q-2} \cdot \frac{p+q}{p \cdot q} \tag{5.6}$$

The number of degrees of freedom, $N = p + q - 2$.

Unequal standard deviations for the two sets of results

With the same notation as before

$$\sigma_m^2 = \frac{\sigma_x^2}{p} + \frac{\sigma_y^2}{q} \tag{5.7}$$

The number of degrees of freedom is approximated by

$$N = \frac{[\sigma_x^2/p + \sigma_y^2/q]^2}{\dfrac{\sigma_x^4}{p^2(p-1)} + \dfrac{\sigma_y^4}{q^2(q-1)}}$$

$$= \sigma_m^4 \bigg/ \left\{ \frac{\sigma_x^4}{p^2(p-1)} + \frac{\sigma_y^4}{q^2(q-1)} \right\} \tag{5.8}$$

If the above gives a non-integral value of N, round it to the nearest whole number.

Difference of a mean from a reference value

This is a special case of equation 5.5 where $\bar{y} = R$, which by definition has zero standard deviation; thus

$$\sigma_m = \sigma(\bar{x}) = \sigma_x/\sqrt{p}$$

and

$$t = \frac{|\bar{x} - R|}{\sigma_x/\sqrt{p}} \quad \text{with} \quad N = p - 1$$

Paired measurements on a series of samples

If each of a series of n samples is analysed by two different methods, or by two different analysts, the difference (with sign) is calculated for each pair $d = x - y$ and the mean, $\bar{d}$, compared with zero, thus

$$t = \frac{|\bar{d}|}{\sigma(\bar{d})} = \frac{|\bar{d}|}{(\sigma_d)/\sqrt{n}} \quad \text{with} \quad N = n - 1$$

One-sided and Two-sided Tests

If we are only concerned with differences between two results and it does not matter which is the larger, the test is two-sided. If we want to know whether one result exceeds the second, i.e. we are concerned with deviations in one direction only, it is a one-sided test. A given value of t is statistically twice as likely to occur in a two-sided test as in a one-sided test and the same table of t-values can be used for both tests, but it is important to note for which type of test the probabilities are defined. Table 5.1 shows values of the t distribution for both types of test and may be used to check how any other tabulation is defined.

Use of the t-test

Analyst X analyses a sample six times and obtains a mean of 4.92 ppm with a standard deviation of a single result of 0.05 ppm, while analyst Y analyses the same

Table 5.1 Percentage points of Student's t-distribution

Degrees of freedom	Probability (%) of a deviation greater than t[a]			
	0.5 1	1 2	2.5 5	5 10
4	4.604	3.747	2.776	2.132
5	4.032	3.365	2.571	2.015
6	3.707	3.143	2.447	1.943

[a] 1-sided, upper row; 2-sided, lower row.

sample 5 times, obtaining 5.06 ppm and 0.09 ppm instead. Are the two sets of results significantly different?

Applying the *F*-test (below) to the standard deviations, we confirm that the standard deviations are unequal and, therefore, from equations 5.7, 5.8 and 5.5

$$\sigma_m = \sqrt{\frac{(0.05)^2}{6} + \frac{(0.09)^2}{5}} = 0.045 \text{ ppm}$$

$$N = (0.045)^4 / \left\{ \frac{(0.05)^4}{6^2 \cdot 5} + \frac{(0.09)^4}{5^2 \cdot 4} \right\} = 5.93 \simeq 6$$

$$t = (5.06 - 4.92)/0.045 = 3.111$$

Consulting Table 5.1 (two-sided) we see that, for $N = 6$, t lies between the values for 2% and 5% probability, i.e. there is less than a 5% chance that the difference of the two means could arise from the random errors associated with the two analysts' work and, therefore, we can be 95% confident that there is a systematic bias between the two sets of results.

If we wish to test that the two results do not differ from the true result (5.00 ppm), we have

$$t_X = \frac{5.00 - 4.92}{0.05/\sqrt{6}} = 3.919, \text{ with } N_X = 5$$

and

$$t_Y = \frac{5.06 - 5.00}{0.09/\sqrt{5}} = 1.491, \text{ with } N_Y = 4$$

Consulting Table 5.1, we see that t_X lies between the values for 1% and 2% probability (two-sided), i.e. there is less than a 5% chance that the difference could have arisen from random errors in X's analyses and there is probably a bias in his result. On the other hand, there is a better than 5% chance that the difference of Y's result from the true value arises from random errors and we would classify the difference as being non-significant.

Suppose we were concerned only that the analytical result were not significantly less than 4.975 ppm. The results of analyst Y clearly meet this criterion, but for those of analyst X

$$t_X = \frac{4.975 - 4.92}{0.05/\sqrt{6}} = 2.694 \text{ with } N_X = 5$$

Consulting Table 5.1, we can see that t_X lies between the 1% and 2.5% values for a one-sided test. The difference between X's result and the reference value is, therefore, significant at the 2.5% level and X's result would be rejected.

Comparison of Standard Deviations — the F-test

The *F*-test is used to test the consistency of the precision of two sets of

Table 5.2 Limits for F, 5 % level (2-way)

Degrees of freedom for denominator	Degrees of freedom for numerator		
	3	5	8
3	15.4	14.9	14.5
5	7.76	7.15	6.76
8	5.42	4.82	4.43

measurements obtained on different occasions or by different analysts or by different methods. The ratio of the variances of the sets of results

$$F = \sigma_x^2 / \sigma_y^2$$

is calculated and compared with tabulated values of the F-distribution (e.g. Table 5.2). F is always calculated so that it is greater than unity, i.e. with the larger standard deviation in the numerator.

It should be noted that Table 5.2 shows values of F corresponding to 5% confidence limits for deciding whether two variances are the same. Such 5% *critical points* are commonly found in tables giving 2.5% *points of the F-distribution*, e.g. Lindley and Miller (1962). This apparent anomaly arises from the convention that numerator and denominator are assigned so that $F > 1$. In the event of confusion when faced with tables labelled as 5% or 2.5% critical points, Table 5.2 may be used as a reference to indicate which level is meaningful for our purposes.

The values in Table 5.2 are also the 2.5% critical points for a one-sided test, i.e. if we are concerned not with whether two standard deviations are different, regardless of which is the larger, but with whether one specified standard deviation is larger than a second specified standard deviation.

Example Two independent analyses of a sample have the same mean but different standard deviations, 0.3 and 0.8 ppm. There were 9 replicates in the first analysis and 4 in the second. Is it reasonable to say that the two sets of measurements belong to the same population?

$$F = (0.8)^2 / (0.3)^2 = 7.11$$

The numerator has $4 - 1 = 3$ degrees of freedom and the denominator $9 - 1 = 8$ degrees of freedom. From Table 5.2, the limit for F at the 5% level is 5.42 for that combination of degrees of freedom, i.e. a ratio of more than 5.42 could be expected on only 5% of all such occasions. The hypothesis that the two sets of results belong to the same population cannot be held with 95% confidence and is rejected.

Criterion and Limit of Detection

It is often difficult to know what is the useful lower limit of an analytical method and in order to investigate this, a properly designed procedure must be followed. The analytical principles have been discussed by Roos (1962).

A *blank* is a portion of pure water to which the reagents are added in the same proportions as for the sample solutions and which is treated in exactly the same way as a real sample. A series of readings taken with blank solutions will be distributed about their mean and the blank will have a certain standard deviation σ_B. If a sample and a blank are analysed and the difference in the readings obtained exceeds 2.326 σ_B, there is less than one chance in twenty that the sample contains the same concentration of determinand as the blank. The *criterion of detection* is defined as the concentration corresponding to this difference of 2.326 σ_B and it is the concentration at which there is only a 5% chance of claiming that the determinand is present when it is not. Note that a sample containing a concentration equal to the criterion of detection will be detected in only 50% of analyses, as half the results will lie below the criterion of detection and half above. To be 95% confident of detecting the determinand, only 5% of the results may lie below the criterion of detection, and this occurs for a sample with a concentration 4.652 σ_B above the blank. This concentration is defined as the *limit of detection*.

The value of σ_B used in the above definitions must be obtained in realistic circumstances: successive replicates of the blank solution will give an underestimate of σ_B. A reasonable value of σ_B will be obtained by including a blank solution along with about four standard solutions in a precision test of the sort described in the following section. The standard solutions chosen should be fairly low in concentration and should not span too wide a range.

Within-batch and Between-batch Errors

It is generally observed that replicate analytical results agree more closely within the same batch of analyses than when the analyses are carried out in a number of separate batches, i.e. there are both *within-batch* and *between-batch* errors giving rise to within-batch and between-batch standard deviations, σ_w and σ_b, respectively. The total standard deviation (σ_t) of any one result in any one batch is

$$\sigma_t = \sqrt{\sigma_w^2 + \sigma_b^2}$$

General conclusions can be drawn from the relative magnitudes of σ_w and σ_b, thus if $\sigma_b \gg \sigma_w$, the calibration slope is probably changing from batch to batch. If $\sigma_w \gg \sigma_b$ contamination and measurement errors are likely important factors. When it is desired to detect small differences between samples or between a sample and a standard solution, the analyses should be performed in one batch, so that only the within-batch standard deviation is relevant.

Determination of within-batch and between-batch standard deviations

Analyse m batches of standard solutions by the procedure to be used for real samples. Each batch should contain four or five solutions, each of which should be analysed twice per batch. Take the solutions in random order, found from tables of random numbers, and never analyse the same solution twice in succession. For each standard solution there will, therefore, be a total of $2m$ e.m.f. readings. At least five batches are necessary. Care should be taken to see that all batches are at the same temperature, or else a between-batch error will be systematically introduced. Use a water bath to bring the standard solutions to the same working temperature.

Suppose a standard solution A gives pairs of millivolt readings (80.0; 80.2), (79.8; 79.6), (80.1; 80.4), (80.5; 80.0), (80.1; 80.2) over five batches. Calculate the within-batch sum of squares

$$M_0 = \Sigma d^2 / 2m$$

where d is the difference between the pairs of readings in the same batch. Thus in the example taken

$$M_0 = (0.2^2 + 0.2^2 + 0.3^2 + 0.5^2 + 0.1^2)/(2 \times 5)$$
$$= 0.043$$

M_0 has m, i.e. 5 degrees of freedom.

It will be seen in the analytical methods section that a standard solution is always taken as a reference solution. Suppose the readings observed with this solution were (90.0; 89.9), (89.9; 89.7), (90.3; 90.3), (90.2; 90.0), (90.0; 89.7). Calculate the batch means for standard solution A and the reference solution and take the difference, x_i, for each batch, e.g. in batch 1, $x_1 = (89.95 - 80.1) = 9.85$. Similarly $x_2 = 10.1$, $x_3 = 10.05$, $x_4 = 9.85$, $x_5 = 9.80$. Calculate the between-batch sum of squares

$$M_1 = \frac{2(\Sigma x^2 - (\Sigma x)^2/m)}{m-1} = \frac{2(493.0975 - 493.0245)}{4}$$
$$= 0.0365$$

M_1 has $m - 1 = 4$ degrees of freedom.

M_0 and M_1 are compared by the F-test to see if there is a significant between-batch standard deviation.

(a) $M_1 > M_0 : F = M_1 / M_0$

F is calculated and compared with the tabulated values ($m - 1$ degrees of freedom across and m down). If $F > F(5\%)$ the between-batch standard deviation is significant and is given by

$$\sigma_b^2 = 0.5(M_1 - M_0)$$

If $F < F(5\%)$ the between-batch standard deviation is non-significant, although it is not justifiable to say that it is zero.

In both cases, the within-batch standard deviation is given by $\sigma_w^2 = M_0$ and the total standard deviation by $\sigma_t^2 = 0.5(M_0 + M_1)$.

(b) $M_1 < M_0 : F = M_0 / M_1$

F is compared with tabulated values (m degrees of freedom across and $m - 1$ down). If $F < F(5\%)$, the best estimate of the between-batch standard deviation is zero and the total standard deviation is the same as the within-batch standard deviation.

$$\sigma_t^2 = \sigma_w^2 = M_0$$

If $F > F(5\%)$ there is probably an abnormal source of error in at least one of the batches and the analytical procedure should be critically examined.

In the example above, $F = M_0/M_1 = 1.18$, which is less then $F(5\%)$; therefore, $\sigma_b = 0$ and $\sigma_t = \sigma_w = 0.21$ mV.

TESTING ANALYTICAL METHODS

The fundamental properties of an analytical method are its accuracy, precision and selectivity. According to circumstances, other properties may also be significant, e.g. cost, the degree of skill required, the time taken for the analysis to be completed and suitability for automatic or *in-situ* analysis.

Precision

A method of determining the precision has been given above (see within-batch and between-batch standard deviations). Synthetic solutions may be prepared fresh for each batch, but a sample solution can only be included if there is a sufficient quantity for all the tests and contamination or depletion of the sample do not occur during storage. Note that the precision obtained with standard solutions may differ from that obtained with real samples, because the latter are more likely to contain interfering substances. The significance of any difference observed between the precision of analysing a sample and a standard solution of the same concentration can be checked by the F-test. If the difference is significant, examine the experimental procedure and check that there was no danger of contamination or depletion of the bulk sample, which may be indicated by a significant between-batch deviation for the sample when there is none for the standard solution.

Accuracy

It is difficult to gauge the accuracy of a method from measurements with standard solutions, as the calibration is defined with respect to the same, or very similar solutions. The results of analysing real samples may be compared with results from a reference method, but this is not always possible and a reference method need not be reliable merely because it is long established. A useful way of testing accuracy in real samples is by means of 'spiking' or recovery tests.

Recovery Tests

The tests are carried out with real samples analysed according to the relevant procedure. Before analysis, each sample is divided into two parts and a spike, a small volume or weight of a concentrated standard solution of determinand, is added to a known volume or weight of one of the portions of sample solution. The spike is chosen so that the concentration of determinand in the spiked sample is approximately twice that in the original sample. If necessary, spiking is delayed until the sample has been analysed.

For example, if a sample is analysed and found to contain 2.3 ppm of determinand, spike V_0 ml of sample with v ml of 2000 ppm standard. If analysis of the spiked sample gives 4.35 ppm, the recovery is given by:

$$R = 100 \left\{ 4.35 - \frac{2.3 V_0}{V_0 + v} \right\} \bigg/ \left\{ \frac{2000 v}{V_0 + v} \right\} \%$$

If $v \ll V_0$,

$$R \cong \frac{100(4.35 - 2.3)}{2000 v / V_0} \%$$

A more detailed analysis of recovery tests is given below.

Suppose n determinations of both spiked and unspiked samples are made in each of m batches. Calculate the recovery R as the difference between the results for n pairs of spiked and unspiked samples in each batch. Calculate the mean recovery, $\bar{R}$, from the mn results and the standard deviation, s, of the m mean recoveries for the m batches. Recovery is complete (with 95% confidence) if

$$\frac{|\bar{R} - 100|}{s/\sqrt{m}} < t_{10}$$

where t_{10} is the tabulated value of Student's t-test at the 10% level (two-sided) for $m - 1$ degrees of freedom. Note that the level to be consulted in the t-test is $2 \times (100 - Q)$ for $Q\%$ confidence.

In carrying out a test as above, attention should be paid to the nature of the sample. (a) If the sample is chemically stable, collect enough at one time to serve for all m batches. (b) If the sample is unstable, take a fresh sample for each batch. If the concentration does not vary much between batches (say, ±20%), add the same spike each time and treat the results as above. (c) If the sample is unstable and varies considerably in concentration from batch to batch, take a large sample and carry out m replicate analyses in a single batch. Calculate the mean of the m recoveries, the standard deviation of the recovery, s, and proceed as before, but note that this is now a within-batch statistic only.

Recovery tests should ideally be carried out over the entire likely range of sample concentrations as the recovery may depend on concentration. Recovery tests are appropriate to direct potentiometry and titrimetric methods, but not to know addition methods, as the assumption of 100% recovery is implicit in this technique.

Interpretation of abnormal recoveries

Abnormal recoveries may be caused by incorrect calibration of the electrode or by the presence in the sample of substances that react with the determinand. In the former case, re-calibration of the electrode should restore the recoveries to normal. The extent of any deviation from normality depends on the way the calibration is performed and particularly the concentration of the standard solution relative to the concentrations in the spiked and unspiked samples.

Incorrect calibration

The following errors may be caused by using incorrectly prepared standard solutions or by calibrating at a different temperature from that of the sample:

(a) Calibration slope overestimated   The recoveries will be low but the effect is worse if at least one of the concentrations in the spiked and unspiked samples is greater than that of the standard solution used for calibration.

(b) Calibration slope underestimated   The recovery is low if the concentrations in both the spiked and unspiked samples are below that of the calibrating solution but high if at least one is greater than the concentration in the calibrating solution. The high recoveries will show larger errors than the low ones.

(c) Standard potential too positive   The recovery will be low for a cation-selective electrode and high for an anion-selective one.

(d) Standard potential too negative   The recovery will be high for a cation-selective electrode and low for an anion-selective electrode.

Cases (*c*) and (*d*) assume the usual polarity for the electrodes, i.e. cation-selective electrode potentials become more positive and anion-selective electrode potentials more negative as the concentration increases. Calibration with incorrectly prepared standard solutions containing more than the nominal concentration of determinand is equivalent to effect (*c*) for a cation-selective electrode and (*d*) for an anion-selective one, while using standards containing less than the nominal concentration gives the opposite effect in each case.

Recoveries in samples containing complexing agents

The presence of substances that form complexes with the determinand generally causes low recoveries, except where there is a large excess of the substance and the electrode has been calibrated on the basis that a fixed proportion of the total determinand is present as free determinand.

Recoveries and interferences

Substances that interfere in the sense that they mask the determinand are dealt with above. Genuine interferences, i.e. substances, other than the determinand, that participate in the electrode reaction, do not affect the results of recovery tests with ion-exchange electrodes unless the selectivity coefficient changes with concentration and, conversely, recovery tests do not give information about interferences with these electrodes.

With solid-state electrodes, the presence of an interferent at concentrations that satisfy the relationship 3.12 can result in low recoveries. If control of the e.m.f. changed from the interferent to the determinand as sharply as predicted, spiking with exactly the indicated concentration of determinand should result in the recovery not of the spike but of the true concentration of determinand originally present. In practice, however, such ideal behaviour cannot be expected.

Interference Tests

Two types of interference must be distinguished: (1) that in which the interfering substance reacts with the determinand, e.g. by forming complexes or by oxidation or reduction, thus reducing the concentration of free determinand and hence giving a low result; (2) that in which a substance other than the determinand can also react at the electrode, with the result that the e.m.f. corresponds to a higher concentration of determinand than is actually present. Interferences of this kind are often described by means of a selectivity coefficient or selectivity constant, K_{MN}, as in equation 5.9

$$E = E^\circ + \frac{k}{z_M} \log (a_M + K_{MN}\, a_N^{z_M/z_N}) \tag{5.9}$$

where a_M and a_N are the activities of determinand and interferent, respectively, and z_M and z_N are their charges.

Such coefficients are most meaningful when applied to glass or liquid ion-exchange electrodes and may be misleading when applied to solid-state membrane electrodes or metal-based electrodes. Selectivity coefficients have to be obtained empirically and, as they often depend on the concentrations of determinand and interferent, their use in conditions other than those in which they were measured is not generally recommended, except as an approximate guide to the possible interference affects.

To determine the extent of an interference in an analytical method, calibrate the electrode with two standard solutions spanning the concentration range of interest and then determine the apparent determinand concentrations in two more standards that contain the same concentrations of determinand as before and also a known concentration of the suspected interferent. If the concentrations of the two standard solutions are c_A and c_B and the apparent concentrations of the solutions containing a concentration c_N of interferent are c'_A and c'_B, the interference effect at each concentration is $c'_A - c_A$ and $c'_B - c_B$. If the differences are small, it will be necessary to repeat the measurements about five times and use the t-test to find if the interference is significant. If the differences are significant and positive ($c'_A > c_A$), the selectivity coefficient may be calculated from

$$K_{MN} = (c'_A - c_A)/c_N \quad \text{and} \quad K_{MN} = (c'_B - c_B)/c_N$$

The test should preferably be repeated with a second concentration of interferent.

In many cases substances present in the sample are not expected to interfere and the above procedure may be unnecessarily elaborate. The interference test may be done at only the lower determinand concentration with a concentration of the suspected interferent greater than is likely to occur in the samples. If no interference effect is found in this case, further tests will be of little practical value.

For even greater convenience, similar substances may often be tested in the same solution, e.g. the alkali metal ions or the alkaline earth ions.

Blanks, Baselines and Reference Solutions

Reagent blanks

These solutions are used in many methods of analysis to correct for the effects of interferences or determinand introduced as impurities in any reagents added to the sample solution. The reagent blank contains the same concentrations of reagents as are added to the sample solution before measurement, but no determinand is added. All operations performed on the sample solutions should be repeated with the blank solution. Provided the quantity being measured is directly proportional to the concentration of determinand, as in titrimetric, photometric and amperometric techniques, the blank is easily applied by subtracting the absorbance, current etc., observed with the blank from those obtained with the standard or sample solutions in the same batch.

In direct potentiometry the blank is generally a misleading concept. At very low concentrations the Nernstian response of an ion-selective electrode is no longer valid and subtracting the e.m.f. for the blank solution from that for a sample solution gives no quantitative information, although if the difference is greater than 120 mV for a univalent electrode or 60 mV for a divalent electrode, the blank is less than 1% of the sample concentration. The concentration of determinand in the blank may be estimated by extrapolating a calibration graph obtained at higher concentrations, but this concentration will not necessarily enable results to be corrected for the effect of the blank. With many electrodes the blank is generated by dissolution of the materials forming the electrodes themselves and as the extent of the dissolution usually increases as the concentration of determinand in the solution decreases, the blank concentration obtained by extrapolation would over-compensate for the effect of dissolution. If the calibration of an electrode is Nernstian there is no need for a blank correction of any sort. Even if the e.m.f. $-\log c$ calibration is curved, applying a blank correction will be no better than using the curved calibration directly and in many cases will introduce further inaccuracy. An important proviso to the above is that the water used to prepare standard solutions should not introduce any determinand or interferences. This may be checked by preparing standards with water that has been evaporated to one-fifth to one-tenth of its original volume: involatile impurities will be concentrated and the curvature will be worse while volatile impurities will be driven off and the calibration will be less curved. In either case, further purification of the water is necessary before an accurate calibration graph can be obtained.

In the great majority of cases, reagent blank measurements are of no use in potentiometry and reagent blanks should not normally be included with batches of samples. An exception to this rule occurs when an electrode is being used near its limit of detection and the e.m.f. is directly proportional to concentration. The only common example of this is the low-level determination of chloride (p. 333).

Electrodes often take a long time to reach equilibrium in reagent blank solutions and inclusion of reagent blanks in batches of samples could take up a disproportionate amount of time.

Baselines

As the reagent blank is generally of little use in potentiometry, so the baseline is not a concept that can usefully be transferred from absorptiometry to potentiometry. Checking for drift in the calibration should always be done by means of measurements with a standard solution and the e.m.f. values observed with sample solutions should always be referred to the e.m.f. corresponding to the standard and not to a baseline.

Reference solutions

The e.m.f. observed when electrodes are immersed in a sample solution is never considered independently in direct potentiometry, but only with respect to a reference e.m.f. established by means of a standard solution. In order to minimize errors arising from inaccuracy of the calibration slope, the reference solution should have a concentration in the middle of the expected concentration range of the sample, but as standard solutions can be prepared more accurately at high concentrations, the reference solution should also be as concentrated as possible. As a good compromise, the reference solution should generally be in the highest 25% of the concentration range of the samples, but if samples are being analysed to check that they are within a certain control level, the reference solution should correspond exactly to that level. Note that recovery tests are best carried out with a reference solution more concentrated than the spiked samples, i.e. at least twice the highest expected sample concentration. In known addition methods and most titrimetric methods, no reference solution is needed.

Control Charts

A control chart is a means of keeping a running check on the performance of a method. First, a control parameter is chosen. This may be the concentration of a standard solution analysed with the samples or the recovery of a spike added to one of the samples. An initial estimate of the total standard deviation for the determination of the control parameter should be found from preliminary precision or recovery tests performed as above. The control chart is constructed by plotting the control parameter on the y-axis against the batch number or the date on the x-axis. The theoretical value of the control parameter, i.e. the concentration of the standard solution or 100% recovery is drawn as a horizontal straight line and a set of parallel lines is drawn ±2 σ and ±3 σ from the theoretical line; as each batch of samples is analysed, the value of the control parameter is found and plotted on the graph.

Five per cent of results should lie between the 2 σ and 3 σ lines on the plot but very few (0.3%) should lie outside the 3 σ lines. A result falling outside the 3 σ lines is an indication that a real change in the analytical conditions is likely to have occurred and should be followed by an investigation of the analytical procedure to reveal the cause of the change, e.g., the malfunction of a particular piece of apparatus, contamination of standard solutions, etc. The ±3 σ lines are sometimes called the *action limits*.

A result's falling between the 2 σ and 3 σ lines does not call for action, provided the next result is within the 2 σ line, but if results frequently lie in this region the chart indicates either (a) that a change has occurred causing a systematic tendency for results to be high or low (if only the +2 σ or −2 σ line, respectively, has been crossed), or (b) that a change has occurred causing the random error to increase (if both 2 σ lines are crossed randomly). The ±2 σ lines are sometimes called the *warning limits*, as action is taken only if they are persistently violated.

The estimates of the standard deviations used to define the action and warning limits should be revised periodically, in practice whenever a new chart has to be drawn up. The results recorded on each completed chart should be used, together with all previous results, to give a better estimate of σ. Results that lie outside the action limits and sets of results that persistently lie outside the warning limits are not included in the revised estimates, as they are by definition atypical of the true performance of the method.

CALIBRATION GRAPHS

Calibration graphs for ion-selective electrodes are obtained, with very few exceptions (see p. 121), by plotting the e.m.f. against the logarithm of the concentration of determinand. In the most useful part of any such calibration a straight-line graph is obtained, corresponding to the Nernst equation

$$E = E^\circ + k \log c$$

In practice, the calibration graph will always show some scatter about a perfect straight line, but for many purposes a line ruled through the points by eye will be sufficiently accurate. If greater accuracy is desired or if the precision of the measurement of E° and k is required, the data should be treated by the method of least squares to find the best straight line.

Linear Regression

If two variables x and y are related by means of the equation

$$y = a + bx$$

where a and b are constants and if the data consists of the set of points (x_1, y_1), $(x_2, y_2), \ldots (x_n, y_n)$, the best values of the intercept, a, and the slope, b, are given by

$$a = [n\Sigma xy - \Sigma x \Sigma y]/[n\Sigma x^2 - (\Sigma x)^2]$$
$$= [(XY)_n - X_n Y_n]/[(X^2)_n - X_n^2]$$

$$b = [\Sigma x^2 \Sigma y - \Sigma x \Sigma xy]/[n\Sigma x^2 - (\Sigma x)^2]$$
$$= [(X^2)_n Y_n - X_n(XY)_n]/[(X^2)_n - X_n^2]$$

where $X_n = \Sigma x/n$, $Y_n = \Sigma y/n$, $(XY)_n = \Sigma xy/n$ and $(X^2)_n = \Sigma x^2/n$, i.e. they are the mean values of x, y, xy and x^2, respectively.

The best estimates to the standard deviations of a and b have $n - 2$ degrees of freedom and are given by

$$\sigma_a = n\sigma_y/\{(n-2)[n\Sigma x^2 - (\Sigma x)^2]\}^{1/2}$$
$$= \sigma_y/\{(n-2)[(X^2)_n - X_n^2]\}^{1/2}$$
$$\sigma_b = n\sigma_y(\Sigma x^2)^{1/2}/\{n(n-2)[n\Sigma x^2 - (\Sigma x)^2]\}^{1/2}$$
$$= \sigma_y(X^2)_n^{1/2}/\{(n-2)[(X^2)_n - X_n^2]\}^{1/2}$$

where σ_y, the standard error of y upon x, is defined by

$$\sigma_y = \frac{1}{n}\{n\Sigma y^2 - (\Sigma y)^2 - [n\Sigma xy - \Sigma x \Sigma y]^2/[n\Sigma x^2 - (\Sigma x)^2]\}^{1/2}$$
$$= \{(Y^2)_n - Y_n^2 - [(XY)_n - X_n Y_n]^2/[(X^2)_n - X_n^2]\}^{1/2}$$

where $(Y^2)_n = \Sigma y^2/n$ is the mean value of y^2. Which of the alternative forms of each of the above equations should be used is a matter of personal convenience. For calculation by computer or programmable electronic calculator the equations in x etc. should be the more useful, but for 'long-hand' arithmetical calculations the forms in X_n etc. will be found less cumbersome.

When the above general formulae are applied to potentiometry,

$$y = E, \quad x = \log c, \quad a = E^\circ \quad \text{and} \quad b = k$$

Correlation Coefficients

The above procedure will calculate the parameters of a straight line whether or not x and y are in fact linearly related. In most cases in analytical chemistry, the linear relationship is obvious from inspection of the graph, but the *correlation coefficient*, r, provides a quantitative indication of the linearity. This may be particularly valuable if data are being processed by calculator or computer without being plotted on a graph. For a perfect straight line, $r = 1$ if the two variables increase together or -1 if one increases while the other decreases. If the two variables are randomly scattered, i.e. are not correlated at all, $r = 0$.

$$r = (n\Sigma xy - \Sigma x \Sigma y)/\sqrt{[n\Sigma x^2 - (\Sigma x)^2][n\Sigma y^2 - (\Sigma y)^2]}$$
$$= (\Sigma xy - nX_n Y_n)/\sqrt{[\Sigma x^2 - nX_n^2][\Sigma y^2 - nY_n^2]}$$

Testing for Linearity

Linearity may be tested by the following procedure, but certain conditions must be fulfilled: (1) There must be an equal number of observations (m) at each value of x.

(2) Duplicate observations at six different levels of x is the minimum amount of data to which the test should be applied. Natrella (1963) has described a more general procedure.

Consider m observations of y_{ij} at each of n equally spaced levels of x_i. The grand mean of y is

$$Y = \frac{1}{mn} \sum_{i=1}^{n} \sum_{j=1}^{m} y_{ij},$$

the mean of x is

$$X = \frac{1}{n} \sum_{i=1}^{n} x_i$$

and the mean of the observations of y at each level of x is

$$Y_i = \frac{1}{m} \sum_{j=1}^{m} y_{ij}.$$

Calculate the total sum of squares,

$$T = \sum_{i=1}^{n} \sum_{j=1}^{m} (y_{ij} - Y)^2,$$

the residual sum of squares,

$$R = \sum_{i=1}^{n} \sum_{j=1}^{m} (y_{ij} - Y_i)^2,$$

the linear regression sum of squares,

$$L = \left[\sum_{i=1}^{n} \sum_{j=1}^{m} (x_i - X)(y_{ij} - Y) \right]^2 \bigg/ m \sum_{i=1}^{n} (x_i - X)^2$$

and the higher terms sum of squares,

$$H = T - R - L.$$

Next calculate the variances

$$s_R = \frac{R}{n(m-1)} \quad \text{and} \quad s_H = \frac{H}{(n-2)}$$

with $n(m-1)$ and $n-2$ degrees of freedom, respectively. If s_H is not significantly greater than s_R a straight line may be used to describe the curve. If $s_H > s_R$, the F-test must be used to test the significance of the difference with $F = s_H/s_R$. As we accept as significant all $s_H \leq s_R$, the significance levels given in tables such as Table 5.2 are half the levels required for this test. Thus Table 5.2 gives the 2.5% level for the linearity test but the 5% level for the general case where there is no knowledge of which should be the higher deviation.

Bibliography

Probably the most useful book on statistical procedures is that by Natrella (1963) and among the very many books on the theory of statistics, Parratt (1961) and Hoel (1954) provide good introductions for the non-specialist. Tables of the statistical functions discussed in this chapter are included in most statistics texts, including those mentioned above, but the compilations of Fisher and Yates (1963), Selby (1970) and Lindley and Miller (1962) may be noted. Lindley and Miller's tables are particularly inexpensive and convenient for laboratory use.

Fisher, R. A., and F. Yates, 1963, *Statistical Tables for Biological, Agricultural and Medical Research*, 6th edn., Oliver and Boyd, Edinburgh.

Hoel, P. G., 1954, *Introduction to Mathematical Statistics*, 2nd edn., Wiley, New York and London.

Lindley, D. V., and J. C. P. Miller, 1962, *Cambridge Elementary Statistical Tables*, Cambridge University Press.

Natrella, M. G., 1963, *Experimental Statistics*, National Bureau of Standards Handbook 91, US Department of Commerce, Washington, D.C.

Parratt, L. G., 1961, *Probability and Experimental Errors in Science*, Wiley, New York and London.

Roos, J. B., 1962, 'The limit of detection of analytical methods', *Analyst*, **87**, 832.

Selby, S. M. (ed.), 1970, *Handbook of Tables for Mathematics*, 4th edn., The Chemical Rubber Co., Cleveland, Ohio.

Chapter 6

Potentiometric Titrations and Related Methods

Titrations offer a number of advantages over direct potentiometry: (a) They can determine the concentration of a species with greater precision, especially at high concentrations. (b) The experimental data are more directly related to the total concentration of determinand than in direct potentiometry, where the measured e.m.f. is a function of the activity of free determinand in solution and can only be related to the total concentration in carefully controlled conditions. (c) They may permit a substance to be determined accurately even in the presence of species that interfere at the electrode, provided that the titrant reacts selectively with the determinand. (d) Substances for which no selective electrodes exist can be determined by titrating them with a suitable electroactive species. (e) In general, less stringent requirements are laid on the electrode as regards stability of calibration slope and standard potential, with the result that electrodes that are unsuitable for direct potentiometric analysis may be adequate for titrimetry.

The titrimetric procedure also has its disadvantages. (a) Substances that do not interfere directly with the electrode may interfere with the titration because they compete with the determinand for the titrant. (b) The time taken for an analysis is much longer. (c) Trace analysis may not be possible. (d) Although titrations may be automated, genuine continuous monitoring is not possible. The problems of (b) and (c) may be considerably reduced by using a known addition or known subtraction procedure instead of a full titration.

Titrations can be classified by the type of reaction involved and by the mathematical treatment applied to the basic data, e.m.f. and volume of titrant added, in order to obtain a result with the desired degree of accuracy or convenience.

TITRIMETRIC PROCEDURES

Acid – base Titrations

These titrations can be subdivided into two categories: those where the acid or base to be determined is fully dissociated and those where it is not. The former can

Table 6.1 Precipitation titrations

Determinand	Titrant	Electrode	Comments
Ag^+, silver	KCl/KBr/KI	Ag	See bromide and iodide
Al^{3+}, aluminium	NaF	F	See *Determination of Aluminium*, p. 248
AsO_3^{3-}, arsenite	$AgNO_3$	Ag	Add NaOH to give pH 11
AsO_4^{3-}, arsenite	{ $AgNO_3$	Ag	Add NaOH to give pH 11
	NaF	F	Add excess of $La(NO_3)_3$ and back-titrate
Ba^{2+}, barium	{ $Pb(ClO_4)_2$	Pb	Add excess of Na_2SO_4 and back-titrate
	Na_2SO_4	Ba	
Br^-, bromide	$AgNO_3$	Ag	Make sample 0.1 mol l^{-1} in KNO_3 or $Ba(NO_3)_2$
Cs^+, caesium	$Ca(\phi_4 B)_2$	K	$\phi_4 B$ = tetraphenylborate, other alkali metals absent
Cl^-, chloride	$AgNO_3$	Ag	
CN^-, cyanide	$AgNO_3$	Ag	Add NaOH to give pH 11–12
CrO_4^{2-}, chromate	$AgNO_3$	Ag	Dilute 1:1 with methanol
F^-, fluoride	$La(NO_3)_3$	F	Dilute 1:1 with methanol
I^-, iodide	$AgNO_3$	Ag	Add KNO_3 or $Ba(NO_3)_2$ to 0.1 mol l^{-1}
K^+, potassium	$Ca(\phi_4 B)_2$	K	See caesium
Li^+, lithium	NH_4F	F	Dilute 1:1 with methanol
MoO_4^{2-}, molybdate	$Pb(ClO_4)_2$	Pb	Dilute 1:1 with methanol (cf. sulphate)
NO_3^-, nitrate	$(\phi_2 Tl)_2 SO_4$	NO_3^-	0.1 mol l^{-1} K_2SO_4 + 0.05 mol l^{-1} H_2SO_4
PO_4^{3-}, phosphate	{ NaF	F	Add excess of $La(NO_3)_3$ and back-titrate
	$Pb(ClO_4)_2$	Pb	pH 8.25–8.75 (NH_4OAc/NH_3 buffer)
Pb^{2+}, lead	Na_2WO_4	Pb	Dilute 1:1 with methanol
Rare earths	NaF	F	Add sample to standard NaF solution + methanol
Rb^+, rubidium	$Ca(\phi_4 B)_2$	K	See caesium
S^{2-}, sulphide	$Pb(NO_3)_2$	S	Add antioxidizing buffer, (p. 362)
SO_4^{2-}, sulphate	{ $Pb(ClO_4)_2$	Pb	Dilute 1:1 with methanol, (p. 366)
	$Ba(NO_3)_2$	Ba	
SCN^-, thiocyanate	$AgNO_3$	Ag	Bromide absent. Add KNO_3 or $Ba(NO_3)_2$
Thiols	$AgNO_3$	Ag	Feasible ?
Th^{4+}, thorium	NaF	F	
WO_4^{2-}, tungstate	$Pb(ClO_4)_2$	Pb	Dilute 1:1 with methanol (cf. sulphate)
ClO_4^-, perchlorate	$\phi_4 AsCl$	ClO_4^-	Tetraphenylarsonium chloride, 2 °C
Hg_2^{2+}, mercurous	KCl/KBr/KI	Hg	
Hg^{2+}, mercuric	$HgCl_2$	Hg or I	Add excess of KI and back-titrate

almost always be determined simply and accurately from the titration curve, but with weak acids or bases more involved calculations may be necessary if the analysis is to be accurate or even practicable. Usually the pH is measured with a glass electrode as the titration proceeds, but other pH-sensitive electrodes may also be used, e.g., the quinhydrone and antimony − antimony oxide electrodes.

Precipitation Titrations

The electrode may be sensitive to either the determinand or the titrant. In many cases, the titration curve is adequate for finding the equivalent volume, but, if the ions forming the precipitate are of unequal charge or if the solubility of the precipitate is not negligibly small, Gran plots or linear titration plots may have to be used. It is often practicable to reduce the solubility of the precipitate by adding a solvent of relatively low polarity to the sample, e.g. methanol, acetone or dioxane. Table 6.1 lists some of the commoner titrations.

Compleximetric Titrations

The electrode is sensitive to a metal ion and the ligand is normally chosen so that the complex formed is sufficiently strong for the titration curve to give the end-point directly. Table 6.2 shows some common titrations, the choice of chelating agents being very wide.

Redox Titrations

A platinum or other noble-metal electrode is used to measure the changes in redox potential as the titrant, which is either an oxidizing or reducing agent, is added to the solution of the determinand, which must be capable of being oxidized or reduced by the titrant. Table 6.3 lists some of the more reliable reactions for potentiometric titrations (Kolthoff and Furman, 1931; Davis, 1964).

Addition Titrations

In this type of titration a standard solution of the determinand is added to the sample; the electrode must be selective for the determinand. The titration curve does not contain an end-point, as no reaction takes place, but Gran plots and linear titration plots can be used to calculate the original concentration of determinand in the sample. The known addition method (p. 98) is a variation of this titration. Addition titrations can be applied to the determination of any species for which there is a suitable electrode, but particularly those, such as the alkali metals, that do not form strong complexes or sparingly soluble salts. Addition titrations can be used when a precipitation or compleximetric titration would be impracticable because of the presence in the sample of a second substance that would react with the titrant, e.g. phosphate in the titration of fluoride with lanthanum ion. It should

Table 6.2 Compleximetric titrations

Determinand	Titrant[a]	Electrode	Comments
Ba^{2+}	DCTA	Ba or Cu	Add known amount of $Cu(NO_3)_2$ or CuDCTA for competitive titration with Cu electrode
Ca^{2+}	EGTA	Ca or Cu	Add known amount of $Cu(NO_3)_2$ or CuEGTA for competitive titration with Cu electrode
Cd^{2+}	EDTA	Cd	Add 1 ml 2 mol l^{-1} acetate buffer per 100 ml sample
Co^{2+}	EDTA	Cu	Add $Cu(NO_3)_2$ or CuEDTA
Cu^{2+}	EDTA	Cu	
K^+	Kryptofix 222	K	Add triethanolamine to give pH 9–10
Li^+	Kryptofix 211	Na	See K^+. Na^+ interferes
Mg^{2+}	EDTA	W.H.[b] or Cu	Add known amount of $Cu(NO_3)_2$ or CuEDTA for competitive titration with Cu electrode
Mn^{2+}	EDTA	Cu	Add $Cu(NO_3)_2$ or CuEDTA
Na^+	DCTA	Na	Add piperidine to give pH 12.6
	Kryptofix 221	Na	Add triethanolamine to give pH 9–10
Ni^{2+}	TEPA	Cu	Add $Cu(NO_3)_2$ or CuTEPA
Sr^{2+}	EDTA	Cu	Add $Cu(NO_3)_2$ or CuEDTA
Zn^{2+}	TEPA	Cu	Add $Cu(NO_3)_2$ or CuTEPA
Chelating agents	$CuSO_4$	Cu	See p. 226

[a]DCTA (=CDTA) = *Trans*-cyclohexane-1,2-diamine-N,N,N′,N′-tetraacetic acid.
EDTA = Ethylenediamine-N,N,N′,N′-tetraacetic acid.
EGTA = 2,2′-Ethylenedioxybis[ethyliminodi(acetic acid)].
Kryptofix are a series of bicyclic diamines or 'cryptates' made by the Merck company (Czerwenka and Scheubeck, 1975).
TEPA = PETA = Pentamethylenediamine-N,N,N′,N′-tetraacetic acid.
[b]W.H. = Water hardness electrode.

be noted that addition titrations are subject to the same interferences as direct potentiometry and cannot be used to avoid them.

The calculation associated with Gran plots and linear titration plots can be eliminated if the titration is carried out in a way known as *null-point potentiometry*. In this technique there is no reference electrode, but two similar sensing electrodes, one immersed in V_0 ml of the sample solution of concentration c_0 and the other in V_s ml of either a standard solution of the determinand of concentration c_s or a blank solution containing no determinand. The two solutions are joined by a salt bridge or a porous frit. The blank or standard solution should be identical to the sample, except for the concentration of determinand; in particular, the pH should be the same in both halves of the system, or liquid junction potentials would cause a bias. Standard determinand solution of concentration c_t is added to the half of the system containing the lower concentration and the e.m.f. noted on each addition. The e.m.f. is plotted against the volume added and the volume, V_e, corresponding to equal concentrations in both halves, read off the

Table 6.3 Redox titration

Oxidant	Reductant	n^a	Titration conditions
$KMnO_4$	Arsenite, AsO_3^{3-}	1.5	Na_2CO_3
	Iodide, I^-	5	0.1 mol l^{-1} H_2SO_4
	Nitrite, NO_2^-	2.5	45 °C
	Nitrite, NO_2^-	2.5	Add nitrite slowly to solution of $KMnO_4$ (10% excess) + 0.75 mol l^{-1} H_2SO_4, then add excess KI and titrate with $KMnO_4$
	Sulphite, SO_3^{2-}	2.5	Add excess of $KMnO_4$ to alkaline sulphite, add H_2SO_4 to give 0.5 mol l^{-1} and excess KI, which is titrated with $KMnO_4$
	Ferrocyanide, $Fe(CN)_6^{4-}$	5	0.75 mol l^{-1} H_2SO_4
	Ferrous iron, Fe^{2+}	5	0.2 mol l^{-1} H_2SO_4
$K_2Cr_2O_7$	Ferrous iron, Fe^{2+}	6	>0.4 mol l^{-1} HCl or >0.2 mol l^{-1} H_2SO_4
$Ce(SO_4)_2$	Ferrocyanide, $Fe(CN)_6^{4-}$	1	>1 mol l^{-1} HCl or H_2SO_4
	Hydrogen peroxide, H_2O_2	0.5	0.5–3 mol l^{-1} HCl or acetic acid
	Iodide, I^-	1	
	Ferrous iron, Fe^{2+}	1	H_2SO_4
	Oxalic acid, $(COOH)_2$	0.5	Add 20 ml conc. HCl + 10 ml 0.005 mol l^{-1} ICl per 70 ml sample
KIO_3	Iodide, I^-	5	H_2SO_4
	Sulphite, SO_3^{2-}	2.5	1st step ⎫ Add sulphite to iodate in
		0.5	2nd step ⎭ 0.5 mol l^{-1} H_2SO_4
$KBrO_3$	Ferrous iron, Fe^{2+}	6	5% HCl
	Arsenious ion, As^{3+}	3	10% HCl
	Antimony, Sb^{3+}	3	5% HCl
	Iodide, I^-	6	1 mol l^{-1} H_2SO_4
KBrO	Thiosulphate, $S_2O_3^{2-}$	0.25	1 mol l^{-1} NaOH
Ferric iron, Fe^{3+}	Stannous tin, Sn^{2+}	0.5	HCl, 75 °C

$^a n$ = Number of moles of reductant reacting per mole of oxidant.

graph. Ideally, V_e should occur at 0 mV, but differences in manufacture may cause a small bias between the electrodes. This bias should be determined by placing both electrodes in the sample solution and noting the e.m.f. Null-point potentiometry does not depend on the slope of the electrode calibration and it may be particularly useful when working near the limit of detection of electrodes, when the slope is often non-Nernstian. Unlike Gran plots or linear-titration plots, no intermediate calculations are required. The concentration of determinand is calculated as follows, depending on the solution to which additions were made:

(a) Additions made to a blank solution

$$c_0 = \frac{c_t \cdot V_e}{V_0(V_s + V_e)}$$

(b) Additions made to the sample solution

$$c_0 = \frac{c_s(V_0 + V_e) - c_t \cdot V_e}{V_0}$$

(c) Additions made to a standard solution

$$c_0 = \frac{V_s c_s + V_e c_t}{V_0(V_s + V_e)}$$

A drawback of the method has been that only low-impedance electrodes could be used, because of the limitations of voltage-measuring instruments, but the development of differential high-impedance amplifiers means that this need no longer be the case.

Competitive Titrations

These are means of determining substances for which there are no ion-selective electrodes. The change in concentration of determinand is followed indirectly by an electrode sensitive to an ion – the indicator ion – that is added to the sample solution and is in equilibrium with both the determinand and the reagent, which may form complexes or sparingly soluble salts with the two other species. The procedure to be followed depends on the relative strengths with which the determinand and the indicator ion react with the common reagent. In practice, only compleximetric competitive titrations have been studied; these have been included in Table 6.2.

If the indicator ion forms the more soluble precipitate or the weaker complex, an excess of reagent is added to the sample solution and the concentration of the unreacted reagent determined by titration with the indicator ion. The difference between the reagent added and that unreacted gives the concentration of the determinand. An example is the determination of phosphate: an excess of lanthanum nitrate is added to the sample and the unprecipitated lanthanum ion titrated with fluoride, using a fluoride-selective electrode.

If the indicator ion forms the less soluble precipitate or the weaker complex, it is added to the sample solution and the mixture titrated with the reagent. One such titration is the determination of calcium using a copper-selective electrode: cupric ions are added to the sample solution, which is then titrated with EDTA. Two end-points appear, the first corresponding to the copper and the second to the calcium.

FINDING THE EQUIVALENT VOLUME

The Titration Curve

The simplest treatment of data is to plot the e.m.f. (or pH or pX) on the y-axis and the volume of titrant added on the x-axis; the curve has the characteristic S-shape and the end-point is the point of maximum slope on the curve. The advantage of this type of plot is its simplicity: no mathematical treatment of the data is required and the standard potential and calibration slope of the electrode do not need to be known.

The point of maximum slope does not, however, correspond exactly to the equivalence point; in acid—base, complexime tric and those precipitation titrations in which the ions are of equal charge it precedes the equivalence point and in precipitation titrations involving ions of unequal charge it occurs on that side of the equivalence point on which the ion of lower charge is in excess. The error increases as the reactants become more dilute and as the strength of the reaction decreases, i.e. the larger the solubility product in a precipitation titration, the weaker the acid or base in an acid—base titration and the weaker the complex in a compleximetric titration. Simultaneously with this increase in the systematic error in the conditions stated, the point of maximum slope itself becomes harder to locate. A theoretical treatment of different types of titration has been given by Meites and Goldman (1963; 1964) and Meites and Meites (1967).

The error caused by taking the point of maximum slope as the equivalence point in precipitation titrations involving ions of equal charge is less than 0.1% if

$$c \geqslant 100 K_s$$

where c = the concentration of determinand and K_s = the solubility product. If the ions are of unequal charge, a Gran titration or a linear titration plot should be used. In strong acid—strong base titrations, the error is negligible and in weak acid titrations is no more that 0.1% if at a concentration of 0.1 mol l^{-1}

$$pK_a \leqslant 9$$

The error increases with dilution, thus at 0.01 mol l^{-1}, the error for $pK_a = 9$ would be 1%. In compleximetric titrations, the error is less than 0.1% if

$$\beta c > 5000$$

where c = the concentration of determinand and β = the stability constant of the complex.

Titration to a Fixed Potential or pH

In this method the concentration of the indicator ion at the equivalence point is calculated from the equilibrium constant involved. The sample is titrated until the concentration of indicator ion is equal to the calculated value, when the volume of titrant is noted. It may also be possible to find the concentration at the equivalence point empirically, by titrating a standard solution of the determinand and noting the value of pH, pIon or e.m.f. at the theoretical equivalence point. It follows from the above that the electrodes must be calibrated as in direct potentiometry.

Strong acid—strong base titrations

The equivalence point occurs when [H$^+$] = [OH$^-$], which can be taken as when pH = ½pK_w, where pK_w is the negative logarithm of the autoprotolysis constant of water.

Weak acid—strong base titrations.

The pH at the equivalence point depends on the concentration of the acid and therefore the procedure is not rigorously accurate. For dilute solutions, pH_e, the pH at the equivalence point, can be approximated by

$$pH_e \simeq \tfrac{1}{2}(pK_w + pK_a + \log c)$$

where c mol l^{-1} is the concentration of the acid and $pK_a = -\log K_a = \log ([HA]/[H][A])$ is the dissociation constant. The approximation works best for moderately weak acids, where pK_a is in the range 3.5–6. The errors have been considered by Ricci (1952) and summarized by Mattock (1961) for this and the other weak acid and weak base tritrations considered below.

Titration of two weak acids (or a dibasic acid) with a strong base

Let the two dissociation constants be K_a and K_a' such that $pK_a' - pK_a > 4$ and the corresponding concentrations be c and c'. For a dibasic acid, K_a' is the second dissociation constant. The two end-point pH values are given by

$$pH_{e_1} \simeq \tfrac{1}{2}(pK_a + pK_a' + \log c - \log c')$$

$$pH_{e_2} \simeq \tfrac{1}{2}[pK_a + pK_a' + pK_w + \log(c' \cdot K_a + c \cdot K_a')]$$

For a dibasic acid, $c = c'$ and the expressions can be simplified; note that pH_{e_1} is independent of the concentration.

Tribasic acid—strong base titration

The titration curve will have three inflections, provided the ratios of the three values of K_{a_1}, K_{a_2}, and K_{a_3} are sufficiently large. The pH values at the end-points are as follows.

$$pH_{e_1} \simeq \tfrac{1}{2}(pK_{a_1} + pK_{a_2})$$

$$pH_{e_2} \simeq \tfrac{1}{2}(pK_{a_2} + pK_{a_3})$$

$$pH_{e_3} \simeq \tfrac{1}{2}(pK_{a_3} + pK_w + \log c)$$

Weak base—strong acid titrations

K_a is the dissociation constant of the conjugate acid of the weak base and $K_b = K_w/K_a = [B][OH]/[BOH]$ is the dissociation constant of the base. At the end-point

$$pH_e \simeq \tfrac{1}{2}(pK_a - \log c)$$
$$= \tfrac{1}{2}(pK_w - pK_b - \log c)$$

Precipitation titrations

If the precipitate formed has the composition A_xB_y and the electrode responds to ion A, which may be anion or cation, determinand or titrant, the activity of A at the equivalence point is given by

$$(x+y) \log [A] = \log K_s + y (\log x - \log y) = -(x+y) \text{pA}$$

where $K_s = [A]^x[B]^y$ is the solubility product. This assumes that there are no side reactions with either A or B.

Derivative Titrations

The point of maximum slope of the titration curve may be found more easily by plotting the first derivative curve, $\partial E/\partial V$ against V. In practice, the increments must be of finite size and the derivative curve is approximated by plotting $(E_j - E_i)/(V_j - V_i)$ against $0.5\,(V_j + V_i)$, where i and j represent successive additions of titrant. Automatic titrators are available that can transcribe the plot directly onto a chart recorder without the need for the analyst to calculate the coordinates. Although this procedure increases the accuracy of finding the point of maximum slope, it is still subject to the systematic errors arising from equating this with the equivalence point.

Gran Titrations

Gran (1952) devised ways of treating titration data that did not use the point of maximum slope of the titration curve. The data are transformed into functions which, when plotted against the volume of titrant added, give straight lines intersecting the V-axis at $V = V_e$, the equivalence point. There are two functions for each titration, one valid before the end-point and one after, except for addition titrations, where, as there is no reaction, only one function is required. The functions are summarized below; in each case the function, F, is plotted against the volume of titrant added, V, and the equivalent volume found from the intercept of the V-axis. V_0 is the volume of solution titrated, i.e. the volume of sample taken plus the volume of any reagent, such as a buffer solution or ionic strength adjustor, added before the titration; E is the e.m.f. observed and k is the calibration slope of the electrode. The functions can also be expressed in terms of a pH or pIon reading.

A. *Strong acid–strong base titrations*

1. Acidic side of the equivalence point

    $F = (V_0 + V)10^{E/k}$ or $(V_0 + V)10^{-\text{pH}}$

2. Alkaline side of the equivalence point

    $F = (V_0 + V)10^{-E/k}$ or $(V_0 + V)10^{\text{pH}}$

B. *Weak monobasic acid – strong base titrations*

1. Acidic side of the equivalence point

   $F = V \times 10^{E/k}$ or $V \times 10^{-pH}$

2. Alkaline side of the equivalence point – as A.2

C. *Weak monoacidic base–strong acid titrations*

1. Acidic side of the equivalence point – as A.1
2. Alkaline side of the equivalence point

   $F = V \times 10^{-E/k}$ or $V \times 10^{pH}$

D. *Precipitation titrations* $xA + yB = A_x B_y$

   (a) *Electrode responding to A (determinand)*

   1. Before the equivalence point

      $F = (V_0 + V)10^{E/k}$ or $(V_0 + V)10^{-pA}$

   2. After the equivalence point

      $F = (V_0 + V)10^{-xE/yk}$ or $(V_0 + V)10^{(x/y)pA}$

   (b) *Electrode responding to B (titrant)*

   1. Before the equivalence point

      $F = (V_0 + V)10^{-yE/xk}$ or $(V_0 + V)10^{(y/x)pB}$

   2. After the equivalence point

      $F = (V_0 + V)10^{E/k}$ or $(V_0 + V)10^{-pB}$

E. *Compleximetric titrations* $xA + yB = A_x B_y$

   (a) *Electrode responding to A (determinand)*

   1. Before the equivalence point – as D(a).1
   2. After the equivalence point

      $F = (V_0 + V)^{1-(1/y)} \times 10^{-xE/yk}$ or $(V_0 + V)^{1-(1/y)} \times 10^{(x/y)pA}$

   (b) *Electrode responding to B (titrant)*

   1. Before the equivalence point

      $F = (V_0 + V)^{1-(1/x)} \times V^{1/x} \times 10^{-yE/xk}$

      or

      $(V_0 + V)^{1-(1/x)} \times V^{1/x} \times 10^{(y/x)pB}$

   2. After the equivalence point – as D(b).2

F. *Redox titrations*

   (a) *Oxidation of determinand* $n_B \cdot A_{red} + n_A \cdot B_{ox} = n_B \cdot A_{ox} + n_A \cdot B_{red}$

   1. Before the equivalence point

   $F = V \times 10^{-n_A E/k}$

   2. After equivalence point

   $F = 10^{n_B E/k}$

   (b) *Reduction of determinand* $n_B \cdot A_{ox} + n_A \cdot B_{red} = n_B \cdot A_{red} + n_A \cdot B_{ox}$

   1. Before equivalence point

   $F = V \times 10^{n_A E/k}$

   2. After equivalence point

   $F = 10^{-n_B E/k}$

G. *Addition titrations*

   $F = (V_0 + V)10^{E/k}$ or $(V_0 + V)10^{-pA}$

In this case the equivalent volume is given by the negative intercept on the V-axis.

Gran plots are often written differently from the above because an arbitrary constant was originally included in the formulae to facilitate slide-rule calculation; thus wherever E or pX appear they may be replaced by $(E + E_c)$ and $(pX + pX_c)$, respectively, where E_c and pX_c are arbitrary constants. If a constant is included in the calculations, a convenient choice is a rounded value of E or pX near the equivalence point. In principle, readings on the direct-activity scale of a pIon meter can be used, so that in the above equations 10^{-pA}, 10^{pA}, $10^{-(x/y)pA}$ etc. can be replaced by $[A]$, $1/[A]$, $[A]^{x/y}$ etc. The disadvantage of using the direct-activity scale is that on some meters it has a restricted range and is not suitable for following the wide-ranging concentration changes observed during a titration. With experience, however, the analyst may come to recognize an e.m.f. as being a good starting point for the Gran plot, in which case the titrant can be slowly added until the starting e.m.f. (± 10 mV) is reached, when the meter is switched to the direct-activity scale and adjusted by means of the calibration control to either its maximum point (if the electrode responds to the determinand) or its minimum point (if the electrode responds to the titrant). Immediately after the equivalence point, which again may be recognized with experience, the scale is reset to its maximum or minimum value as before. The direct-activity scale is not recommended for titration of samples varying widely in composition, but it may be useful if similar samples are analysed routinely enough for the analyst to become familiar with the course of the titration.

A simpler way of treating the data for Gran plots of types A and D is to plot the results directly on semi-antilogarithmic graph paper, which is made by Orion Research Inc. One type of paper is not corrected for dilution as the titration proceeds, which will result in a curved plot. The other type is corrected for a 10% dilution at the equivalence point, which is commonly found in practice, but as the degree of dilution becomes increasingly different from 10%, the more curved the plot will be. It should also be noted that the antilogarithmic axis is strictly valid only for one value of the calibration slope (±58 mV per decade for univalent electrodes; ±29 mV per decade for divalent electrodes) and that error will be introduced if the actual calibration slope does not coincide with this value. The 10% corrected paper is nevertheless very convenient if the accuracy achieved meets the analyst's requirements and, if the sample concentration has an expected value, the titrant can be prepared so that the dilution will be close to 10%.

Although each point on a Gran plot requires more calculation than in the titration curve or derivative curve, far fewer points are needed and the titration can be finished more quickly, as six points are usually enough for one branch, i.e. before or after the equivalence point, of a titration plot. Gran plots are not subject to the errors involved in taking the point of maximum slope of the titration curve as the equivalence point, as they are fully corrected for dilution effects and are valid for precipitation titrations even when the precipitate is formed of ions of unequal charge. The derivations of the Gran functions are not completely rigorous and in certain circumstances the plots may be curved for the following reasons:

(1) Gran plots neglect variations in activity coefficients. McCallum and Midgley (1973) calculated the variations in activity coefficients for strong acid–strong base and precipitation titrations and showed that for most practical titrations the variation has a negligible effect if $V < 1.3 V_e$.

(2) Gran plots neglect minor components in the equilibria involved in the titration reactions, resulting in curved plots in very dilute solutions or when the reaction between determinand and titrant is comparatively weak. McCallum and Midgley (1973) calculated ranges for precipitation titrations within which the Gran function deviated from linearity, as summarized in Table 6.4. The range within which a given percentage error, δ, is found for a titration $xA + yB = A_x B_y$ in which the electrode responds to ion A is given by the formula

$$-\log [A] = \frac{-1}{x+y} \log K_s - \frac{y}{x+y} \log \frac{x}{y} \pm \frac{y}{x+y} \log \frac{\delta}{100}$$

where K_s is the solubility product. For strong acid–strong base titrations and precipitation titrations in which $-\log K_s \geqslant 5(x+y)$, Gran functions are generally reliable and in other cases a good answer can often be obtained by extrapolating the linear portion of the plot. The Gran plots for the titration of weak acids are more prone to curvature than the others, particularly function B.1 if $pK_a \leqslant 3$ and B.2 if $pK_a \geqslant 8$. Midgley and McCallum (1974; 1976) have discussed weak acid titrations and proposed better functions.

Table 6.4 Ranges of $-\log[A]$ for which the Gran function deviates by more than 1% from linearity in precipitation titrations $x A + y B = A_x B_y$

			$-\log K_s$	
x	y	14	10	6
1	1	6.0–8.0	4.0–6.0	2.0–5.0
1	2	3.1–6.2	2.0–5.1	0.7–3.8
1	3	2.3–5.4	1.3–4.4	0.3–3.4
2	1	4.0–5.1	2.7–3.8	1.4–2.5
2	3	1.6–4.3	0.8–3.5	0.0–2.7
3	1	2.8–3.9	1.8–2.9	0.8–1.9
3	2	2.0–3.5	1.2–2.7	0.4–1.9

Linear Titration Plots

By rigorously considering the charge and mass balance equations for all the components in the equilibrium system during the titration, linear functions can be derived that do not rely on the assumptions made by Gran and therefore deviate from linearity only because of experimental error. Such functions require more calculation than Gran's functions, but computers and electronic calculators have made this a less burdensome problem than was formerly the case. Ingman and Still (1966) developed some functions, although no allowance was made for activity coefficients. Midgley and McCallum (1974; 1976) and McCallum and Midgley (1973) have derived general formulae for acid–base and precipitation titrations, including titrations of polyfunctional weak acids and weak bases, and allowing for activity coefficients. This approach requires the knowledge of both the calibration slope and standard potential of the electrode, as well as all the equilibrium constants involved.

The following functions, when plotted against V, the volume of titrant added, intersect the V-axis at $V = V_e$, the equivalence point. Activity coefficients are shown in the functions, but an interative procedure is needed if they are included in the calculations. In strong acid–strong base titrations and precipitation titrations, neglecting the activity coefficients causes little error and they may be eliminated from the calculations by adding a relatively high concentration of an indifferent electrolyte to the sample, so that the ionic strength is kept constant. In the latter case, set all the activity coefficients in the following functions to unity, but note that the values of the standard potential E° and the equilibrium constants refer to the same ionic background. In contrast to Gran titrations, there is only one function for each type of titration.

Strong-acid–strong-base titrations

$$F = (V_0 + V)(10^{-pH} - K_w/10^{-pH})/f_H$$

where K_w = the autoprotolysis constant for water and f_H = the univalent ion activity coefficient.

Weak acid–strong base titrations

$$F = \frac{h(V_0 + V)(10^{-pH} - K_w/10^{-pH})/f_H + C_t \cdot V \cdot R}{h - R}$$

In the above, h is the number of protons per molecule of acid, e.g. for citric acid $h = 3$, for sodium dihydrogen citrate $h = 2$, for ammonium chloride $h = 1$; C_t is the concentration of the titrant and R is a function of the pH and the acid *association* constants

$$R = \frac{\sum_{1}^{n} i \cdot Q_i(10^{-pH})^i f_A/f_i}{1 + \sum_{1}^{n} Q_i(10^{-pH})^i f_A/f_i}$$

where n is the number of dissociation steps of the acid, $Q_i = ([H_iA] \, f_i/[A][H]^i f_A)$ is the ith overall *association* constant of the acid and f_A and f_i are the activity coefficients of the unprotonated form of the acid and the ith protonated form, respectively. The association constants Q_i are related to the stepwise dissociation constants as follows

$$\log Q_1 = pK_n, \log Q_2 = pK_n + pK_{n-1} \text{ etc.,}$$

where $pK_i = -\log K_i = -\log ([H][H_{i-1}A] \, f_{i-1})$ and $[H] = 10^{-pH}$.

Weak base–strong acid titrations

$$F = \frac{(n - h)(V_0 + V)(10^{-pH} - K_w/10^{-pH})/f_H + C_t \cdot V(R - n)}{h - R}$$

C_t, R, n and h are defined as for weak acid titrations, e.g. for ammonia $n = 1, h = 0$, for dipotassium hydrogen phosphate $n = 3$, $h = 1$. The terms Q_i are defined as before as the overall association, i.e. protonation, constants and are related to the stepwise dissociation constants of the conjugate acid of the base being determined, e.g. for phosphate solutions $\log Q_1 = pK_3$ for phosphoric acid, $\log Q_2 = pK_3 + pK_2$ etc. Sometimes the equilibria are expressed as hydrolysis reactions

$$AH_{i-1}^{(a+1-i)-} + H_2O = AH_i^{(a-i)-} + OH^-,$$

$$K_i^H = [OH^-][AH_i^{(a-i)-}]/[AH_{i-1}^{(a+1-i)-}]$$

where a is the negative charge on the base A^{a-}. The terms Q_i can now be related to the hydrolysis constants by

$$\log Q_i = ipK_w - \sum_{1}^{i} pK_i$$

Precipitation titrations

$$xA + yB = A_xB_y; K_s = [A]^x[B]^y$$

If the electrode responds to the ion A, whether it is the determinand or the titrant, the function is

$$F = (V_0 + V)\left[\frac{10^{-pA}}{f_A} - \frac{xK_s^{1/y}(10^{-pA})^{-x/y}}{yf_B}\right]$$

where $-pA = (E - E°)/k$ and f_A and f_B are the activity coefficients of the ions A and B.

Experimental Limitations

Acid–base titrations

Carbon dioxide Absorption of carbon dioxide from the atmosphere is a problem in all acid–base titrations, whether carried out potentiometrically or not. In the pH range 6–8 this absorption may cause drifting readings which make it difficult to obtain accurate e.m.f. values. In the titration of a strong acid with a strong base, the end-point may be so sharp that absorption of carbon dioxide is not important, otherwise a Gran plot or linear titration plot obtained in a pH range where absorption is negligible may be extrapolated to give the equivalence point. Provided that the acid or base being titrated is non-volatile, nitrogen may be bubbled through the solution during the titration, thus keeping the absorption of carbon dioxide to a minimum.

Non-ideal electrode response The quinhydrone and antimony–antimony oxide electrodes both show deviations from ideality above pH 8 and titration curves measured with these electrodes may be distorted. The hydrogen electrode is not suitable for the titration of aromatic acids or bases as they may be reduced at the electrode by the action of hydrogen gas. All these electrodes are also affected by oxidizing and reducing agents to some extent, with a loss of symmetry in the titration curve. All these difficulties can be avoided by using a glass electrode.

Precipitation titrations

Adsorption The shape of the titration curve is often distorted by adsorption of whichever ion is in excess. The extent of the adsorption can be reduced by increasing the ionic strength of the solution with an indifferent electrolyte, and potassium nitrate and barium nitrate are often used for this purpose, especially in the titration of halide ions with silver ions. Vigorous stirring can help to overcome this problem and sufficient time must be allowed for a steady e.m.f. to be observed. A Gran plot of the type D(a).2 or D(b).1 (see p. 90) i.e. when the indicator ion is not in excess, will usually give a more accurate result than other methods.

Co-precipitation The formation of some precipitates in the presence of other ions may result in the premature partial precipitation of these ions to an unpredictable extent giving a spuriously large value for the equivalent volume of titrant. This problem can be reduced by stirring the solution vigorously and adding the titrant as slowly as is conveniently possible. In some cases the addition of an indifferent electrolyte such as potassium nitrate may help, e.g. in the titration of mixtures of halides with silver ion. Usually the problem can only be assessed empirically by carrying out a titration with a synthetic standard solution containing all the ions present in the sample.

Moderately soluble precipitates In the titration $xA + yB = A_xB_y$, a reasonably sharp end-point will be observed in the titration curve if the concentration of determinand is greater than $100^{x+y}\sqrt{K_s}$. Gran plots show increasing errors as the concentration of determinand falls. Linear titration plots, however, are not affected by this problem, provided the solubility product is known accurately. The solubility can often be reduced by adding up to twice the volume of an organic solvent of relatively low dielectric constant, e.g. acetone, methanol, or by reducing the temperature at which the titration is performed.

Masking effects The formation of soluble complexes between either the determinand or the titrant and other ions present in the sample will affect the shape of the titration curve and may obscure the location of the equivalence point by reducing the size of the step in the titration curve and causing curvature in Gran and linear titration plots. It is sometimes possible to de-mask the determinand or titrant by adding a substance that will selectively precipitate or complex the masking agent without reacting with either the determinand or the titrant. The pH of a solution can often affect the result of the titration by changing the proportion of free anion present, e.g. in phosphate titrations, or of free metal, by the formation of metal hydroxy complexes, or by changing the composition of the substance precipitated, e.g. the formation of basic salts. The addition of an excess of strong acid or strong base, as appropriate, or of a suitable buffer solution may be required.

Interference effects During a precipitation titration, the free concentration of the ion to which the electrode responds will generally be very low on one side of the equivalence point. In such circumstances an interfering ion, i.e. one that gives an electrode response rather than one that takes part in a competing precipitation reaction, may dominate the potential. If this happens, the titration curve will be distorted, the potential jump at the end-point reduced and the point of inflection made more difficult to locate. Gran plots or linear titration plots would probably be useless on this side of the equivalence point, but could be used with data from the half of the titration where the indicator ion is present in reasonable concentrations. Liquid ion-exchange electrodes are particularly prone to errors arising from this cause, which, as Schultz (1971), has calculated can be of the order of several per cent. The error, which gives an underestimate of the equivalent volume, increases as the concentration of the indicator ion decreases, as that of the interfering ion increases, as the solubility product increases and as the degree of dilution during the titration increases. Similar problems are encountered in compleximetric titrations.

Limit of detection At very low free concentrations of indicator ion, dissolution of the ion-selective membrane of the electrode may contribute significantly to the concentration of free indicator ions in solution, in which case the e.m.f. will not be governed solely, or perhaps primarily, by the precipitation reaction. The effect on the titration curve is much the same as that of an interfering substance and the same considerations apply. In practice, the two effects will often act simultaneously.

Compleximetric titrations

Influence of pH As most ligands are acids or bases, the shape of the titration curve will depend on the pH of the solution and whether it varies during the titration. A more symmetrical curve will be obtained if the pH is constant, which may be achieved by adding a buffer that does not itself form strong complexes with the metal ion. The pH should generally be as high as possible without causing precipitation of the metal as its hydroxide or the formation of metal hydroxy complexes.

Interference effects The same considerations apply as for precipitation titrations. The errors involved have been discussed theoretically by Schultz (1971) and Carr (1972). Whitfield and co-workers (1969) treated calcium and magnesium titrations with a liquid ion-exchange electrode both practically and theoretically. The most practical approach has been to use the Gran function $D(a).1$ or $D(b).2$ (see p. 90) on *both* sides of the equivalence point. The resultant plot shows a break at the equivalence point but a more accurate answer is obtained from the intersect of the two lines drawn through the linear portions of the two arms of the plot (and *not* the intersect of the straight line through one arm of the plot with the V-axis, as in a normal Gran plot).

Limit of detection The problem is essentially the same as in precipitation titrations and will often be associated with interference effects also. The use of the Gran functions $D(a).1$ or $D(b).2$ on both sides of the equivalence point is recommended (Whitfield *et al.*, 1969).

Redox titrations

Effect of pH Many redox reactions are affected by the pH of the solution and the shape of the titration curve may therefore depend on the pH, which can usually be adjusted to a suitable value.

Effect of atmospheric oxygen Oxygen in the atmosphere may be capable of reacting with the reducing species; in this case, careful handling of the sample may be required, or the titrant will need to be restandardized at fairly frequent intervals, depending on which contains the reducing agent.

Irreversible redox couples. If the reduction or oxidation of one of the reactants occurs irreversibly the titration curve will have a different shape from that predicted theoretically by consideration of the standard redox potentials and the concentrations of the species present. The potential jump at the end-point will be smaller than if both the oxidation and reduction steps were reversible and the titration will therefore be less sensitive. Whether a particular titration is practicable can only be found empirically. An example of an irreversible redox couple is the sulphite−sulphate reaction.

SINGLE-POINT TITRATIONS

The number of readings taken during a titration increases the time of analysis compared with direct potentiometry, but by making only one addition of titrant and calculating the concentration of determinand from the observed change in e.m.f. the increase in time can be kept to almost negligible proportions. Single-point titrations are classified as *known* (or *standard*) *addition methods* and *known* (or *standard*) *subtraction methods*. The former corresponds to the earlier addition titration and the latter to the compleximetric or precipitation titrations. Although they are theoretically possible, single-point techniques are not normally applied to acid−base or redox titrations.

The conditions required for a successful application of single-point methods are as follows:

(1) The activity coefficient of the indicator ion should not vary significantly in the course of the titration.
(2) If the determinand forms complexes with other substances in the sample solution, the ratio of the concentration of free determinand to the sum of the concentrations of these complexes should not change significantly during the titration.
(2a) In the known subtraction method, the titrant must react much more strongly with the determinand than any substance originally in the sample
(3) In the subtraction method, the reaction equilibrium must so favour the product that each mole of titrant added can be considered to remove the same number of moles of determinand on all occasions, i.e. the reaction must be stoichiometric.
(4) The calibration slope, but not necessarily the standard potential, of the electrode must be known.

Known Addition

This technique consists of measuring the potential in a sample before and after the addition of a known concentration of the substance that is being determined. The practical advantage of this procedure is that it can be used to obtain concentrations of a determinand without reference to a calibration curve. In addition, it is especially suitable for measurements in samples that contain large amounts of complexing agents in equilibrium with the determinand or in a series of samples whose ionic strengths are unknown. In order to make a suitable addition, only a rough knowledge of the concentration of the determinand is required. Although a calibration curve is not required, it is still necessary for work of the greatest accuracy to obtain an accurate value of the slope constant, k, in the analytical conditions.

In any sample solution containing an unknown total concentration of determinand, c_T, the electrode will respond to the activity of the free ionic form of the determinand and its potential can be expressed by

$$E_1 = E° + k \log (f c_T \alpha) \tag{6.1}$$

where f is the activity coefficient and α is the fraction of the total concentration of the determinand present as free ions. If a volume, V_A, of a solution containing a known concentration, c_A, of determinand is added to the original volume, V_S, of the sample in which the electrodes are immersed, the new potential is

$$E_2 = E° + k' \log \left[f'\alpha' \frac{(V_S c_T + V_A c_A)}{(V_S + V_A)} \right] \tag{6.2}$$

where the primed terms k', f' and α' denote values of the previously defined terms in the second solution. If the electrode is responding in its Nernstian range, k and k' are equal and the difference between the potentials observed in the two solutions is

$$E_2 - E_1 = k \log \left[\frac{f'\alpha'(V_S c_T + V_A c_A)}{f \alpha c_T (V_S + V_A)} \right] \tag{6.3}$$

If the conditions are arranged so that $f = f'$ and $\alpha = \alpha'$, equation 6.3 can be simplified to

$$E_2 - E_1 = k \log \frac{(V_S c_T + V_A c_A)}{c_T (V_S + V_A)} \tag{6.4}$$

which can be rearranged to give an expression for the total concentration of determinand present in the sample

$$c_T = \frac{V_A c_A}{V_S \left[\left\{ \text{antilog} \left[\frac{E_2 - E_1}{k} \right] \right\} \left\{ \frac{V_A}{V_S} + 1 \right\} - 1 \right]} \tag{6.5}$$

The concentration change brought about by the known addition should be such that the difference in potential $(E_2 - E_1)$ is large enough to be determined accurately. On the other hand, if $(E_2 - E_1)$ is very large there is a greater possibility that f will not remain constant in both solutions and the best practical compromise is made by an addition which will approximately double the concentration of determinand in the sample. The initial concentration can often be estimated from E_1 and an existing calibration graph with sufficient accuracy for the volume and concentration of the known addition to be decided. If the volume of the reagent added is less than or equal to approximately 1% of that of the sample, the volume change can be ignored and equation 6.5 simplified to

$$c_T = \frac{V_A c_A}{V_S \left\{ \text{antilog}\left[\frac{E_2 - E_1}{k}\right] - 1 \right\}} \qquad (6.6)$$

The calculation involved in solving equation 6.6 for c_T can be simplified by the use of Appendix 4, in which values of the function

$$\frac{1}{\text{antilog}\left[\dfrac{E_2 - E_1}{k}\right] - 1}$$

are tabulated for a range of millivolt differences $|E_2 - E_1|$ and specified values of $|k|$ for singly and doubly charged ions. The total concentration, c_T, is obtained by multiplying the function obtained from the tables by the nominal change in the determinand concentration, i.e. by

$$\frac{c_A V_A}{V_S}$$

The success of the known addition method depends on the validity of the assumption that the terms α, f and k are constant throughout. In solutions containing no complexing agent, α is unity in both solutions and there is no problem. In general, however, $\alpha = 1/(1 + \Sigma \beta_i L_i)$, where L_i is the free concentration of the ith ligand and β_i is the stability constant of the complex between the determinand and the ith ligand, if α is to remain effectively constant, so must the values of L_i. In order to meet this condition, the dilution on making the known addition should be negligible, i.e., $V_S \geqslant 100 V_A$, and the quantity of ligand removed in forming a complex with the added determinand should also be negligible, i.e. $\beta c^* \leqslant 10^{-3}$, where c^* is the *molar* concentration of free determinand in the original solution and where the total concentration of determinand is approximately doubled by the addition. If the stability constant is large, $\beta \geqslant 10^3$, the latter condition implies that there is a thousandfold excess of ligand over total determinand. If dilution is negligible, $\beta c^* = 10^{-3}$ and $c_T \cdot V_S = c_A \cdot V_A$, c_T will be underestimated by 0.2% compared with 3% if $\beta c^* = 10^{-2}$ in otherwise identical conditions. An estimate of c^* may be obtained, by extrapolation if necessary, from a calibration graph prepared in the absence of ligand. If insufficient ligand is present in the sample for α to remain constant, more should be added before E_1 is measured.

An alternative procedure, which is particularly useful when complexation takes place simultaneously with several ligands, is to add to the sample an excess of a complexing species which forms stronger complexes than those naturally present in the sample; α will then depend predominantly on the free concentration of the added ligand. If V_R is the volume of the complexing agent added to the sample, the equation 6.5 is modified as follows

$$c_T = \frac{V_A c_A}{V_S \left[\left\{ \text{antilog} \left[\frac{E_2 - E_1}{k} \right] \right\} \left\{ \frac{V_S + V_A + V_R}{V_S + V_R} \right\} - 1 \right]} \quad (6.7)$$

Where the volume of reagent added, V_R, is less than approximately 1% of V_S, equation 6.7 simplifies to equation 6.5.

When an excess of ligand is added, it is possible to reduce the concentration of free determinand below the working range of the electrode and no useful results will be obtained: this is more of a problem with liquid ion-exchange electrodes than with solid-state membrane electrodes.

The activity coefficient f is dependent on the total ionic strength, I, of the sample (see Chapter 2) which will be increased by the known addition. However, the nature of the relationship is such that only in those samples where I is totally dependent on the determinand salt and also greater than approximately 10^{-3} mol l^{-1} will the errors be appreciable. The effect is greater the higher the charge on the ion, errors of approximately 15% occurring for divalent ions when the initial value of I in the sample is 10^{-2} mol l^{-1}. If the ionic strength of the sample is dominated by ions other than those in the salt of the determinand, f can be considered to be constant provided the addition is made with negligible dilution.

In order to solve equations 6.5, 6.6 or 6.7, it is necessary to use a value of k that is representative of the performance of the electrode in the analytical conditions. Theoretical values of k are not sufficiently accurate for most purposes and better values can be obtained from standard solutions covering the same concentration range as the samples and containing any additional ionic background that is common to the samples. The calibration slope may be obtained as part of the known addition procedure by measuring the e.m.f. after a dilution step following the known addition.

If the known addition technique has been followed and the solution in which the potential E_2 was recorded is diluted by a known volume, V_D, the potential is

$$E_3 = E^\circ + k \log \left[f \alpha \frac{(V_S c_T + V_A c_A)}{(V_S + V_A + V_D)} \right] \quad (6.8)$$

The change in potential dilution is

$$E_2 - E_3 = k \log \left[\frac{(V_S + V_A + V_D)}{(V_S + V_A)} \right] \quad (6.9)$$

assuming that α and f have the same values in both solutions. By rearrangement of equation 6.9.

$$k = \frac{E_2 - E_3}{\log \left[\frac{(V_S + V_A + V_D)}{(V_S + V_A)} \right]} \quad (6.10)$$

If the dilution volume V_D equals $V_S + V_A$, $k = (E_2 - E_3)/0.301$. This is a particularly useful dilution to use if the known addition approximately doubled the initial concentration of the determinand in the sample. A subsequent 1:1 dilution returns the concentration to its starting level, so that k is measured over the concentration range directly applying to the analysis; this is particularly advantageous if the calibration is non-linear in the concentration range of the sample. If reagent solutions were added to the sample before measuring E_1, the diluent must contain these in the same proportion as they were added to the original sample so that f and α will not change significantly on dilution. For the same reason, if other substances are present in samples at approximately constant concentrations, the diluent should be prepared so that it contains similar concentrations of these substances.

Known Subtraction

The technique is analogous to the known addition method described above except that a reagent added to the sample reduces the concentration of the ion being determined by a known amount either by precipitation or complexation. The theoretical expressions are similar to those for the known addition technique except that a negative sign prefixes the concentration term for the reagent added (see equation 6.2). If the values of α and α', f and f' are considered the same and the reaction is such that n moles of added reactant remove one mole of determinand, the change in potential following addition of reagent is

$$E_2 - E_1 = k \log \frac{(V_S c_T - n V_A c_A)}{(V_S + V_A) c_T} \qquad (6.11)$$

which can be rearranged to give an expression for the total concentration of the determinand present in the sample

$$c_T = \frac{-n V_A c_A}{V_S \left[\left\{ \mathrm{antilog} \left[\frac{E_2 - E_1}{k} \right] \right\} \left\{ 1 + \frac{V_A}{V_S} \right\} - 1 \right]} \qquad (6.12)$$

If the volume of the reagent added is less than or equal to approximately 1% of that of the sample, the volume change can be ignored and equation 6.12 simplified to

$$c_T = \frac{-n V_A c_A}{V_S \left(\mathrm{antilog} \left[\frac{E_2 - E_1}{k} \right] - 1 \right)} \qquad (6.13)$$

The extent of the concentration change brought about by the addition of the reagent should be such that the original concentration in the sample is approximately halved. The limitations brought about by changes in α, f and k are the same as those previously discussed with the addition that the reagent must bind the determinand more strongly than any species already in the sample. If k is not known, it is possible to obtain a value by a known dilution procedure following the known subtraction but this step is not as applicable as in the known addition technique, because each step involves a reduction in concentration of the ion being determined. The expression for k, following addition of a volume V_D of diluent and the measurement of the potential E_3, is

$$k = \frac{E_2 - E_3}{\log\left[\frac{(V_S + V_A + V_D)}{(V_S + V_A)}\right]}$$

Both the known addition and the known subtraction methods are subject to the same interferences as direct potentiometry. These interferences can be eliminated, if at all, only by a titrimetric method, unless there is a means of removing the interferent from the solution. Both methods theoretically require k to be constant over the range of concentrations involved in the procedures, but if the changes in concentration are kept to a factor of about two, the error in using an average value of k over that range will often be acceptable.

Bibliography

Carr, P. W., 1972, 'Intrinsic end-point errors in titration with ion selective electrodes', *Anal. Chem.*, **44**, 452.

Czerwenka, G., and E. Scheubeck, 1975, 'Zum analytischen Einsatz von Kryptaten für die massanalytischen Erfassung von Lithium, Natrium und Kalium in wässerigen Lösungen', *Z. Anal. Chem.* **276**, 37.

Davis, D. G., 1964, 'Potentiometric titrations', in *Comprehensive Analytical Chemistry, Vol. IIA*, C. L. Wilson and D. W. Wilson (eds.), Elsevier, Amsterdam, London, New York, Ch. 3.

Gran, G., 1952, 'Determination of the equivalence point in potentiometric titrations. Part II', *Analyst*, **77**, 661.

Ingman, F., and E. Still, 1966, 'Graphic method for the determination of titration end-points', *Talanta*, **13**, 1431.

Kolthoff, I. M., and N. H. Furman, 1931, *Potentiometric Titrations*, 2nd ed., Wiley, New York.

McCallum, C., and D. Midgley, 1973, 'Improved linear titration plots for potentiometric precipitation and strong acid–strong base titrations', *Anal. Chim. Acta*, **65**, 155.

Mattock, G., 1961, *pH Measurement and Titration*, Heywood & Co., London.

Meites, L., and J. A. Goldman, 1963, 'Theory of titration curves. Part I. The location of inflection points on acid–base and related titration curves', *Anal. Chim. Acta*, **29**, 472.

Meites, L., and J. A. Goldman, 1964, 'Theory of titration curves. Part II. Location of points of maximum slope on potentiometric heterovalent ("asymmetrical") precipitation titration curves', *Anal. Chim. Acta*, **30**, 18.

Meites, L., and T. Meites, 1967, 'Theory of titration curves. Part VI. The slopes and inflection points of potentiometric chelometric titration curves', *Anal. Chim. Acta*, **37**, 1.

Midgley, D., and C. McCallum, 1974, 'Improved linear titration plots for weak acid titrations', *Talanta*, **21**, 723.

Midgley, D., and C. McCallum, 1976, 'Linear titration plots for polyfunctional weak acids and weak bases', *Talanta*, **23**, 320.

Ricci, J. E., 1952, *Hydrogen Ion Concentration*, Princeton University Press, Princeton, N.J.

Schultz, F. A., 'Titration errors and curve shapes in potentiometric titrations employing ion-selective indicator electrodes', *Anal. Chem.*, **43**, 502.

Whitfield, M., and J. V. Leyendekkers, 1969, 'Liquid ion-exchange electrodes as end-point detectors in compleximetric titrations. Determination of calcium and magnesium in the presence of sodium. Part I. Theoretical considerations', *Anal. Chim. Acta*, **45**, 383.

Whitfield, M., J. V. Leyendekkers, and J. D. Kerr, 1969, 'Liquid ion-exchange electrodes as end-point detectors in compleximetric titrations. Part II. Determination of calcium and magnesium in the presence of sodium', *Anal. Chim. Acta*, **45**, 399.

Chapter 7

Potentiometric Analytical Practice

This chapter is intended as a general introduction to the individual analytical methods in Part II and its structure corresponds as far as possible to the arrangement of those methods.

ION-SELECTIVE ELECTRODES

No preference is stated for the products of any particular manufacturer, but in some cases it is clear that some electrodes have advantages over others for particular types of analyses and the results are allowed to speak for themselves. We have not been able to find any results in the literature for the products of several of the manufacturers listed in Appendix 3 and it is idle to suppress the fact that Orion electrodes account for most of the citations of commercial liquid ion-exchange and solid-state membrane electrodes. Sodium- and pH-sensitive glass electrodes are made by so many companies that a survey of the market is almost impossible. As a further complication, identical electrodes are often sold under different names and some are unavailable in certain countries because of patent laws or licensing agreements.

APPARATUS

Apart from the electrodes themselves and a pH meter, most of the methods require little or no special apparatus. Reference electrodes are discussed in Chapter 3, but it should be emphasized that the choice of reference electrode can have major influence on the precision and accuracy achieved in the analysis. pH meters are dealt with in Chapter 4 and the choice is usually a compromise between accuracy, convenience and cost. For the best precision and accuracy in the laboratory, a meter that can discriminate 0.1 mV is required for direct and known addition potentiometry, but for titrimetric methods 1 mV discrimination is usually sufficient. Portable meters are suitable for field trials, but their millivolt ranges are often insufficiently sensitive to yield a performance as good as can normally be obtained with a laboratory pH meter. With the use of an expanded millivolt or pIon range or a direct concentration range, however, portable pH meters can give good results, although they will need frequent recalibration if the concentration range of the samples is too wide to be accommodated on the expanded scale.

Stirrers, flow cells and pumps

It will be apparent from later sections that magnetic stirrers can cause temperature drift and electrical noise. A properly maintained stirrer should not create noise and one that does should be repaired or changed. As to the generation of heat, the larger stirrers with well-ventilated metal casings are the most satisfactory.

At concentrations approaching an electrode's limit of detection, better results will be obtained by using the electrodes in a flow cell and pumping the sample solution past them. The following features are required of a good flow cell: (a) When pumping ceases, the flow cell should not drain dry, leaving the electrodes exposed. (b) Solution from the reference electrode should not be able to reach the sensing electrode, even when the flow has stopped. (c) A new sample solution should rapidly displace the previous one. (d) The flow rates required should not be so high as to necessitate taking large volumes of sample or preparing large quantities of reagents. A flow cell with the desired characteristics can be machined out of Perspex (Figure 7.1) or can be obtained commercially, e.g. the Electronic Instruments Ltd model 24 8990 240. The latter has a water jacket so that the temperature can be controlled by means of a thermocirculator. Special flow-through end-caps are available for some kinds of ion-selective electrodes, so that no flow cell as such is required. Except for gas-sensing membrane electrodes, however, there remains the problem of housing the reference eletrode. One suggestion is

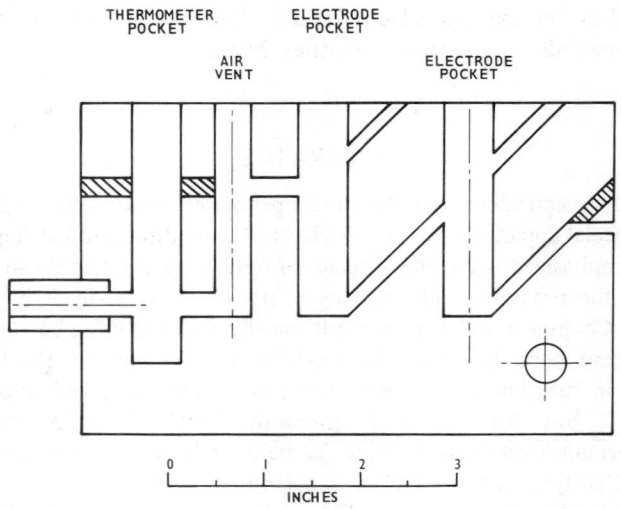

Figure 7.1 Flow cell for ion-selective electrodes

that the reference electrode be kept in a beaker containing the standard (or sample) and reagent solutions in the correct proportions and that the outlet tube of the flow-cap be immersed in the same solution, so that electrical contact is made. There is no need to change the solution in the beaker, it is merely displaced by each successive sample. The beaker should have as small a capacity as possible and be allowed to overflow into a catchpot or to a drain. Flow-caps of this sort require small (<1 ml min^{-1}) flow rates, but they are prone to noise caused by the passage of air bubbles across the membrane or through the narrow tube connecting the sensing and reference electrodes, whereas in the immersion type of flow cell these problems can be largely avoided. The flow-cap may be of advantage if contact with the atmosphere is undesirable.

Many pumps are suitable for use with flow cells, but one that can pump at least two channels simultaneously at different rates — the greater for the sample and the lesser for the reagent solution — is very convenient. Peristaltic pumps are most frequently used. The EIL series 89 Laboratory Monitor incorporates a flow cell and two peristaltic pumps in the same housing.

Laboratory ware

Most of the methods in Part II require only common laboratory glassware such as beakers, flasks and pipettes, although in some cases plastic containers are necessary. The use of pipettes for measuring volumes of sample and standard solutions is essential in titrimetric and known addition methods and is recommended for the best precision and accuracy in direct potentiometry, but in the latter case measuring cylinders are sufficiently accurate for most purposes. Suppose 5 ml of ionic strength adjustment solution are added to 49 ml of sample instead of 50 ml: the difference in concentration of determinand in the final solution caused by the 2% error in sample volume is only 0.17%. If a smaller volume of more concentrated reagent had been added, the final error would have been even smaller. The discrimination of the best pH meters is 0.1 mV, which is equivalent to about 0.4% error in concentration for a univalent ion and 0.8% for a divalent ion or, for 10:1 mixtures of sample and reagent solutions, 5% and 10% errors in the volume of sample dispensed. There should be no difficulty in avoiding errors of this size. If a particular analysis is carried out regularly, the reagents added can often be delivered more conveniently by an automatic dispenser. Many dispensers are capable of excellent reproducibility (0.1% at 5 ml) but some may need calibrating before a specified volume can be delivered accurately. In direct potentiometry the reproducibility of the additions is generally more important than their accuracy and dispensers should be satisfactory for most purposes.

In a number of methods, precise additions of small volumes ($\leqslant 1$ ml) are recommended, e.g. in a known addition procedure. For volumes $\geqslant 0.1$ ml a grade A micro-burette with 0.02 ml divisions should be used and for volumes <0.1 ml a micrometer syringe is recommended. The latter consists of a modified engineering micrometer whose scale measures the distance of travel of the piston of a syringe which is attached to the moveable stem of the micrometer. It is essential to

Table 7.1　Density of water from 10 to 29 °C

°C	Density g ml^{-1}	°C	Density g ml^{-1}	°C	Density g ml^{-1}	°C	Density g ml^{-1}
10	0.99 973	15	0.99 913	20	0.99 823	25	0.99 707
11	0.99 963	16	0.99 897	21	0.99 802	26	0.99 681
12	0.99 952	17	0.99 880	22	0.99 780	27	0.99 654
13	0.99 940	18	0.99 862	23	0.99 756	28	0.99 626
14	0.99 927	19	0.99 843	24	0.99 732	29	0.99 597

calibrate the micrometer scale by weighing the water expelled from the syringe for a given distance of travel and using the density (Table 7.1) to calculate the volume.

SAMPLE COLLECTION

The general aims of sampling have been summarized by Wilson (1974) as follows:

(a) To obtain a sample in which the concentrations of determinands are identical to those in the water of interest at the time of sampling.
(b) To repeat this sampling operation at times such that the required information or any changes in water quality is obtained.
(c) To ensure that the concentrations of the determinands in the samples do not change between sampling and analysis.

The major problem in obtaining a sample that is representative of the water of interest is in identifying the correct location of the sampling point. For example, if it is necessary to know the levels of a substance being discharged from a known source into a river, the sampling point can only be located in the discharge stream. Should it be necessary to know the level of this substance in the river, the correct siting of the sampling point becomes much more difficult, as characteristics of the river such as velocity profiles and turbulence patterns must also be known (FWQA, 1970). Similar problems can arise within complex industrial plant, with the added difficulty that it may be necessary to withdraw the sample from a closed vessel or pipe requiring a specially engineered outlet. Sampling procedures for pressurized systems such as steam-raising plant are highly specialized and require careful consideration (ASTM, 1973; BSI, 1969).

The technique used to collect the sample can be as important as the location of the sampling point in obtaining representative samples. In every case it is essential to have a fixed sampling procedure which can be easily understood, as it is common for samples to be collected by personnel other than trained laboratory staff.

Samples can be taken from open water by simply filling a wide-mouthed container from just under the surface, but special apparatus (Hutchinson, 1957; BSI, 1969; DoE, 1972) is needed if samples are required from present depths. For determinands that can be gained from, or lost to, the atmosphere, e.g. carbon dioxide and ammonia, or that can be oxidized, e.g. sulphide, it is advisable to take the sample from well below the surface, as the surface layers will not be

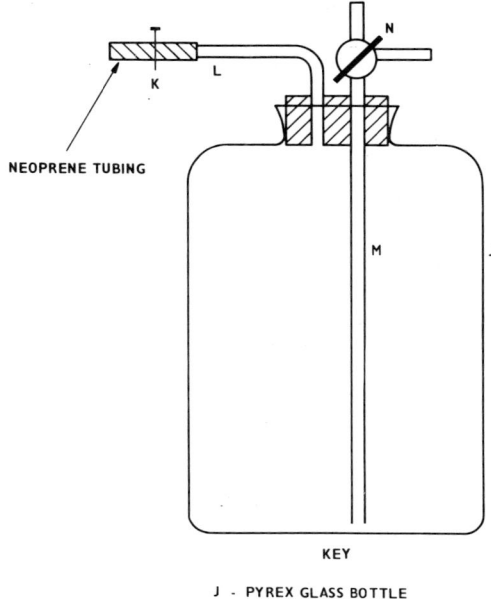

Figure 7.2 Sampling apparatus for volatile determinands

KEY

J - PYREX GLASS BOTTLE
K - SCREW CLIP
L - OUTLET TUBE
M - INLET TUBE
N - 3-WAY STOPCOCK

representative of the bulk; the sample container should be completely filled and preferably sealed under water.

When samples are taken from closed systems such as industrial plant, consideration should be given to the length of sampling line. If the system is being sampled continuously few problems need arise, but before occasional samples are taken, the line should be flushed with about five times its own volume of solution before any is collected, so that nay stagnant and, therefore, unrepresentative solution resident in the line since the last sample was taken is not collected. If the sample has to be protected from the atmosphere, the apparatus shown Figure 7.2 should be used as follows.

Purge the empty bottle with a stream of nitrogen for 10 minutes, then close the outlet screw clip against the nitrogen flow and turn the inlet three-way tap to bypass the nitrogen. Disconnect the bottle from the nitrogen supply and connect to the sampling point by the shortest possible length of neoprene or other suitable tubing. Allow the water to flow to waste via the three-way tap until a volume of water approximately five times the capacity of the sampling line has passed. Rotate the tap to connect the bottle to the sampling line and release the screw clip on the outlet tube, but do not open it fully. Fill the bottle and allow about 100 ml (for a

1-litre bottle) to flow to waste from the outlet. Rotate the three-way tap to the bypass position and close the outlet screw clip.

Samples should always be analysed as quickly as possible, but this is especially important in the case of determinands that are volatile, liable to oxidation, subject to biological production or degradation or easily adsorbed on the walls of the container. It is nevertheless impossible always to analyse immediately and ways of storage should be considered. In the first three cases above, storage in a deep-freeze will often be effective (plastic containers are essential) or preservative solutions may be added, e.g. acid to prevent the volatilization of ammonia, ascorbic acid to prevent the oxidation of sulphide and biocides to inhibit bacterial activity, as with nitrate solutions. Adsorption can sometimes be prevented by adding a complexing agent to the sample; this usually means that the analysis has to be completed by a known addition method. Before the storage of samples is accepted as routine, a test should be performed in which typical samples are analysed immediately and then after various periods of storage, so that the effects of storing them are known.

The volume of sample required for each analytical method is largely determined by the bulk of the electrodes and the convenience of handling by standard pipettes. The volume recommended is usually 50 ml, but it should be possible to scale this down easily to 10 ml by using either a combination sensing and reference electrode or a normal sensing electrode in conjunction with a remote reference electrode fitted with a micro-junction tip. The volume required may be reduced to less that 1 ml by the use of special dishes (Orion Research Inc., model OR920014), but no stirring is possible and any problems of volatilization or oxidation will be multiplied; these dishes are therefore most suitable for fairly concentrated solutions, but even then are not to be preferred to normal methods of working.

CONDITIONING AND STORAGE OF ELECTRODES

In the instructions supplied with electrodes, manufacturers recommend methods of storage and state whether the electrode should be conditioned before use. It is advisable to follow these instructions, although it may be found that they can be modified to suit individual circumstances. Some general rules apply, however, according the nature of the electrode:

1. No glass electrode should be allowed to dry out if possible as reconditioning may not be successful.

2. Solid-state electrodes can either be stored dry, with their membranes covered by protective plastic caps, or in deionized water. For long periods of storage (weeks) the former method is preferable. Once these electrodes have been rinsed with water following a period of storage they are generally ready for use, although some manufacturers recommend a short conditioning period in a solution ($\sim 10^{-3}$ mol l^{-1}) of the ion to which the electrode responds. Impregnated graphite electrodes (p. 27) are slightly different in that all but the silver electrodes have to be conditioned before use and the divalent metal electrodes should not be stored dry.

3. The instructions for storing liquid ion-exchange electrodes (including plastic

membrane electrodes) vary according to the manufacturer. Most electrodes can be stored dry overnight, but there is a danger that the membrane will lose its conducting properties because of evaporation of the solvent if it is stored in this manner for long periods (or for short periods at high temperature). The sensing modules of Orion series 93 electrodes should be detached from their electrode bodies and returned to their storage capsules between use. Orion series 92 electrodes can be stored for a number of days in a small volume of deionized water or in a 10^{-3} mol l^{-1} solution of the ion to which the electrode responds. The performance of liquid ion-exchange electrodes is improved after a period of dry storage if they are immersed in a standard solution ($\sim 10^{-3}$ mol l^{-1}) for about 10 min before being used.

4. Gas-sensing membrane electrodes should be assembled according to the manufacturer's instructions and conditioned as described in the relevant analytical procedures. As their membranes should never be allowed to dry out, these electrodes should be stored with their tips immersed in a portion of internal filling solution or in deionized water plus any reagents added during the appropriate analytical procedure.

5. The fact that the condition of the reference electrode is as important as that of the sensing electrode is often overlooked. Fortunately, the design of most laboratory reference electrodes is such that no pretreatment is necessary if the electrodes have been correctly stored. Reference electrodes are received from their manufacturers filled with an internal reference solution and are ready for use following a rinse with deionized water to remove any electrolyte that has accumulated on the outer surface of the liquid junction. In electrodes having a refillable reservoir, the filler hole will be sealed with a rubber bung or cap to prevent loss of solution during transit. It is necessary either to remove the bung or to pierce the cap with a pin so that the flow of solution through the liquid junction is not restricted.

6. The commonest fault with reference electrodes is having insufficient solution in the reservoir; the level should be inspected frequently and topped up. If the level of the reference solution falls below that of the element there is a danger of the element's drying out. This is not serious if it is a silver—silver chloride electrode since it will recover in minutes when re-immersed in solution. In calomel or mercury—mercurous sulphate electrodes, however, the mercurous salt can dry to form a hard layer which is difficult to rehydrate. The best treatment is to fill the reservoir and stand the electrode in water at 40—50 °C for a few hours. In many cases the electrodes never recover and have to be replaced and even if the treatment is successful the electrodes may take some time to recover from the high temperature. Check the e.m.f. of the electrode against another reference electrode to see that the treatment has worked and that stability has been established.

7. Reference electrodes can be stored for a few days in deionized water. The main precaution is to make sure that the element is covered with reference solution at all times to avoid drying out. It is also advisable to have the level of solution outside the electrode below that in the electrode reservoir or else diffusion of the external solution into the reference solution will occur. Filler ports should be sealed

even for short periods of storage, particularly for electrodes with large leakage rates, and for longer periods of storage, junctions should be kept wet by covering them with small rubber or plastic teats containing deionized water. When the electrode is returned to service, the filler port should be opened and the junction rinsed with water; a few drops of reference solution should be allowed to flow through sleeve-type junctions before re-use.

CONCENTRATION RANGE AND UNITS

Estimates are given of the range of concentration over which each ion-selective electrode is expected to have a Nernstian response when it is in good condition and also of the useful non-Nernstian range. The Nernstian range for any individual electrode should be found by preparing a calibration graph, as even electrodes of the same manufacture may have considerably different Nernstian ranges. It is also fairly common for the Nernstian range of liquid ion-exchange electrodes to change with time. The lower limit of response of ion-selective electrodes is a rather ill-defined concept: as the methods for determining chloride show, if sufficient trouble is taken to control the temperature very precisely in a flowing system, it is possible to determine concentrations that would be unattainable by the usual technique involving beakers of solution at room temperature. The upper limit is sometimes set by the solubility of the determinand, but the membrane may tend to dissolve in concentrated solutions of determinand, e.g. silver chloride in strong chloride solutions. A linear relationship between the e.m.f. and the logarithm of the concentration will not be observed at high concentrations, as the assumptions in Chapter 2 about the constancy of activity coefficients will be invalid. Deviations from linearity can be expected at 0.1 mol l^{-1} or even lower, depending on the nature of the determinand. Concentrated solutions are usually best analysed after dilution.

The units adopted are those in common analytical use – ppm, ppb and % – rather than the more formal ones. Molar concentrations are used where appropriate, e.g. in discussing activity coefficients, solubility products, equilibrium constants, selectivity coefficients etc. The relationships between the units are given below, where W is the molecular weight of the determinand.

$$1\% \equiv 10 \text{ g l}^{-1} \equiv 10 \text{ mg ml}^{-1} \equiv 10^4 \text{ ppm} \equiv 10/W \text{ mol l}^{-1}$$
$$1 \text{ ppm} \equiv 1 \text{ mg l}^{-1} \equiv 1 \text{ } \mu\text{g ml}^{-1} \equiv 10^3 \text{ ppb} \equiv 10^{-3}/W \text{ mol l}^{-1}$$
$$1 \text{ ppb} \equiv 1 \text{ } \mu\text{g l}^{-1} \equiv 1 \text{ ng ml}^{-1} \equiv 10^{-3} \text{ ppm} \equiv 10^{-6}/W \text{ mol l}^{-1}$$
$$1 \text{ mol l}^{-1} \equiv 10^3 \text{ } W \text{ ppm} \equiv 10^6 \text{ } W \text{ ppb} \equiv 0.1 \text{ } W\%$$

ANALYTICAL PROCEDURES

The analytical procedures in Part II are based on published analytical results where possible and any exceptions are clearly indicated. It cannot be hoped that the procedures will suit every type of water sample, nor that all electrodes with a selectivity for a particular ion will give equally good results in the conditions laid down. The reliability of any method will be improved by adopting a more rigorous control of temperature, by including more standard solutions in each batch of analyses and by carrying out replicate analyses on each sample. The analyst himself must decide on the balance between accuracy and precision, on the one hand, and cost and convenience on the other. Only experience can show what proportion of the analyst's time should be spent testing and calibrating the analytical system, but one solution in five is not too many to devote to this purpose (including all measurements with standard solutions and spiked samples) and one in ten is probably too few, especially during the first period of a method's application. The methods will always work best when the determinand is dissolved in what is otherwise virtually pure water and when concentrations are neither very high nor very low. The following discussion is restricted to general points of technique of relevance to almost all the methods in Part II. Individual problems are discussed in the methods themselves.

Temperature

Control of the temperature of both the solutions being analysed and the electrode assembly is essential for accurate potentiometry. Ideally, all analyses should be carried out at the same temperature, but this is not generally possible and it is useful to consider methods of reducing temperature-induced errors in a typical analytical laboratory. Standard and sample solutions can readily be brought to a constant temperature by immersing them in a thermostatically controlled water bath. It is undesirable to have the water-bath at a temperature greatly different from ambient, as thermal gradients will be set up when the electrodes are immersed in the solution. Preferably, the bath should be at the mean ambient temperature, but unless the bath has a cooling circuit this is impracticable; a temperature about 2 °C above the mean is a good compromise. The above precautions may not be adequate if the ambient temperature is subject to fairly large and rapid fluctuations, even though the mean temperature is steady; such conditions may occur close to large air-conditioning units and their effects can be reduced by housing the electrode assembly in an insulated cabinet. In laboratories in which there is a large change in temperature during the day, electrodes should be recalibrated fairly frequently to correct for any changes in standard potential and calibration slope, even if all the above measures are applied. The care taken in handling solutions is largely wasted, however, if the magnetic stirrer produces enough heat to change the temperature during the measurement of e.m.f. (see *Apparatus*). Temperature problems may arise from the heat generated by treating the sample with reagents prior to analysis, but for most purposes only the neutralization of acids or bases more concentrated than ~0.05 N need be considered.

The size of any errors depends not only on the differences in temperature but on the analytical procedure and the nature of the sample:

1. When a permanent calibration graph is used, the sample and standard solutions ahould not only be at the same temperature as one another but at the temperature at which the calibration graph was obtained.

2. The error caused by a given temperature difference, whether between sample and standard solutions or between both solutions and the calibration graph, will increase as the concentration difference between sample and standard solutions increases.

3. The various makes of ion-selective electrodes may have different temperature coefficients because of internal differences.

4. The various types of reference electrodes have different temperature coefficients and are affected to differing extents by temperature hysteresis (see Chapter 3). The reference electrode can be made almost immune to the effects of temperature changes in the test solution by using it with a remote liquid junction, but this will not affect its susceptibility to changes in ambient temperature.

5. All pH meters have some means of correcting for changes in calibration slope, but only on the pH, pIon or direct activity scales and not on the millivolt, relative millivolt or expanded millivolt scales. Relatively few pH meters have the ability to correct for the change in standard potential with temperature and some of these may work only for specific combinations of sensing and reference electrodes; so far this type of correction has been applied only to pH measurements. For accurate work using a pIon or direct activity scale, the slope/temperature control on the pH meter should be used only to achieve the correct calibration slope with standard solutions (see the section on calibration for the procedure) and it cannot generally be used to compensate for any variation from the temperature of calibration.

6. Temperature changes affect equilibria and so may cause changes in the concentration of free determinand in solution. Even an ideal correction for the changes in standard potential and calibration slope would not allow for this effect, which occurs most commonly when the determinand participates in protonation equilibria, e.g. fluoride, cyanide and most importantly in the determination of pH.

7. It is particularly unwise to attempt to compensate for temperature differences when operating in the non-Nernstian range of an electrode, as effects 1, 2 and 6 are invariably larger than for analysis in the Nernstian range.

8. Temperature effects in titrimetry are less important than in direct potentiometry, but they may still be significant in Gran plots; linear titration plots and titrations to a fixed e.m.f. If the end-point is determined from the point of inflection of the titration curve, temperature will not generally change the position of the end-point, but may affect its sharpness.

Stirring

Stirring is necessary in a beaker to ensure that constant solution conditions are maintained in the immediate vicinity of the sensing surface of the indicator electrode, otherwise local conditions may develop such that the e.m.f. of the cell is not representative of the concentration of the determinand in the bulk of the solution. This is particularly true of electrodes operating near their limits of detection, where materials dissolved from the membranes are of greater importance. Insufficient stirring at a liquid junction can lead to an accumulation of concentrated electrolyte from the reference electrode such that continuously changing ionic conditions at the end of the junction give rise to a variable liquid junction potential. In most analyses this is not serious, as the variations will be very small in the short time that the electrodes are immersed in the sample.

Stirring rates are important, but precise recommendations cannot be made, as the shape of the beaker, the volume of solution it contains, the depth of immersion of the electrodes and the size of the stirrer bar are only the main factors affecting the correct choice. In practice, there is a relatively wide range of suitable rates: one that is too slow will result in unnecessarily long response times and one that is too fast may cause a noisy signal (especially with liquid ion-exchange electrodes). Temperature changes due to the heat generated by magnetic stirrers are a source of error and should be minimized by using as slow a speed as is consistent with adequate performance in other respects and by separating the beaker from the body of the stirrer with a 1-cm thick layer of insulating material such as plastic foam. To find the best stirring speed, carry out the appropriate analytical procedure with a standard solution of as low a concentration as is likely occur in the samples, starting with a steady slow stirring speed. When the potential is steady, or only changing slowly, increase the stirring speed and observe any shift in the potential or change in the level of noise (a chart recorder is of great assistance here). The speed above which there is no significant shift in potential should be noted and used for all subsequent e.m.f. measurements with those electrodes, provided the level of noise is acceptable. Another relevant test is the time taken for an electrode to reach a steady e.m.f. after transfer from one solution to another in different stirring conditions; allow the electrode to reach a steady e.m.f. in a dilute standard solution, inject a spike of concentrated solution so that the original concentration is at least doubled and record the e.m.f. as a function of time. Repeat for a number of stirring speeds, always starting with the same concentration in the dilute solution. Alternating the electrode between two standard solutions is another way of performing the same test, but it takes longer and care must be taken to use exactly the same rinsing procedure on each transfer.

Light

Photo-electric e.m.f.s have been observed from several types of solid-state membrane electrodes. In all cases this should not constitute a serious limitation in their use and the effects are negligible when the analyses are carried out in opaque beakers or in a laboratory that is screened from direct sunlight. Should a solid-state silver halide membrane be exposed to strong light it may be necessary to renew the membrane surface by polishing it. The presence of silver sulphide in silver halide membranes is thought to decrease the photosensitivity of the halide electrode. The reported effects of light on glass electrodes are considered by Bates (1973) to be the result of either temperature changes or the photosensitivity of the silver–silver chloride internal reference electrode.

Flow-cell analysis

The majority of the analytical methods in Part II are carried out by immersing the electrodes in a stirred sample contained in a beaker, but there are a few methods which specify the use of flow cells. These analyses require special conditions which cannot be produced in a beaker, the details of which are discussed in the individual methods. Although beaker analyses are simpler and require less apparatus, most of the direct potentiometric methods could be adapted to flow-cell procedures with the following advantages:

(a) When the electrode is operating near its limit of detection it is easier to obtain reproducible equilibrium conditions in flow system. This applies equally to solid-state electrodes, where the material of the membrane is involved in solution equilibria with the determinand, and to liquid ion-exchange electrodes where there is partition of the exchanger between the organic and aqueous phases. As a result, it is possible to lower the limit of detection, relative to analysis in a beaker, by as much as two orders of magnitude.

(b) There is less risk of contamination when the analysis is carried out with few manipulative steps; this is particularly important for common species such as sodium and chloride. The comparatively enclosed analytical system used with a flow cell is suited for analyses where the sample can be contaminated by the atmosphere or where an interference, such as carbon dioxide, will be introduced.

(c) Once the flow system is set up, it is simple to operate and has fewer opportunities for error and, therefore, there is less need for skilled laboratory staff.

(d) The precision and accuracy of the analysis are usually improved.

(e) Flow cells with water jackets may provide more accurate and convenient control of temperature than can be achieved otherwise.

In adapting beaker procedures to flow cell procedures, pump the sample and reagent solutions at rates such that they are mixed in the same proportions as in the beaker, although small differences can be tolerated. If the pumps available cannot give the required ratio of flow rates, the reagent solution may be prepared in a more concentrated or dilute form, so that the concentration of reagent in the final mixture is practically the same as it would have been in the beaker. A change to a flowing system may also bring problems and some general faults are discussed later (p. 129).

The flow cells mentioned in the apparatus section were designed for flow rates of 8 ± 4 ml min^{-1}, but an optimum rate cannot be specified, as it will depend on the electrodes used, the volume of sample available and the response time of the electrodes. Small-bore plastic tubing, either PVC or PTFE, is used for transmission lines between sample and reagent bottles, the pumps and the flow cell. Silicone rubber tubing is too permeable to be used for this purpose in analyses where contact with air is to be avoided.

The temperature in the flow cell has to be controlled as carefully as in a beaker, but in some ways flow systems are less prone to temperature variations because the electrodes are not handled during the analysis. The simplest method of controlling the temperature is to carry out the analysis with sample, standard and reagent bottles standing in a water bath whose temperature is set about 2 °C above ambient. If the room temperature is not steady, it is advisable to use, in addition to the water-bath, a flow cell fitted with a water jacket through which water can be circulated at a controlled temperature.

CALIBRATION

Except when they are applied to some titrimetric procedures, ion-selective electrodes always need some form of calibration, i.e. the relationship between the e.m.f. and the logarithm of the concentration must be found empirically. The standard potential, E°, and the calibration slope, k, in the Nernst equation

$$E = E^\circ + \frac{k}{z}\log f + \frac{k}{z}\log c$$

cannot be predicted with sufficient accuracy for analytical purposes, although k is generally close to the theoretical value (Appendix 1). The activity coefficient, f, may also be difficult to predict accurately and it is generally better to keep it constant by making measurements in a medium of constant ionic strength. An empirically determined value of E° in this medium will include activity coefficient effects and give an equation

$$E = E^\circ_{\mathrm{emp}} + \frac{k}{z}\log c$$

It is not valid to determine E°_{emp} and k once and assume that they will be the same for the rest of the electrode's life. Both may change with time, especially E°_{emp}, and both will certainly change with temperature. It is, therefore, unthinkable not to check at least the E°_{emp} value with every batch of samples.

There are a number of approaches to calibration, the best for any particular analysis depending on the number of samples involved, the frequency with which the analysis is carried out, the nature of the sample and the performance of the electrode. In the methods described in Part II, recommendations for calibration are based on the last two factors. A precision test (p. 69) will help in the choice of a calibration procedure, particularly if the electrode is to be used in or near its non-Nernstian range (see p. 78 for checks on linearity) or if a permanent calibration graph is wanted.

Direct Potentiometry

In controlled conditions in the laboratory, a permanent calibration graph prepared as below may be the only one required, but considerable time is taken in obtaining such a calibration and circumstances may demand a more rapid procedure, either because the conditions cannot be controlled or because the electrode does not have stable characteristics. With liquid ion-exchange electrodes it is common for both the standard potential and the calibration slope to change considerably with time, so that a permanent calibration graph cannot be used for accurate analyses.

Preparation of a permanent calibration graph

A permanent calibration graph should not be used for accurate work unless the following conditions are met:
 1. The calibration graph should be prepared from standard solutions in

conditions that are, as nearly as possible, identical with those applying to the samples, i.e. any reagents added to the samples should also be added to the standards and the standard solutions should cover the expected concentration range of the samples.

2. The temperature of every batch of samples must be within 1 °C of the temperature at which the calibration was obtained.

3. A precision test of the sort described on p. 69 should have shown the electrode to have zero or non-significant between-batch standard deviations.

4. A standard solution should be introduced with each batch of samples and the e.m.f. values observed with the samples should be referred to the e.m.f. observed with the standard. This procedure compensates for changes in the standard potential of the electrode.

5. The calibration graph should be prepared with all e.m.f. values referred to the e.m.f. of the standard solution mentioned above.

6. A second standard solution should be analysed with each batch of samples and the result plotted in a control chart (p. 76). If a significant bias appears in the analysis of this standard solution, either the calibration graph must be redefined or the analytical procedure should be examined for a source of error.

The results from the precision test in condition 3 can be used to form the calibration graph itself; consider the example below;

Five batches of duplicate measurements are made with five standard solutions, 10, 5, 1, 0.5 and 0.1 ppm. The readings observed with the 10 ppm standard were (90.0; 89.9), (89.9; 89.7), (90.3; 90.3), (90.2; 90.0), (90.0; 89.7) and the corresponding readings with the 5 ppm standard were (72.5; 72.7), (72.3; 72.1), (72.6; 72.9), (73.0; 72.5), (72.6; 72.7). The mean e.m.f. for each solution in each batch is calculated and subtracted from the corresponding mean for the 10 ppm solution, which is chosen as the reference standard to be introduced with each batch of samples. Thus for the 5 ppm solution, $\Delta_1 = 89.95 - 72.6 = 17.35$ mV and similarly $\Delta_2 = 17.6$, $\Delta_3 = 17.55$, $\Delta_4 = 17.35$, $\Delta_5 = 17.3$ and the mean of these five e.m.f. differences is calculated $\bar{\Delta} = 17.43$ mV. Similarly-calculated mean differences for the 1, 0.5 and 0.1 ppm standards are 57.9, 73.3 and 110.0 mV, respectively, and for the 10 ppm standard we have, by convention, $\bar{\Delta} = \Delta_1 =$ etc. $= 0.0$ mV. The mean e.m.f. differences are plotted on the y-axis against the logarithms of the concentrations on the x-axis, as in Figure 7.3. The upper horizontal scale of Figure 7.3 gives the logarithm of the concentration, i.e. log 10 = 1, etc. and shows how the graph would be plotted on ordinary squared graph paper. The lower horizontal axis has a logarithmic scale, enabling the concentrations to be plotted directly without being converted to logarithms. The two-cycle scale shown allows results over all hundredfold concentration range to be plotted. If the concentration range were wider, semi-logarithmic graph paper with more cycles would be required; this is readily available. Semi-logarithmic graph paper is very convenient, but its use can lead to errors with inexperienced operatives, especially when only two standard solutions are used to calibrate the electrode (see below) or all the standards are one decade apart. To avoid error, follow the calibration graph from low to high and check that it crosses the lines marked 2, 3 . . . , 9 in ascending order.

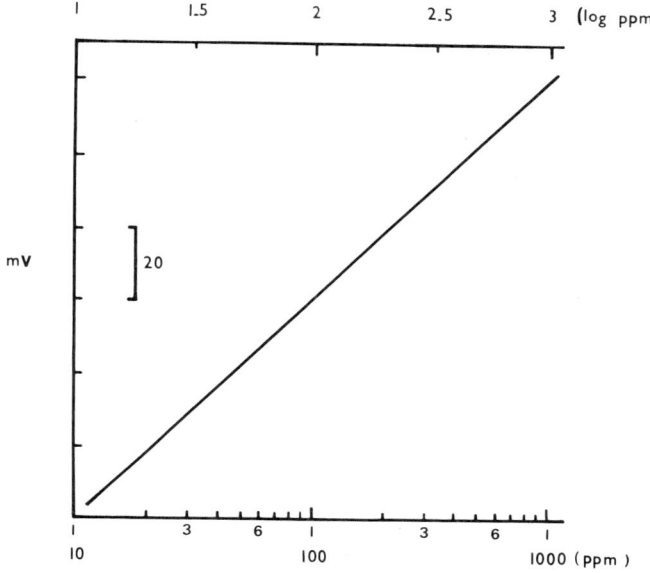

Figure 7.3 Calibration graph for a univalent determinand

Nernstian calibration with two standard solutions

This is the most frequently recommended mode of calibration in Part II. It is appropriate for any analysis by direct potentiometry in the Nernstian range of an electrode, particularly for analyses carried out in conditions of varying temperature, e.g. during field trials. The calibration is performed with two standard solutions in each batch of samples: the first to give (implicitly) the value of $E°$ and the second so that the calibration slope, k, can be found. Measurements with one or both standards may be repeated at intervals during the batch, depending on the constancy of the temperature and the stability of the electrode. If the temperature varies, both standards will be needed fairly frequently, but otherwise only the first ($E°$) standard should be necessary in most cases.

The two standards should be chosen so that they span the concentration range expected in the samples, as any errors are exaggerated if the calibration is extrapolated. If k is to be determined with reasonable accuracy, the two standards should differ in concentration by a factor of at least two or (for divalent electrodes) three. It is normally recommended that the more concentrated and, therefore, more accurate standard be used to check the standard potential, but if the purpose of the analysis is to check that samples are within a control limit, a standard solution corresponding exactly to that limit should be used. Similarly, if the samples are to be between two limits, the two standard solutions should be chosen accordingly. If the e.m.f. values E_1 and E_2 mV are observed with standards of concentration s_1

and s_2, respectively, the calibration slope is given by

$$k = \frac{E_1 - E_2}{\log s_1 - \log s_2} \text{ mV per decade}$$

Note that k is a signed quantity, conventionally negative for anions and positive for cations. With most pH meters, the above convention is maintained when the ion-selective electrode is connected to the high-impedance input terminal, but with some it is reversed; no error should arise from this if measurements are made in a consistent fashion. If the value of k differs from the theoretical (Appendix 1), the calibration should not be used unless its linearity has been checked with more standard solutions.

In calculating concentrations from this type of calibration, the e.m.f., E_x, observed with the sample is always related to that, E_1, of the standard used to check the standard potential. If E_1 changes during the batch, E_x, may be referred to the value of E_1 obtained with the least separation in time, or to a value of E_1 interpolated between those obtained before and after E_x. The concentration is calculated from

$$c = s_1 \times \text{antilog}\left[\frac{E_x - E_1}{k}\right] = s_1 / \text{antilog}\left[\frac{E_1 - E_x}{k}\right]$$

The concentration may also be obtained from a calibration graph formed by drawing a straight line through the points $(\log s_1, 0)$ and $(\log s_2, E_1 - E_2)$ or from the direct activity scale of a pIon meter, as below.

Nernstian calibration using the direct activity scale of a pIon meter

The direct activity scale should only be used if the electrode is operating within its Nernstian range, which should be confirmed from a calibration plot made on the millivolt scale. Two standard solutions are used to calibrate the direct activity scale and the procedure is, therefore, exactly equivalent to that given immediately above. The methods in Part II do not describe the use of direct activity scales, but they can easily be adapted to this type of calibration, as the readings R_1 and R_2 below are obtained at the same stages as E_1 and E_2 in the methods and all other parts of the procedure such as the conditioning of the electrodes, the addition of reagents and control of temperature and stirring are the same, except that the result is read directly from the meter.

Direct readings of concentration or activity are made from a logarithmic scale which usually spans two or three decades. All pIon meters have function switches or controls which have to be selected according to the charge (with sign) on the determinand. If the meter lacks a setting for a particular charge, it cannot be used on the direct activity scale for determinands of that charge-type.

Immerse the electrodes in the first standard solution of concentration s_1 and when the reading is steady, adjust the calibration or buffer control so that the reading R_1 corresponds to that point on the scale which is independent of the slope or temperature control. This point is usually specified in the instruction manual for the meter; for example, it is the mid-point (100) on the logarithmic scale of the Orion 401, 404 and 407 meters and on the EIL 7030 and 7050 meters it occurs at the points 1.2 and 1.8 on the p scale for cations and anions, respectively. Alternatively, it can be found empirically by shorting the input terminals and observing the effect of varying the slope control at various points selected by means of the buffer control. If altering the slope control from 100% to 90% causes the reading to increase, the point selected is too high. The opposite change indicates that the setting is too low and no change means that the setting is correct.

Immerse the electrodes in a second standard solution whose concentration, s_2, is preferably one tenth of s_1. When the reading has stabilized, adjust the slope control until the reading corresponds to $R_2 = R_1 \cdot s_2/s_1$.

Rinse the electrodes and re-immerse them in a second portion of the first standard solution: the reading should return R_1 ($\pm 2\%$). If this is not so, possible causes are contamination of one or other of the standard solutions, a non-equilibrium reading in the first portion of the first standard solution or an incorrect selection of the initial scale position, R_1. Confirm the position of R_1 and repeat the calibration procedure using fresh portions of both standard solutions. When the calibration has been completed, immerse the electrodes in the sample solution and note the steady reading, R_x, from which the corresponding concentration can be calculated,

$$c = s_1 \cdot R_x/R_1.$$

It is obviously convenient to have the indicated reading numerically equal to the concentration and this can be achieved with most meters by using a value of s_1 that is an exact power of ten. Alternatively, the calibration procedure can be extended, following the stage at which the electrodes are re-immersed in the first standard solution to confirm that the reading returns to R_1, by adjusting the calibration control until the indicator reads the required numerical value of s_1.

The reading with the first standard solution should be checked from time to time and returned to its initial setting R_1 or s_1 by adjustment of the calibration control, so that the meter is corrected for changes in the standard potential of the electrodes. The frequency with which this should be done depends on the type of electrode used and the accuracy required.

Non-Nernstian calibration

In the non-Nernstian region, the relationship between E and $\log c$ is no longer linear and two standards do not suffice to define the calibration graph, except in certain special circumstances where other linear functions can be derived (see the methods for determining chloride). Concentrations are obtained from a calibration graph rather than by calculation and because the graph is curved at least four standard solutions are needed to define it. As with the Nernstian calibration, one standard is chosen as a frequent check on the stability of E° and all other e.m.f. values are referred to the e.m.f. obtained with this standard. It is even more important than in the Nernstian calibration that the standard solutions span the range of sample concentrations and that the calibration be renewed with every change of temperature, as the errors will increase as the calibration becomes less linear.

Known Addition and Known Subtraction Potentiometry

Both of these techniques require a knowledge of the calibration slope, but not of the standard potential. A calibration graph of the type prepared for direct potentiometry is useful, however, as it defines the linear range of the electrode, outside which these methods are no longer accurate. The calibration graph also provides a value of the calibration slope, but this can only be used for subsequent known addition or subtraction calculations if a precision test (p. 69) revealed no significant between-batch standard deviation and if all solutions are analysed at the temperature at which the calibration graph was prepared. The calibration slope may be determined with each batch of samples by making measurements with two standard solutions, as in direct potentiometry. An alternative method is the dilution technique described on p. 101; this is more suitable for known addition than for known subtraction and care should be taken that any reagents added to the sample are added to the diluent in the same proportions and that the temperature of the diluent is within 1 °C of that of the sample. The dilution method is more convenient than the separate determination of the calibration slope, but is less accurate. Errors arise if either the ionic strength or, in the presence of a complexing agent, the ratio of free to bound determinand changes on dilution.

As a general rule, the best results are obtained if the original concentration is approximately doubled in known addition or halved in known subtraction, but before the correct quantity of determinand or reagent can be added, a reasonable estimate of the determinand concentration is required. In some analyses, the concentration may be known by experience to be within a certain range and so an addition that permits an accurate determination can be made without difficulty. In general, however, it is convenient to be able to estimate the concentration roughly from the e.m.f. (E_1 in equation 6.1, p 99) observed with the sample solution. In the absence of complexing agents, a calibration graph obtained as in direct potentiometry either from precision test data or from measurements with two standard solutions could be used and for this purpose it is permissible to extrapolate an existing graph or use one obtained at a slightly different temperature.

The most important use of the known addition method is for determining the total concentration of determinand in the presence of a complexing agent, but this is the most difficult case in which to make a good estimate of the quantity of determinand to be added. Because the known addition method requires the total concentration to be approximately doubled, adding the quantity of determinand indicated by a calibration prepared in the absence of complexing agents will produce only a small change in e.m.f. Provided the identities and approximate concentrations of the main complexing agents present in the samples are known, a working calibration graph for indicating the correct addition may be prepared as follows:

1. Immerse the electrodes in a stirred portion, V_S ml, of solution containing no determinand but approximately the same concentrations of complexing agents as would be present in the samples.
2. Add V_R ml of any reagents that would be added in the known-addition procedure. The ratio of V_R to V_S should be the same as in the known-addition procedure.
3. Add a small volume (less that 2% of V_S) of a standard solution of determinand (concentration c_0) to the solution in the beaker. A burette or graduated pipette will be sufficiently accurate. Allow the electrodes to reach a steady potential.
4. Note the total volume added, V_i, and the steady e.m.f., E_i.
5. Repeat steps 3 and 4 for five or six additions of standard solution.
6. Calculate the concentration, $c_i = c_0 \cdot V_i/(V_S + V_i)$, corresponding to each e.m.f.
7. Plot the e.m.f. values on the y-axis against the logarithms of the concentrations on the x-axis.

A calibration of the above sort will be useful not only for calculating the additions required, but also because of the information it gives on the performance of the electrode at low free concentrations of determinand. It is observed, particularly with solid-state membrane electrodes, that they can respond to very low concentrations of free determinand in equilibrium with fairly large concentrations of complexed determinand, even though it is not possible to obtain a good response with very dilute solutions of uncomplexed determinand containing similar concentrations of free determinand. A linear calibration confirms that the electrode can be used for the known addition method in the conditions tested and this calibration should give a reasonable estimate of the quantity of determinand to be added. As this is an approximate procedure, larger variations in temperature and standard potential can be tolerated than with a calibration for direct potentiometry.

Titrimetry

Each type of titration procedure has its own calibration requirements. If the equivalence point is obtained from the point of inflection of a plot of e.m.f. against volume added or from the corresponding derivative plot, there is no need to know either the standard potential or the calibration slope of the electrode, but a Gran plot (p. 89) uses the calibration slope and the more refined linear titration plots require both the slope and the standard potential. The calibration procedures are the same as for direct potentiometry in these cases and the same conditions apply with regard to the temperature and the constancy of other factors. For titration to a fixed potential, the calibration slope is not needed, but the correct potential must be chosen by some means of calibration. In pH titrations, this can be by means of a buffer solution, especially if titrating to pH 7, but a more general method is to carry out the titration with a standard solution of the determinand and note the e.m.f. or pH at the theoretical equivalence point; this e.m.f. or pH will again be valid only for titrations carried out in the same conditions, but the errors involved will be much smaller than in direct potentiometry.

SOURCES OF ERROR

Sources of error associated with the electrode system itself usually involve faulty calibration or, what often amounts to the same thing, poor control of temperature. These matters are subject to the control of the analyst and can be checked during the analytical procedure. Faults which, being manifest, are unlikely to lead the analyst into error are dealt with in the final section of this chapter. Errors that originate in the sample solution are more likely to go undetected and therefore to lead to an incorrect decision, but if the general nature of the samples being analysed is known, it is often possible to modify the procedure so as to reduce or even eliminate the error.

Ionic strength effects

If samples contain variable and high concentrations of background electrolyte, the ionic strength adjustment solution normally added to dilute samples may not be concentrated enough to swamp the variability of the samples and low results will be obtained. The concentration of the ionic strength adjustment solution may be increased to cope with this problem or the usual solution may be added in greater proportions, e.g. 1 + 1 instead of 1 + 10, or the sample may be diluted. The two latter procedures should be avoided if the determinand concentration is near the limit of response of the electrode. Particularly with liquid ion-exchange electrodes, there is a danger that increasing the concentration of electrolyte added may also increase the separate problem of direct interference at the electrode by substances in the ionic strength adjustment solution. If the ionic strength adjustment solution contains a buffer so that it also controls the pH, the effect of the above procedures on the pH control should be considered. An alternative approach is to use the known addition technique, as this is little affected by the ionic strength except in concentrated solutions of determinand, which can easily be measured after dilution. Note, however, that the calibration slope should not be found by the dilution method when known additions are made to solutions of high ionic strength.

Masking

The presence in the sample of substances that form complexes or ion pairs with the determinand will reduce the concentration of free determinand and so lead to a low result. This problem can arise with any determinand at high enough concentrations, even with alkali metal ions, and can be detected by doing a recovery test. If the complex formed is weak, so that the masking effect is significant only at high concentrations of determinand, diluting the sample will remove the problem, e.g. determination of sodium ions. If the complex is strong the known addition method can be used, unless the concentration of free determinand is depressed below the limit of detection of the electrode. In some cases the masking can be removed by adding a substance that forms a stronger complex than the determinand, but this substance must not itself interfere; most frequently, this means controlling the pH.

Interferences

If other substances in the sample apart from the determinand can react at the electrode, a high result will be obtained. Interferences are not necessarily detected by recovery tests, but their occurrence may be marked by drift or long response times. Interferents can sometimes be masked by adding substances that form complexes with them, e.g., acetylacetone masks magnesium in the determination of calcium, or that precipitate them, e.g., the removal of chloride as silver chloride in the determination of nitrate. Volatile interferents may be removed by heating or gas-stripping, provided the determinand is not itself volatile, e.g. the removal of carbon dioxide from nitrite solutions. In designing procedures for removing interferences, the chemistry of the electrochemical cell as a whole must be considered, i.e. the reagents added should not introduce substances that can react with the determinand as well as the interferent, nor substances that themselves interfere. As an example, chloride may be removed from nitrate solutions — but not from copper solutions — by adding silver ions, because the silver ions interfere with the copper electrode. Interferences in direct potentiometry may sometimes be avoided by using a titrimetric method in which the end-point is obtained from the point of inflection of the titration curve, although the sharpness of the end-point will be reduced. Gran plot methods may sometimes be applicable, but not linear titration plots, titration to a fixed potential or the known addition method.

As a last resort, if the concentration of the interferent is known the result may be corrected for its presence by means of a selectivity coefficient. Selectivity coefficients are quoted throughout Part II, but mainly as a guide to the relative importance of different interfering substances. Inspection of some of the results shows that the coefficients reported may vary by a factor of ten or more and therefore it would be unwise to use them for accurate work. If the analyst wishes to try this procedure, he should determine his own selectivity coefficients by making measurements with a range of mock sample solutions containing the determinand and the interferent at concentrations likely to occur in reality (p. 74).

PRECISION

Precision is quoted when possible as within-batch, between-batch and total standard deviations, but usually only within-batch figures are available and these should be assumed unless it is stated otherwise. Carrying out a precision test with at least five batches of five solutions, as described in Chapter 5, is a most valuable exercise before any method is used extensively over a long period, as it allows the analyst to know with some confidence what variations are likely to occur by chance and what may be due to some new factor affecting the analysis. The use of a control chart will be of assistance in detecting the occurrence of such factors and a knowledge of the precision is essential for this. An incidental benefit of the precision test is that it also provides a good calibration graph.

ACCURACY

The accuracy of the methods depends in part on factors under the control of the analyst, e.g., temperature and the frequency of calibration, and also on largely imponderable factors such as the concentration of interferents in the sample solution. Tests with synthetic solutions can indicate likely sources of error, but the extent of such errors can only be known if the sample is very thoroughly characterized. Recovery tests carried out on sample solutions as described in Chapter 5 will show up some kinds of error and should always be done when a new type of sample is being analysed for the first time. Even when one type of sample is analysed routinely, the inclusion of a recovery test on one sample out of each batch can be a valuable control. Before a method is adopted, the analyst may wish to compare the results it yields with those given by another method for the same samples. Agreement between the two sets of results will increase the confidence in the method, but it is not a proof of accuracy, however well established the reference method may be. Such comparisons may show up errors due to interferences, but poor technique in sampling or in preparing standard solutions will generally affect both methods equally. McFarren, Lishka and Parker (1970) proposed that a method was 'excellent' if 95% of the results from inter-laboratory trials were within 25% of the true answer and 'acceptable' if the same proportion were within 50% of the answer. Such figures show the importance of controls as set out in Chapter 5 and of frequently checking an electrode for changes in standard potential or calibration slope.

RESPONSE TIME

The time taken to reach a steady e.m.f. depends on many factors: the nature of the membrane, the concentration of determinand, the temperature of the sample, the temperature and concentration of the solution to which the electrode was previously exposed, the presence of interferents and the rate of stirring. In the absence of temperature variations and interferences, most electrodes operating within their Nernstian ranges will reach equilibrium in less than 2 min for an increase in concentration, but as their limits of detection are approached, much slower responses are observed. Times as long as 10—15 min are required for some liquid ion-exchange electrodes at concentrations in the range $10^{-4}-10^{-5}$ mol l^{-1}. One method of reducing the time per sample is to take the reading of e.m.f. after a fixed time in the solution. If this procedure is followed consistently for both sample and standard solutions, acceptable precision can be obtained (Torrance, 1974). It is also possible to reduce the time taken to reach a steady e.m.f. by increasing the temperature at which the analysis is carried out, but care must be taken to avoid drift if there is a large difference (~10 °C) between the sample and ambient temperatures. This technique is more suited to continuous analysis, where steady thermal conditions are more easily achieved. In all cases, the decrease in response time will be accompanied by an increase in the limit of detection.

The response time can depend on the concentration of the solution previously analysed, particularly when the concentration changes from high to low. In manual analyses it is possible to reduce this 'memory' effect by immersing the electrodes in deionized water (plus reagents in the usual proportions) between each sample, as in the determination of ammonia.

The responses of many electrodes are slower in the presence of an interferent and it is possible for such a substance to have a permanent effect, so that the subsequent response becomes sluggish and sometimes non-theoretical. As an example, traces of sulphide in a sample whose chloride content is determined by a silver chloride electrode will contaminate the surface of the membrane with silver sulphide. A solid-state membrane electrode can often be restored by polishing its surface and so removing the layer of contamination; many manufacturers supply polishing kits for this purpose. An equivalent effect may be observed with liquid ion-exchange electrodes that have been exposed to interferents; in this case immerse the electrode in a fairly concentrated ($\sim$100 ppm or 10^{-3} mol l^{-1}) solution of the determinand to restore the performance.

TRACING FAULTS

Some faults are caused by specific reactions of the determinand or the materials composing the electrode and these are considered individually in the appropriate analytical methods. The most frequently occurring faults, however, are common to all electrodes or to whole classes of electrodes and are discussed below.

Off-scale readings

When no readings can be obtained on any scale of the pH meter there is a faulty electrical contact causing an open-circuit condition. Another symptom of this trouble is the occurrence of large changes in readings caused by even slight movements of the operator. Check that the electrodes are firmly plugged into the terminals on the pH meter and that any other connections are properly made, e.g. between an amplifier and a separate indicator, recorder or transmitter. A common source of trouble is the soldering of the plug onto the lead from the electrode, especially if the plug has been changed from that originally fitted to the electrode. Extra care is necessary with gas-sensing membrane electrodes and combination electrodes, as the signal from the reference electrode is sometimes taken along the screening of the cable. Some pH meters will not accept a signal of this kind; disconnect the screening from the plug and solder an extension lead onto it. Cover the soldered connection with a rubber or plastic sheath and attach the appropriate reference pin to the end of the lead.

Reference electrodes that have drained dry are another possible cause; They should be regularly inspected and topped up with their correct filling solutions.

Calomel and mercury—mercurous sulphate reference electrodes may be permanently damaged by drying out (see *Conditioning and Storage of Electrodes*). Gas-sensing membrane electrodes and some makes of solid-state and liquid ion-exchange electrodes have to be filled with an internal reference solution before use; if this is omitted, or simply not added in sufficient volume, there will be no contact with the internal reference electrode. Bubbles of air trapped in capillaries can also break the electrical contact; this is a problem particularly of reference electrodes with ground-glass sleeve liquid junctions and some types of liquid ion-exchange electrodes.

Invariant readings

If the e.m.f. is always zero (or close to it, depending on the adjustment of the pH meter), even if the electrodes are not immersed in the solution, there is a short circuit. The most likely cause is a stray piece of wire or graphite from the screening round the cable of the ion-selective electrode making contact with the lead itself: dismantle the plug and remove any extraneous conducting material. This is especially a problem of plugs that have been changed in the laboratory.

A constant but non-zero reading amounts to a total loss of sensitivity and selectivity. This may be caused by rupture of the ion-selective membrane or by the cap holding the membrane not being screwed on properly, so that current flows directly through the solution between the internal and external reference electrodes and not through the relatively high resistance membrane. Removing the electrodes from solution produces an open circuit.

A related problem occurs when the polymer membrane of a gas-sensing electrode is ruptured. The highly concentrated strong acid or base added to the sample makes direct contact with the internal glass electrode and in this situation variations in the concentration of the determinand have a negligible effect on the pH and therefore on the e.m.f. Removal of the electrode from the solution would not cause an open circuit, but the e.m.f. would drift.

Loss of sensitivity

If the electrode no longer gives a near-Nernstian response to changes of determinand concentration, the membrane has probably been damaged in some way. Before taking any action, check that the analytical procedure has been carried through correctly and that the standard solutions have been prepared properly. Glass electrodes that have been allowed to dry out will generally have a poor performance, although it may improve with time. Glass pH electrodes can often be restored by soaking them overnight in 0.01 mol l^{-1} hydrochloric acid. Solid-state membranes may become tarnished or badly scratched, but in many cases the performance can be restored by polishing the membranes according to the manufacturer's instructions. Liquid ion-exchange electrodes often deteriorate with age and the membrane has to be replaced.

Fouling of the membrane can impede the access of fresh solution to the electrode and so cause a loss of sensitivity. This is rare in laboratory analysis, but is not uncommon with electrodes used for continuous monitoring of natural or industrial waters. Once the fouling has been removed (see Table 4.1, p. 58), the electrode will in most cases work normally again. Changes in temperature will affect the sensitivity, but increases and decreases are equally likely and the effects are small compared with those above.

Slow response

All the causes of low sensitivity apply here also. If the response is normal in standard solutions but slow in sample solutions, the latter probably contain an interfering substance. Air bubbles at the surface of the membrane can be a cause of slow response with gas-sensing electrodes. The slow response may be an adjustment to a new temperature equilibrium if the temperature of the solutions being analysed is not kept constant. Inadequate stirring may result in long response times.

Response times in flow systems are partially dependent on the time taken for a solution to displace the previous one from the system, a factor that is directly proportional to the flow rate. A common cause for the deterioration of the response time is the flattening of peristaltic pump tubes, with a consequent reduction in the flow rate of the solutions passing through them.

Drift

A slow and steady movement of the e.m.f. in one direction is most often caused by changes in temperature, although inadequate stirring is another possibility. Small differences in temperature between successive solutions are unlikely to result in persistent drift, although they are undesirable for other reasons. The main source of drift is probably the heat generated by the motors of magnetic stirrers. Some electrically driven stirrers are adequately ventilated and may cause little or no trouble, but it is advisable always to separate the beaker containing the solution from the body of the stirrer by a layer of plastic foam or similar insulating material.

Noise

Noise describes a number of types of short-lived disturbance of the steady electrical signal wished for from an electrode and it may take the form of random oscillations, regular pulses or spasmodic and very brief jumps in potential. One of the commonest sources of noise is electrical interference by nearby apparatus — magnetic stirrers and water-baths seem to be among the commonest offenders, but all electric motors and any apparatus containing relay switches, e.g. ovens and hot-plates, are suspect. It may be possible to isolate the electrodes from most of this apparatus, but the only cure lies in proper maintenance. Some noise however, arises from the measuring or recording equipment itself. Both pH meters and recorders may exhibit noisy signals even in the absence of other apparatus, but it should be possible to have the amplifier adjusted so that this is not a problem. Some digital pH meters have displays that pulse and produce 'blips' on chart recorders connected to the meter; connecting a capacitor across the recorder output terminals of the meter will usually cure this.

In flow cells, it is possible to have a signal with a noise pattern imposed on it that is associated with the pump. This can be recognized by stopping the pump or, where appropriate, comparing the frequency of the noise with the action of the rollers in a peristaltic pump. Noise of this type is often electrostatically generated by the action of the pump on the plastic pump tubes and can sometimes be eliminated by the following procedure: interpose a 2 cm length of stainless-steel tube in the sample line just before the flow cell and connect the exterior surface to a suitable earth point such as electrical mains earth or a mains-water pipe, provided that this is not contrary to the instructions provided with the pH meter. Observe any change in the signal when the earth lead is connected, and if the noise is improved leave the lead in place and change the concentration of the solution by a known amount. Confirm that the expected change in e.m.f. takes place. This is necessary because, in the attempt to eliminate electrostatic noise, earth loops can be introduced, leading to spurious readings. One frequently found earth loop is formed by the effluent from the cell running to drain in a continuous stream — arrange the cell so that the effluent is collected in a catchpot or drips into a drain.

Another cause of noise in flowing systems is poor mixing of the sample and reagent streams. This can arise from the practice of making the ratio of the sample to reagent flow rates high to avoid excessive dilution of the sample and to reduce the storage and consumption of reagents. If the solution reaches the sensing electrode in pulses of varying composition, the e.m.f. may show sudden deviations of a few millivolts lasting for several seconds or several minutes, depending on the mixing characteristics of the flow cell and the response time of the electrode. The existence of such a fault can be confirmed by pumping a mixture of the sample and reagent solutions (added in the usual proportions) simultaneously along both the sample and reagent supply lines and observing the e.m.f. If the noise is reduced, poor mixing was a contributory factor and should be improved by inserting a mixing coil in the flow stream before the cell or by using a cell with a mixing chamber containing a magnetic stirrer bar. The use of a mixing coil will increase the overall response time of the system and it is therefore desirable that the coils be no longer than is necessary. At flow rates of $4-10$ ml min^{-1} coils of $10-15$ turns should be sufficient.

Noise may be associated more directly with the electrodes, especially in flowing systems. Pressure surges originating in the action of a pump may produce an oscillation in the signal from the electrodes; some forms of liquid ion-exchange electrodes and reference electrodes are particularly susceptible to pressure changes. Placing the sample and reagent solutions about 50 cm above the pump and flow cell, so that the solutions flow by a syphoning mechanism and are only metered by the pump, may alleviate this problem. Stirring that is too vigorous may also cause noise with some electrodes. Gas-sensing membrane electrodes can suffer from noise if the pressure of the sensing electrode on the polymer membrane is too great, but improved design has reduced the occurrence of this fault. Some solid-state membrane electrodes are photosensitive and if placed in an exposed position near a window may show shifts of 1–5 mV on changes in the intensity of sunlight.

Bibliography

ASTM, 1973, *Part 23, Water; Atmospheric Analysis*, American Society for Testing and Materials, Philadelphia.

Bates, R. G., 1973, *Determination of pH. Theory and Practice*, 2nd edn., John Wiley and Sons, London and New York.

BSI, 1969, *Methods of Sampling Water Used in Industry*, B.S. 1328: 1969, British Standards Institution, London.

DoE, 1972, *Analysis of Raw, Potable and Waste Waters*, Department of the Environment, H.M. Stationery Office, London.

FWQA, 1970, *Design of Water Quality Surveillance Systems*, Federal Water Quality Administration, US Government Printing Office.

Hutchinson, G. E., 1957, *A Treatise on Limnology, Vol. 1*, John Wiley and Sons, New York.

McFarren, E. A., J. R. Lishka, and J. H. Parker, 1970, Criterion for judging the acceptability of analytical methods, *Anal. Chem.*, **44**, 878.

Torrance, K., 1974, A potentiometric method for the determination of chloride in boiler waters in the range 0.1 to 10 $\mu g\ ml^{-1}$ of chloride, *Analyst*, **99**, 203.

Wilson, A. L., 1974, *The Chemical Analysis of Water, General Principles and Techniques*, Society for Analytical Chemistry, London.

Part II

ANALYTICAL METHODS

Acidity and Alkalinity	147	Lead	243
Aluminium	248	Lithium	155
Ammonia	279	Metals by Compleximetry	226
Boron	383	Nitrate	374
Bromide and Iodide	343	Nitrite and Nitrogen Oxides	263
Cadmium	239	Nitrogen	290
Calcium	186	Perchlorate	390
Carbon Dioxide	254	pH	135
Chelating Agents	226	Potassium	171
Chloride	323	Silver	205
Chloride (Low level)	333	Sodium	160
Compleximetric Titrations	226	Sulphate	366
Copper	214	Sulphide	355
Cyanide (Cyanide electrode)	298	Sulphur Dioxide	271
Cyanide ($Ag_2 S$ electrode)	306	Thiocyanate	352
Fluoride	313	Water Hardness	194
Iodide	343		

The Measurement of pH

As almost all pH measurements are best made with glass electrodes, other electrodes will not be considered here. In solutions containing fluoride ions, the quinhydrone electrode (p. 22) may have advantages and in conditions of very severe fouling, the antimony–antimony oxide electrode (p. 20) may be more economical if the fairly poor precision of that electrode is tolerable. Glass electrodes may be divided broadly into three classes. (1) General purpose types suitable for wide ranges of temperature and pH. (2) Low-resistance types for work at low temperatures (<10 °C). These electrodes may have special low-resistance glasses, which are unsuitable for use above pH 9–10 or above about 60 °C, or may have the same glass as the general purpose types, but blown into a thinner and, therefore, more fragile membrane. Examples of low-resistance electrodes are the Leeds & Northrup 117224, Activion 003 12 002, Radiometer G2027C, Electrofact 7G 212 and 7G 172, Schott 9202, Pye LoT. 102, Metrohm ANA-EA 107-T, Philips GT. 130 and GAT. 130. (3) High-resistance types for work at high temperatures or at high (>12) pH values. Some of these electrodes are not suitable for low pH values and most are sluggish at temperatures below 10–15 °C. High-resistance electrodes include the Radiometer G. 202CH, Electrofact 7G 411, Schott 9301, 9401 and 9501, Metrohm ANA-EA 107-H and the Philips GH. 210 and GAH. 110. Modern general purpose glass electrodes such as the EIL 1070-1 and 1072-1 and the Corning 4760 are very versatile and the need for special electrodes is diminishing.

Glass electrodes are made in a wide variety of shapes and sizes by many manufacturers. The most popular type is still that with a spherical membrane, but 'toughened' electrodes with hemispherical or conical membranes are increasingly used where extensive handling and, therefore, more frequent breakage is a problem. Electrodes with flat membranes are mainly used for special purposes such as the measurement of the pH of skin or leather; they have no operational advantage for normal use and are more expensive. Micro-electrodes are also available, but their fragility makes them relatively unsuitable for general use. Combination glass and reference electrodes are widely available, but usually only with general-purpose glasses. The reference half of the combination electrode invariably has a silver–silver chloride element and usually a ceramic frit junction, although Beckman, Philips and Pye can supply electrodes with ground-glass sleeve junctions.

Other electrodes can be steam sterilized. Most glass electrodes have silver–silver chloride internal reference electrodes, but both Schott and Electrofact make electrodes with Thalamid internal elements. The commonest internal filling solution is a pH 7 buffer solution containing potassium or sodium chloride for the reference electrodes, but a few manufacturers provide a choice of pH 4.6 or pH 2 fillings. In one or two cases, the internal filling takes the form of a gel, which enables the electrode to be used in any attitude. Some manufacturers will supply specially-modified electrodes at little extra cost.

APPARATUS

Glass electrode; reference electrode; pH meter; magnetic stirrer

The standard pH values assigned by the National Bureau of Standards assume the use of a calomel electrode with a saturated or 3.5 mol l^{-1} potassium chloride filling solution, but any reference electrode can be used, provided it allows a consistent calibration to be obtained. Silver–silver chloride electrodes should be as suitable as calomel types, but Thalamid reference electrodes should be used only if the glass electrode has a Thalamid internal element. Any of the common filling solutions, 3.0, 3.5, 3.8, 4.0 mol l^{-1} or saturated potassium chloride, should be suitable. Sealed reference electrodes are often unsuitable for pH measurements (Midgley and Torrance, 1976). If the sample is clear, satisfactory results can be obtained for most purposes with all types of liquid junction, but emulsions and samples with high suspended solid contents tend to foul the electrodes and a type of junction that can easily be cleaned should be used (see p. 43).

REAGENTS

Water

The specific conductance of the water used to prepare buffer solutions should not exceed 2 μS cm^{-1}. For the preparation of borax and phosphate buffers, the water should also be free of carbon dioxide (its pH should be in the range 6.5–7.5). Water taken directly from the outlet of a mixed-bed deionization unit should meet all requirements. Distilled water, or water that has been stored, should be boiled for 15 min or purged with nitrogen or air free of carbon dioxide and should be protected from contamination by a soda-lime or equivalent guard tube during cooling and subsequent storage (see *Determination of Acidity and Alkalinity*).

Buffer solutions

Instructions are given below for the preparation of NBS standard buffers. Other buffer solutions may be used, but their characteristics (Bates, 1973; Mattock, 1961; Perrin and Dempsey, 1974) are less well defined. Buffers for pH 4, pH 7 and pH 9.2 are available commercially, either as pre-weighed tablets or sachets of powder or in solution form. The pH 4 and pH 9.2 buffers are nominally the same as the phthalate and borax buffers given below, but the pH 7 buffer is neither of the phosphate buffers below.

For the most accurate work, buffers should be freshly prepared, but for many purposes storage periods of 2–3 or even 6 weeks may be tolerated. Borax and carbonate buffers are less amenable to storage, because of contamination by carbon dioxide. Solutions should be discarded if mould is evident.

For most purposes, analytical reagent grade materials are adequate, but for the greatest accuracy use NBS standard reference materials. As the buffers are specified in *molal* units (m, moles of solute per kg of solvent) but prepared volumetrically, the water used should be at $25 \pm 2\ °C$. The pH values of the buffers are given in Table 1 for a range of temperatures.

Citrate buffer (0.05 m) Dissolve 11.41 g of potassium dihydrogen citrate in water and dilute to 1 litre. Before weighing, dry the crystals at 80 °C for 1 h and cool in a desiccator.

Table 1 pH values of NBS buffers

°C	Tetra-oxalate (0.05 m)	Citrate (0.05 m)	Phthalate (0.05 m)	Phosphate (1 + 1)	Phosphate (1 + 3.5)	Borax (0.01 m)	Carbonate (1 + 1)
0	1.666	3.863	4.003	6.984	7.534	9.464	10.317
5	1.668	3.840	3.999	6.951	7.500	9.395	10.245
10	1.670	3.820	3.998	6.923	7.472	9.332	10.179
15	1.672	3.802	3.999	6.900	7.448	9.276	10.118
20	1.675	3.788	4.002	6.881	7.429	9.225	10.062
25	1.679	3.776	4.008	6.865	7.413	9.180	10.012
30	1.683	3.766	4.015	6.853	7.400	9.139	9.966
35	1.688	3.759	4.024	6.844	7.389	9.102	9.925
38	1.691		4.030	6.840	7.384	9.081	
40	1.694	3.753	4.035	6.838	7.380	9.068	9.889
45	1.700	3.750	4.047	6.834	7.373	9.038	9.856
50	1.707	3.749	4.060	6.833	7.367	9.011	9.828
55	1.715		4.075	6.834		8.985	
60	1.723		4.091	6.836		8.962	
70	1.743		4.126	6.845		8.921	
80	1.766		4.164	6.859		8.885	
90	1.792		4.205	6.877		8.850	
95	1.806		4.227	6.886		8.833	

Phthalate buffer (0.05 m)   Dissolve 10.12 g of potassium hydrogen phthalate in water and dilute to 1 litre.

Phosphate buffer (1 + 1)   Dissolve 3.388 g of potassium dihydrogen phosphate and 3.533 g of anhydrous disodium hydrogen phosphate in water and make up to 1 litre. Before weighing, dry each salt for 2 h at 110–130 °C and allow them to cool in a desiccator. The concentration of each salt is 0.025 m.

Phosphate buffer (1 + 3.5)   Dissolve 1.179 g of potassium dihydrogen phosphate and 4.302 g of anhydrous disodium hydrogen phosphate in water and dilute to 1 litre. Each salt should be dried as before. The concentrations are 0.008695 m KH_2PO_4 and 0.03043 m Na_2HPO_4.

Borax buffer (0.01 m)   Dissolve 3.80 g of sodium tetraborate decahydrate (borax) in carbon dioxide-free water and dilute to 1 litre.

Carbonate buffer (1 + 1)   Dissolve 2.092 g of sodium hydrogen carbonate and 2.640 g of sodium carbonate in carbon dioxide-free water and dilute to 1 litre. Ignite the sodium carbonate at 270 °C for 1 h before use and cool in a desiccator. Do not heat the sodium hydrogen carbonate. The concentration of each salt is 0.025 m.

The above are primary standards. The following secondary standard may also be useful:

Tetroxalate buffer (0.05 m)   Dissolve 12.61 g of potassium tetroxalate dihydrate in water and dilute to 1 litre. Do not heat the tetroxalate salt above 60 °C.

CONDITIONING AND STORAGE OF ELECTRODES

If a new electrode is supplied with a dry membrane, it should be conditioned by leaving it overnight in 0.1 mol l^{-1} hydrochloric acid, or as the manufacturer recommends. Even if the electrode is supplied with the membrane wet, the above procedure is a good one to follow. In general, it is undesirable to allow membranes to dry out and the electrodes should normally be stored in distilled or deionized water. High-alkalinity types of glass electrode are probably best conditioned and stored in borax buffer. When combination electrodes with ceramic-frit liquid junctions are being conditioned, the frit should preferably not be immersed in the concentrated acid or buffer solution for long periods. For very accurate measurements, it is sometimes recommended that the electrodes be conditioned in a solution similar to those being measured. Electrodes should always be stored at temperatures close to those at which they are to be used. Some electrodes may be stored dry, after having been thoroughly washed with water; check the manufacturer's instructions before doing this. As the electrodes have to be re-conditioned before use, dry storage is not recommended unless a prolonged period of inactivity is expected.

CONCENTRATION RANGE AND UNITS

The pH range that can be measured depends on the type of glass electrode being used. In the range pH 2–10 almost all electrodes should be satisfactory at room temperature, but at higher pH values and more extreme temperatures some will be found wanting. Below pH 2, liquid junction potentials may cause deviations from a Nernstian response.

A pH scale may be defined in a number of ways. The definition pH = $-\log c_H$, where c_H is the hydrogen ion concentration, may be useful in certain circumstances, but the analyst would have to set up his own empirically based scale. More generally, pH = $-\log a_H$, where a_H is the hydrogen ion activity, but some assumption about individual ionic activity coefficients is implicit in any such definition. The NBS buffers in Table 1 involve the assignation of a conventional individual ionic activity coefficient for chloride ion and are the most widely accepted. Bates (1973) has discussed the problem at length. As the pH values of most solutions are temperature dependent, the result should include the temperature at which the pH was measured.

ANALYTICAL PROCEDURES

A. Routine pH measurements without temperature compensation

Use two buffer solutions spanning the expected pH range of the samples. The temperatures of the sample and buffer solutions should not differ by more than 2 °C.

Step 1A Remove the electrodes from the solution in which they had been stored and rinse them with deionized water. If possible, make a fresh liquid junction and rinse the reference electrode again. Remove surplus water from the reference electrode and the stem of the glass electrode with a tissue. The bulb of the glass electrode may be wiped with a tissue, but it is better to rinse it with a portion of the solution in which measurements are to be made.

Step 2A Immerse the electrodes in the first buffer solution, turn the temperature-control knob to the temperature of the buffer and adjust the buffer or calibration control until the meter reads the theoretical buffer pH. After a minute, check that the reading is steady and, if it is not, adjust the control again. Carry on in this way until a steady reading is obtained (Note 1, below). If the greatest accuracy is required, repeat the procedure with further portions of buffer until no further adjustment is necessary (Note 2).

Step 3A Repeat the rinsing prodedure (Step 1A) and immerse the electrodes in the second buffer solution. The steady reading should be within the desired accuracy of the theoretical pH. If not, repeat with a second portion of the same buffer and so on (Note 2) until successive solutions agree within the accuracy desired (Note 3).

Step 4A For the best accuracy, repeat Step 1A and again immerse the electrodes in the first buffer solution. The reading observed in Step 2A should be obtained again. If this is not the case, consider the possible causes in Note 1.

Step 5A Repeat Step 1A before immersing the electrodes in the sample solution. Note the steady pH reading and, if greater accuracy is required repeat with further portions of solution until agreement is reached between sucessive solutions.

Note 1 In a buffer solution an electrode should attain a steady reading in less than five minutes. If this does not happen, either the glass electrode has deteriorated or one or both of the electrodes has been stored at a different temperature from the buffer and is taking time to reach thermal equilibrium, or the liquid junction of the reference electrode has not been properly formed.

Note 2 No more than three portions of buffer solution should be needed before a reproducible value is attained; if this is not so, see Note 1.

Note 3 If the desired accuracy is not obtained, the glass electrode may have lost its Nernstian response or the buffer solutions may have deteriorated. If the problem occurs with freshly-prepared buffer solutions, re-condition or replace the glass electrode. Alternatively, calibrate the electrode on the millivolt scale (Method C, below).

B. pH measurements with temperature compensation

Errors are least likely to arise if the glass and reference electrodes are of the same manufacture as the pH meter. Meters having a variable isopotential control, however, can be used with a variety of glass and reference electrode pairs. Note that some pH meters have an isopotential setting without indicating it on the meter itself. The instruction manual of any pH meter should be carefully consulted before temperature compensation is attempted; meters without an isopotential setting are not suitable for this procedure (see *Sources of Error*).

The purpose of the isopotential setting is to enable the pH meter to correct for temperature-dependent changes in the standard potential of the electrode pair. Such changes are approximately linearly related to temperature over a range of 10 °C either side of the temperature of the calibrating buffer solutions. Mattock (1961) has discussed the isopotential concept at length. The correction enables a pH system calibrated at one temperature to read the true pH at a different temperature: it does not, with certain exceptions, correct the pH of a sample solution at one temperature back to what it would have been at the calibration temperature.

Step 1B As Step 1A, but also rinse and wipe the temperature sensor if automatic compensation is used.

Step 2B Immerse the electrodes in the first buffer solution, turn the temperature control knob to the temperature of the buffer (unnecessary with automatic compensation) and the isopotential control knob, if any, to the correct isopotential setting (Note 4). Adjust the buffer or calibration control until the meter reads the theoretical buffer pH. After a minute, check that the reading is steady and, if it is not, adjust the control again. Carry on in this way until a steady reading is obtained (Note 1). If the greatest accuracy is required, repeat the procedure with further portions of buffer until no further adjustment is necessary (Note 2).

Step 3B As Step 3A.

Step 4B Repeat Step 1B and immerse the electrodes in the sample solution. Unless automatic compensation is being used, measure the temperature and adjust the temperature control knob on the pH meter accordingly. Note the steady pH reading and, if greater accuracy is required, repeat with further portions of solution until agreement is reached between successive measurements.

Note 4 Either use the manufacturer's value or find the isopotential setting empirically, following the operating instructions for the pH meter.

C. Calibration on the millivolt scale

Use a series of buffer solutions at the same temperature. Samples should be brought to this temperature before analysis.

Step 1C Carry out Step 1A.

Step 2C Immerse the electrodes in the first buffer solution and when a steady potential has been reached note the e.m.f., E_1. If working on an expanded or relative millivolt scale, adjust the calibration or buffer control so that the e.m.f. E_1 is a convenient round figure.

Step 3C Repeat Step 1C and then immerse the electrodes in the next buffer solution. When a steady reading has been obtained, note the e.m.f. E_i. (Do not adjust the calibration or buffer controls if using expanded or relative millivolt ranges.)

Step 4C Repeat Step 3C for each buffer solution until the required pH range has been fully covered.

Step 5C Calculate for each buffer the e.m.f. difference $\Delta_i = E_i - E_1$.

Step 6C Plot the values of Δ_i on the y-axis against the pH values on the x-axis and calculate the slope factor $k \simeq 59$ mV per pH unit.

The calibration graph prepared in the above way should be linear, although some deviations may be observed at the extremes. A glass electrode in good condition should have an almost theoretical Nernstian response (Appendix 1); if the slope of the calibration graph differs greatly from the theoretical value, the electrode should be re-conditioned or discarded.

For routine calibrations, a buffer having a pH close to that of the sample is used in Step 2C and a second buffer is used in Steps 3C, 5C and 6C to check that the slope factor has not changed.

Step 7C Carry out Step 1C and immerse the electrodes in the sample solution. Note the steady e.m.f., E_x, and, if greater accuracy is required, repeat with further portions of solution until successive readings are the same.

Step 8C Calculate the e.m.f. difference $\Delta_x = E_x - E_1$.

Step 9C Read the pH corresponding to Δ_x from the calibration graph or calculate it from

$$pH_x = pH_1 - \Delta_x/k$$

SOURCES OF ERROR

Temperature

The calibration slope and standard potential of a pH electrode are affected by temperature in the same way as for other ion-selective electrodes. If the pH is read directly off the pH scale of the meter, some form of temperature compensation will be available, but often only for the calibration slope and not for the standard potential (see Method B, above). If the electrode is calibrated at temperature T_B and measurements in the sample are made at a different temperature, T, the error will depend on the type of temperature correction applied. If the meter has an isopotential setting, the error will be small within the range $T_B \pm 10$ K. If the meter has no isopotential setting, but a correction is made for the change in slope factor, the error is independent of pH

$$\Delta pH = pH_i(T - T_B)/T$$

where pH_i is the isopotential pH for the electrode pair and T and T_B are expressed in degrees Kelvin. If no compensation of any kind is applied, the error is given by

$$\Delta pH = (pH' - pH_i)(T_B - T)/T$$

where pH' is the apparent pH in the sample. For the most accurate work, the sample and buffer solutions should be at the same temperature, even if an isopotential correction is possible.

Stirring

There is a tendency to neglect stirring when making pH measurements. In well buffered solutions this may be justifiable, but in poorly buffered ones it can lead to error.

Liquid junction potential

At pH values below 3 or above 11, the calibration of the electrodes may deviate from linearity because of the liquid junction potential (Chapter 2) and so cause an error in the pH reading. The size and even the direction of the error depend on the type of reference electrode used, on the composition of the sample and on the

Table 2 Sodium ion corrections for various makes of glass electrodes at 20–25°C. (Corrections in pH units to be added to apparent pH reading)

pH[a]	Beckman[b] GP	Beckman[b] E-2	Corning[c]	EIL[d]	Electrofact[d,e] LT	Electrofact[d,e] CR	Electrofact[d,e] U	Philips/Pye[d] LoT	Philips/Pye[d] A41	Philips/Pye[d] HA	Radiometer[b] B	Radiometer[b] C	Schott[d] U	Schott[d] HA	Schott[d] HTA
0.1 mol l⁻¹ Na⁺															
10.5	0.03											0.02			
11.0	0.05											0.04			
11.5	0.08				0.07							0.06	0.01		
12.0	0.15											0.10	0.01		
12.5	0.25	0.02			0.23	0.03	0.01		0.09	0.04		0.02	0.16	0.02		
13.0	0.48	0.03			0.46	0.11	0.07		0.27	0.18	0.02	0.03	0.26	0.05	0.02	0.02
13.5		0.05										0.05	0.40	0.08	0.05	0.05
14.0			0.02	0.03	0.82	0.22	0.17			0.10				0.09	0.07	0.09
1.0 mol l⁻¹ Na⁺															
10.0	0.05											0.05			
10.5	0.09											0.08			
11.0	0.15											0.14	0.03		
11.5	0.25	0.02							0.16	0.13		0.02	0.22	0.06	0.02	
12.0	0.48	0.03							0.26	0.19	0.01	0.03	0.35	0.10	0.03	0.02
12.5		0.05										0.05	0.50	0.14	0.06	0.05
13.0		0.10	0.02	0.03					0.57		0.13		0.08	0.18	0.08	0.07
13.5		0.18	0.07	0.08									0.12	0.23	0.12	0.11
14.0		0.20	0.17								0.18			0.28	0.17	0.14

[a] pH reading on meter; [b] Bates (1973); [c] calculated from nomogram in Corning literature for triple-purpose glass; [d] taken from graphs in the manufacturer's technical literature; [e] low-temperature, corrosion-resistant and universal glasses.

buffers used to calibrate the electrodes. Pressure can also affect the liquid junction and this may be a problem when the electrodes are used in flow-cells (p. 42).

Contamination

A problem of measurements above pH 5 is absorption of carbon dioxide from the atmosphere. With well-buffered samples the errors caused may be negligible, but with poorly buffered solutions it may be necessary to use a flow cell to obtain accurate results. Volatile acids and bases may cause errors by being lost from, or absorbed by, the sample. Stirring should be as gentle as possible and the apparatus should be kept away from obvious sources of contamination such as bottles of concentrated hydrochloric acid or concentrated ammonia solution.

Effect of other substances

High concentrations of sodium ion may interfere in alkaline solutions; the pH at which interference starts to be significant depends on the composition of the glass. This interference is the *alkali error* and causes the pH to be underestimated. A similar interference could be expected from lithium ions, while potassium, ammonium and alkaline earth ions have negligible effects. The interference could be defined by a selectivity coefficient, K_{HM}, as follows

$$E = E^\circ + k \log (a_H + K_{HM} a_M)$$

where a_H is the hydrogen ion activity and a_M is the activity of the interfering univalent ion. It is more common, however, to express the interference as a correction to be added to the observed pH reading at a given metal concentration. Table 2 shows the corrections for some glasses at 20–25 °C and 0.1 and 1.0 mol l^{-1} sodium (each glass may be used in several types of electrode). These corrections depend on temperature and especially at high pH values may refer to the reading after a time of about 5 min immersion, as the readings may drift considerably. Corning supply a nomogram with their electrodes, enabling corrections to be applied over a wide range of temperatures and sodium concentrations.

Fluoride ions will attack the glass membrane, especially at pH values below 6, and constitute a serious interference.

PRECISION AND ACCURACY

Readings reproducible to 0.05 pH units may reasonably be expected in the pH range 3–10 at room temperature in well buffered solutions. At more extreme pH values and temperatures, poorer precision and accuracy will be obtained, but these will depend on the types of glass and reference electrodes used. It is difficult to improve on a reproducibility of ±0.01 pH units for routine measurements — great precautions as to temperature control and prevention of contamination by carbon dioxide are necessary. The quality of measurements in poorly buffered solutions cannot easily be predicted; often they will be no better than ±0.1 pH units. Tests in thirty laboratories on the same pH 7.3 buffer solution produced a standard deviation of 0.13 pH units (APHA, 1971).

Accuracy may be lost through absorption of carbon dioxide or by the presence of suspensions, sols and gels. In many cases the measured pH may be acceptable as a control parameter, even though its theoretical accuracy is in doubt.

RESPONSE TIME

The time for an electrode that has been rinsed to reach equilibrium in a well buffered solution should be only a few seconds. In poorly buffered solutions or at very high or low pH values the response will be slower, possibly extending to several minutes. Low temperatures increase the response time, especially for electrodes with a moderately high resistance. In many cases the cause of the slow response may be the reference electrode (Midgley and Torrance, 1976).

TRACING FAULTS

The measurement of pH is subject to much the same faults as are found for other ion-selective electrode methods (p. 127), although as the temperature is generally less rigorously controlled, errors from that source are more likely. The following faults are worthy of special consideration:

Drift

In poorly buffered alkaline solutions, absorption of carbon dioxide from the atmosphere is a major cause of the downward drift of pH readings. The drift may be minimized by immersing the electrode deep in a large volume of sample and stirring only gently; a flow-cell technique can eliminate the problem. Other volatile acidic and basic substances should be kept well away from the pH equipment, as their absorption can cause similar drifts, e.g. ammonia and concentrated hydrochloric acid.

Loss of response

Fouling of the electrodes may reduce their sensitivity by impeding the access of each fresh solution to the glas membrane. Remove the fouling according to its nature (see Chapter 4, Table 4.1) and recalibrate the electrodes before making any more measurements.

If the glass membrane is scratched, or has been allowed to dry out, there may be some loss of response. Condition the electrode for 24 h, using the treatment applied when the electrode was new.

A glass electrode with a cracked or broken membrane will indicate the same pH in all solutions, but will not go open circuit.

Slow response

Fouling, scratching and drying out are likely causes, but note that many pH systems participate in slow equilibria and the electrodes may not be at fault.

Bibliography

The theory of glass electrodes is best covered by the book edited by Eisenman (1967). The books by Bates (1973) and Mattock (1961) can be highly recommended also.

APHA, 1971, *Standard Methods for the Examination of Water and Wastewater,* 13th edn., American Public Health Association, Washington, D.C.

Bates, R. G., 1973, *Determination of pH, Theory and Practice,* 2nd edn., John Wiley and Sons, New York and London.

Eisenman, G., 1967, *Glass Electrodes for Hydrogen and Other Cations,* Edward Arnold, London/Marcel Dekker, New York.

Mattock, G., 1961, *pH Measurement and Titration,* Heywood, London.

Midgley, D., and K. Torrance, 1976, 'An assessment of various types of reference electrode for use in continuous potentiometric analysis with particular reference to highly pure waters', *Analyst,* **101**, 833.

Perrin, D. D., and B. Dempsey, 1974, *Buffers for pH and Metal Ion Control,* Chapman and Hall, London.

Determination of Acidity and Alkalinity

Potentiometric titrations can be carried out with a wide variety of pH-responsive electrodes, but in almost all cases the glass electrode is the most accurate and most convenient. The information that can be extracted from a pH titration depends on the nature of the acids or bases present in the sample and their relative concentrations. Samples containing only one acid or base can almost always be analysed by using one of the methods described in Chapter 6, but if several species are present it may be difficult to interpret the titration curve at all.

APPARATUS

Glass electrode; reference electrode; pH meter; burette; stirrer or nitrogen cylinder

There are very few restrictions on the choice of glass or reference electrode to be used. Unless there are particular reasons for titrating hot ($>50\ °C$) or cold ($<10\ °C$) samples, a general-purpose glass electrode with a toughened membrane, such as is available from a wide range of manufacturers, will be suitable. Any of the usual reference electrodes may be used, unless the sample contains ions that react with chloride or sulphate ions. If this is the case, use a double-junction reference electrode with a suitable bridging electrolyte. For most purposes a combination pH and reference electrode is suitable.

The simplest kinds of pH meters are suitable for titrations that show sharp end-points of the step-curve type. Temperature compensation, calibration slope-adjustment controls and similar refinements are not necessary. For Gran titrations, a slope control is desirable if some accuracy is not to be lost and for linear titration plots the electrodes and the pH meter need to be accurately calibrated. Meters reading to 0.01 pH units are adequate for linear titration plots.

Provided none of the acids or bases present are volatile, nitrogen may be bubbled through the sample to prevent the absorption of carbon dioxide at high pH values and also to mix the solution. Otherwise, a magnetic or mechanical stirrer should be used.

REAGENTS

Standard solutions of hydrochloric acid, sulphuric acid and sodium hydroxide can be purchased either as ready-prepared bulk solutions or as vials of concentrated solution for dilution to 500 or 1000 ml. Alternatively, the acids may be prepared by dilution of the concentrated acids and standardized against weighed amounts of sodium carbonate (dried at 140 °C) and the sodium hydroxide solution can be prepared from washed pellets of sodium hydroxide and then standardized against weighed amounts of potassium hydrogen phthalate (ground and then dried at 120 °C). For the most accurate work primary standard grades of anhydrous sodium carbonate and potassium hydrogen phthalate are available. Woodward and Redman (1973) describe the preparation of pure sodium carbonate and discuss many other aspects of standardization.

Water that is free of carbon dioxide is best prepared by taking it directly from the outlet of a mixed-bed deionization unit and collecting it by displacement of nitrogen. Dissolved carbon dioxide may be removed from distilled or deionized water by boiling for about 15 min and cooling rapidly to room temperature with a slightly oversized beaker inverted over the neck of the flask. For better results, insert a wash-bottle head in the neck of the flask and pass nitrogen during the cooling period or exclude carbon dioxide by fitting a tube containing soda-lime or an equivalent proprietary absorbent (Ascarite; Carbosorb; Caroxite) – do not merely stopper the flask.

Standardizations should be carried out in the same manner as the titrations of unknown samples. Store all standard solutions well away from concentrated solutions of hydrochloric acid, ammonia or other volatile acids or alkalies.

SAMPLE COLLECTION

Samples containing volatile species should be collected by the method described on p. 109 and analysed as soon as possible. Do not store the samples near concentrated hydrochloric acid or ammonia or any other volatile acid or alkali.

CONDITIONING AND STORAGE OF ELECTRODES

New glass electrodes, especially if supplied with the bulb dry, should be allowed to soak overnight in approximately 0.01 mol l^{-1} hydrochloric acid solution. Thereafter they should be stored with the bulb immersed in deionized water and should not be allowed to dry out. Electrodes that have dried out can often be restored by immersion in dilute hydrochloric acid. If titration with sodium hydroxide results in the precipitation of a metal hydroxide the electrode should be rinsed with dilute hydrochloric acid before being stored or used again. Conversely, if titration with an acid results in precipitation, the electrode is rinsed with sodium hydroxide or ammonia solution ($0.01-0.1$ mol l^{-1}).

Table 1 Normality and concentration of acids and bases

Acid or Base		10^{-3} N	1 ppm
monobasic acids, 1N ≡ 1 mol l^{-1}			
hydrochloric	as HCl	36.46 ppm	2.743×10^{-5} N
nitric	as HNO$_3$	63.01 ppm	1.587×10^{-5} N
perchloric	as HClO$_4$	100.46 ppm	9.954×10^{-6} N
boric, HBO$_3$	as boron	10.81 ppm	9.251×10^{-5} N
bicarbonate	as NaHCO$_3$	84.01 ppm	1.190×10^{-5} N
acetic	as CH$_3$COOH	60.05 ppm	1.665×10^{-5} N
potassium hydrogen phthalate	as KHC$_8$H$_4$O$_4$	204.23 ppm	4.896×10^{-6} N
Dibasic acids, 1N ≡ 0.5 mol l^{-1}			
sulphuric	as H$_2$SO$_4$	49.04 ppm	2.039×10^{-5} N
hydrogen sulphide	as H$_2$S	17.04 ppm	5.869×10^{-5} N
carbonic	as CO$_2$	22.00 ppm	4.544×10^{-5} N
Monoacidic bases, 1N ≡ 1 mol l^{-1}			
sodium hydroxide	as NaOH	40.00 ppm	2.500×10^{-5} N
bicarbonate	as NaHCO$_3$	84.01 ppm	1.190×10^{-5} N
ammonia	as NH$_3$	17.03 ppm	5.872×10^{-5} N
Diacidic bases, 1N ≡ 0.5 mol l^{-1}			
carbonate	as Na$_2$CO$_3$	52.99 ppm	1.887×10^{-5} N
	as CaCO$_3$	50.04 ppm	1.998×10^{-5} N

UNITS

The acidity or alkalinity of a mixture of substances can most conveniently be expressed in units of normality or equivalents per litre. The relationship between normality and concentration is given in Table 1 for some common acids and bases.

1N ≡ 1 equivalent per litre ≡ 1000 meq l^{-1}

1 meq l^{-1} ≡ 1 epm (equivalents per cubic metre)

ANALYTICAL PROCEDURE

Step 1 Calibrate the electrodes according to the procedure described in *Determination of pH*. If a step curve or derivative curve is being plotted, this step may be omitted.

Step 2 Pipette a suitable volume of sample, V_s, into the titration vessel.

Step 3 Remove the electrodes from the solution in which they have been stored, rinse them with deionized water and then remove the surplus of water with a paper tissue. The bulb of the glass electrode should not be wiped in this process: a drop of water hanging from the bulb may be soaked up on the tip of the paper tissue.

Step 4 If necessary, add carbon dioxide-free water until the bulb of the glass electrode and the junction of the reference electrode are completely covered. If the results are to be obtained from a Gran plot or linear titration plot, this water should be added by pipette and the volume, V_d, noted.

Step 5 Start the stirrer or the flow of nitrogen, as appropriate. Do not stir so violently as to create a vortex, nor pass nitrogen so vigorously as to cause splashing of the sample.

Step 6 Fill the burette with the titrant of concentration N meq l^{-1}, check that there are no air bubbles in the tip and note the volume reading V_0.

Step 7 Add a portion of titrant. Note the new reading on the burette, V_x, and when a steady potential or pH has been reached note the value, E_x

or

Start the automatic titrator and omit Steps 8 and 9.

Step 8 Repeat Step 7 until enough data has been collected, i.e. until the desired pH has been attained or the pH or e.m.f. has shown a sharp increase corresponding to the end-point.

Step 9 Plot the data as a step curve, derivative curve, Gran plot or linear titration plot, as desired.

Step 10 Find the equivalent volume, V_e, and calculate the normality of the sample, $c = N \cdot V_e / V_s$ meq l^{-1}.

Procedural variations

The analytical method outlined above can only be a general guide to the procedure to be followed in any particular case. Further hints are given below and before adopting a titrimetric method for routine use it should be tested with model solutions containing known concentrations of the substances likely to be found in the sample. The shape of the titration curve will often indicate that the conditions have changed and that a procedure may need to be modified.

1. Stirring and mixing

A stream of nitrogen or a similarly inert gas may be used to mix the solution during the titration only if the sample does not contain volatile acids or bases, e.g. carbon dioxide, sulphur dioxide, hydrogen sulphide or ammonia. Gas mixing has the advantage of reducing the absorption of carbon dioxide from the atmosphere.

2. Titration vessels

Open beakers are not satisfactory for precise work. A tall-form spoutless beaker fitted with a clean rubber bung can be used. The bung should have holes for the pH and reference electrodes, for the nozzle of the burette and possibly for a gas-inlet tube. The use of a combination pH and reference electrode will make this arrangement more convenient, as the electrodes require large holes in the bung and often leave little room for the burette. Alternatively, a three-necked flask may be used, or a two-necked flask if a combination electrode is available. Electrodes fitted with standard ground-glass cones are available from a number of manufacturers, including Radiometer, Pye and Philips, and these are particularly convenient for this application. Flasks tend to require larger volumes of sample than beakers of the same capacity because it is more difficult to cover the electrodes in a wide vessel.

3. Titration to a fixed pH

If only strong acids or bases are present, titration to a pH of 7.00 will give an accurate result with properly calibrated apparatus, but if weak acids or bases are present the equivalence point will not usually occur at pH 7 and the pH at the equivalence point will depend on the concentration of determinand. It is recommended that model sample solutions be titrated so that the pH values at the theoretical equivalence points for a range of concentrations can be found empirically.

4. Gran plots

Gran plots are particularly useful when the titration curve is asymmetrical and the point of inflection difficult to locate. They may also be applied when the end-point would otherwise be obscured, e.g. interactions between transition metal ions and hydroxide ion may distort an acid–base titration curve at high pH values, but the data obtained at low pH values may be sufficient to give an end-point by Gran's method.

5. Titrant concentration

The concentration of the titrant should be at least ten times that of the determinand in the solution being titrated. In theory, the best results should be obtained when the titrant is very concentrated, but the practical difficulties of making very small additions and of handling concentrated solutions make titrant concentrations of 10–1000 meq l^{-1} the most popular.

SOURCES OF ERROR

Temperature effects

A significant change in temperature during the titration may lead to some distortion of the titration curve, with a consequent loss of precision. This is least important when the end-point is determined from the point of inflection, as an accurate calibration is not required.

Effects of other substances

In mixtures of acids or bases, the point of inflection may be obscure and other methods of finding the equivalence point inapplicable, depending on the equilibrium constants involved.

Metal ions can affect titration curves by forming complexes with weak acids or bases or by forming hydroxo complexes. The effect is to make the pH lower than it would be in the absence of the metal. Metals that form insoluble hydroxides may precipitate on the surface of the glass electrode. If the coating is not removed by washing the electrode with dilute hydrochloric acid, subsequent titrations may be impaired by a loss of sensitivity by the electrode.

As fluoride ions attack glass, the quinhydrone electrode should be used for the titration of solutions containing fluoride and the titration vessel should be made of plastic.

PRECISION

The great variation in the characteristics of different waters makes it impossible to predict the precision in a particular case. In the titration of standard hydrochloric acid with tetramethylammonium hydroxide solution, Dunsmore and Midgley (1972) obtained relative standard deviations of 0.07–0.15% for approximately 70 meq l^{-1} hydroxide solution by using Gran plots. 0.180 ppm boron was determined with a relative standard deviation of 8.3% by the method described below (APHA, 1971).

ACCURACY

Great accuracy is possible when titrating moderately concentrated (0.01–0.1 meq l^{-1}) solutions of strong acids or bases, but it declines as the concentration decreases and as the acids or bases become weaker. Solutions of a single very weak acid or base may best be analysed by means of a linear titration plot. The accuracy of determining acids or bases in mixtures is particularly hard to predict: in many cases it will be found that the best way of finding the total alkalinity or acidity may be from the Gran plot obtained when the titrant has been added in excess.

APPLICATIONS

1. Total acidity or alkalinity by titration to a fixed pH

In solutions containing complex mixtures of acids or bases, a sharp end-point may not occur, or it may not coincide with the equivalence point. For control purposes it may be better to titrate to a pH fixed by convention. If the full titration curve is plotted, results may be obtained at a number of pH values, with a consequent increase in the information provided. Because the acid–base equilibria are temperature dependent, the quantity of titrant needed for the attainment of a given pH will vary with the temperature, which should always be noted. Results should be reported as 'The acidity (or alkalinity) to pH A at B °C is C meq l^{-1}'.

2. Waters containing ferrous iron

Mine drainage waters and some industrial waste waters may contain substantial amounts of polyvalent metal ions in a reduced state — usually ferrous iron. The oxidation of ferrous iron and the subsequent hydrolysis of ferric iron generate acidity and a reliable measure of the acidity or alkalinity is achieved only if these processes are driven to completion: Pipette 50 ml of sample into a 250 ml beaker and measure the pH with a glass electrode. If necessary, add 20 meq l^{-1} sulphuric or hydrochloric acid in 5 ml increments until the pH is 4.0 or less. Add only 5 drops of 30% hydrogen perioxide solution and then boil the sample for 2–4 min. Cool the sample to room temperature and titrate with standard sodium hydroxide solution to pH 8.2. Report the result as 'The acidity (boiled and oxidized) to pH 8.2 is C meq l^{-1}'. Note that the result may be negative, corresponding to a solution containing excess alkalinity. The results are calculated as follows:

(a) pH $\leqslant 4.0$, no acid added

$\quad$ acidity = $V_e \cdot N_b / V_s$ meq l^{-1}

(b) pH > 4.0, acid added

$\quad$ acidity = $(V_e \cdot N_b - V_a \cdot N_a)/V_s$ meq l^{-1}

where V_a = the volume of N_a meq l^{-1} standard acid added, V_e = the volume of N_b meq l^{-1} standard sodium hydroxide solution required and V_s = the volume of the sample.

3. Determination of boron

Boric acid is so weak that it cannot be accurately titrated, except perhaps by a linear titration plot method. When the solution of boric acid is treated with mannitol, a complex acid is produced which can be titrated in a conventional way.

Pipette a portion of sample containing not more than 1 mg boron into a 400 ml tall-form beaker and dilute to 250 ml. Add a few drops of bromothymol blue indicator solution (1 g of bromothymol blue sodium salt in 100 ml water) and acidify with 0.5 mol l^{-1} sulphuric acid, adding 0.5–1 ml in excess. Bring to the boil and stir to expel carbon dioxide; this should be done cautiously at first, but then more vigorously when most of the carbon dioxide has been lost. Cover the beaker with a clock glass and allow it to cool to room temperature.

Standardize the pH meter and glass electrode with pH 7.00 buffer, rinse the electrodes with deionized water and immerse them in the sample solution. Add 0.5 mol l^{-1} sodium hydroxide until the pH is about 5.0 and then adjust the pH to exactly 7.00 by adding standard (0.02–0.025 mol l^{-1}) sodium hydroxide solution. Add 5.0 ± 0.1 g of boron-free mannitol; if the sample contains boron, the pH will drop. Titrate the solution with standard sodium hydroxide solution, noting the volume required to restore the pH to 7.00. Repeat the procedure with 250 ml boiled deionized water instead of the sample, the blank titre should be no more than about 0.1 ml of 0.025 mol l^{-1} sodium hydroxide. Calculate the boron concentration from

$$c = 10823 \, m \, (V_e - V_b)/V_s \text{ ppm boron}$$

where V_e = the titre of the sample, V_b = the titre of the blank, V_s = the volume of the sample and m mol l^{-1} = the concentration of the titrant.

It is important to store the titrant solution in a boron-free container. Pyrex beakers may be used, but they should first be cleaned by filling them with dilute acid and heating them on a steam bath. Germanium and vanadium(IV) react like boron, but these are fairly rare substances. Phosphate will react, but not quantitatively: remove phosphate if present at more than 10 ppm by precipitation with lead nitrate. Remove the excess of lead by precipitation with sodium bicarbonate.

Bibliography

APHA, 1971, *Standard Methods for the Examination of Water and Wastewater*, 13th edn., American Public Health Association, Washington, D.C.

Dunsmore, H. S., and D. Midgley, 1972, 'The preparation of tetramethylammonium hydroxide solution for use in pH titrations', *Lab. Pract.*, **21**, 791.

Woodward, C., and H. N. Redman, 1973, *High Precision Titrimetry*, Analytical Sciences Monograph No. 1, Society for Analytical Chemistry, London.

Determination of Lithium

GLASS ELECTRODES

There are no lithium-selective glass electrodes made as such, but some sodium-sensitive glass electrodes are almost equally sensitive to lithium ions. Electrodes made of lithium aluminosilicate glass appear to be more lithium-selective than those of sodium aluminosilicate glass and the three most selective electrodes have the former composition. The Beckman 39278, the Electronic Instruments Ltd. GEA 33 and the Leeds & Northrup 117201 electrodes have selectivities for lithium over sodium of 5, 0.1–1 and 0.1–0.2, respectively. As the electrodes are the same, the method is identical to that for the determination of sodium, except that the electrodes are standardized with lithium solution.

Apparatus

See *Determination of Sodium*. As sodium ions interfere with lithium measurements the same precautions to avoid contamination with sodium must be observed.

Reagents

Water

See *Determination of Sodium*. Sodium rather than lithium will be the significant impurity.

Alkaline additive

See *Determination of Sodium*.

Standard Lithium solution A (100 ppm)

Dry anhydrous lithium chloride (analytical reagent grade) in an oven at 105 °C overnight. Rapidly weigh 0.6109 g of the salt, dissolve it in water and make up to the mark with deionized water in a 1-litre calibrated flask. Store in a polyethylene bottle.

$$1 \text{ ml} \equiv 100 \text{ } \mu\text{g lithium}$$

Standard lithium solution B (10 ppm)

Pipette 50 ml of standard solution A into a 500 ml calibrated flask and make up to the mark with deionized water. Store in a polyethylene bottle.

$$1 \text{ ml} \equiv 10 \text{ } \mu\text{g lithium}$$

Standard lithium solution C (1 ppm)

Weigh 4.950 kg of water into a weighed 5-litre polyethylene aspirator. Add 50 ml of standard solution A and mix.

$$1 \text{ ml} \equiv 1 \text{ } \mu\text{g lithium}$$

Standard solution D (100 ppb)

Weigh 4.950 kg of water into a weighed 5-litre polyethylene aspirator. Add 50 ml of standard solution B and mix.

$$1 \text{ ml} \equiv 0.1 \text{ } \mu\text{g lithium}$$

Sample Collection

See *Determination of Sodium*.

Conditioning and Storage

See *Determination of Sodium*.

Concentration Range and Units

The EIL GEA 33 and Beckman 39278 electrodes have linear responses over the range 14 ppm–100 ppb and have useful but curved responses down to 5 ppb.

$$10^{-3} \text{ mol l}^{-1} \equiv 6.939 \text{ ppm}$$
$$1.44 \times 10^{-4} \text{ mol l}^{-1} \equiv 1 \text{ ppm}$$

Analytical Procedure

As for *Determination of Sodium*, except that lithium standards are used and the high-level calibration is used above 100 ppb.

Sources of Error

See *Determination of Sodium*. The effects of other substances are greater or smaller, depending on the selectivity of the electrodes for lithium over sodium. Sodium is the most likely interferent and it may be necessary to correct for its presence as in the determination of potassium by glass electrode (Friedman, 1967). The sodium could be determined separately by flame photometry or with an electrode that was less sensitive to lithium, e.g. the Corning, Orion or Metrohm electrodes (see Table 1 of *Determination of Sodium*).

Precision

The relative within-batch standard deviations obtained at various lithium concentrations with two types of electrode are shown in Table 1. Electrodes of the NAS 11-18 type, which are less lithium-selective that those in Table 1, have given within-batch relative standard deviations of about 0.5% in concentrated (7000–49000 ppm) lithium solutions (Eisenman, 1965).

Table 1 Precision of lithium-selective glass electrodes

Electrode	Relative standard deviation (%) at lithium concentrations of:				
	140 ppm	14 ppm	1.4 ppm	140 ppb	14 ppb
EIL GEA 33	1.4	1.6	3.4	3.5	12.5
Beckman 39278	3	0.8	1.8	3.9	1.7

Response Time

The time to reach a steady potential after a tenfold change in concentration in the range 140 ppm–14 ppb was 10–15 min for the EIL and Beckman electrodes in an EIL 8900 series flow cell with a sample flow rate of 4 ml min^{-1}. NAS 11-18 glass electrodes have a response time of about 15 min for changes between 700 and 70 ppm lithium in unbuffered solutions (Eisenman, 1965).

Comparison with Other Methods

No direct comparisons between methods have been made. Flame photometry is a more sensitive technique, but is subject to interferences from alkaline earth ions. Atomic absorption spectroscopy has about the same sensitivity as the electrodes, but is virtually free of interferences. For the analysis of discrete samples, especially if sodium is present, the spectroscopic methods would generally be preferred. The potentiometric method, however, has advantages for on-stream analysis. The neutral carrier membrane electrode has better selectivity than glass electrodes, but its limit of detection is higher.

Tracing Faults

See *Determination of Sodium*.

NEUTRAL CARRIER MEMBRANE ELECTRODE

A lithium-selective electrode of this type has been described (Güggi *et al.*, 1975) and one is produced commercially (Philips IS 56-Li), but no analytical applications have so far been described. The electrode has a linear response in the range 70 ppm–700 ppb and may be used down to 70 ppb. Each plastic membrane containing the neutral carrier has an operational life of 4–6 months and possibly longer. The response time for a change from 7 to 70 ppm is less than 30 s. The main advantage of the neutral carrier membrane electrode over the glass electrode is its greater selectivity for lithium over sodium and hydrogen ions. Interference by potassium and ammonium ions is approximately the same for both kinds of electrode. Table 2 shows the selectivity coefficients, K_{LiM}, for several commonly-occurring ions according to the equation

$$E = E^\circ + k \log (c_{Li} + K_{LiM}\, c_M^{1/m})$$

where m is the charge on the ion M and c_{Li} and c_M are the molar concentrations of lithium and interferent, respectively.

Table 2   Selectivity coefficients for the Philips IS 561-Li lithium electrode

Interfering ion	H^+	NH_4^+	Na^+	K^+	Rb^+	Cs^+	Mg^{2+}	Ca^{2+}
Selectivity coefficient	1	5×10^{-2}	5×10^{-2}	7×10^{-3}	4×10^{-3}	3×10^{-3}	2×10^{-4}	6×10^{-4}

The electrode should not be used below pH 3 or above pH 11, and prolonged use below pH 4 is inadvisable.

Analytical Procedure

In the absence of analytical results, it is recommended that the procedure for using potassium-selective neutral carrier membranes be followed, with the following exceptions:

(i) The reference electrode should be of the calomel type with a low rate of electrolyte outflow.

(ii) The high-level procedure is tentatively assigned the range 10–100 ppm lithium and the low-level procedure the range 0.1–10 ppm. Prepare the low-level calibration with standard lithium solutions B and C. Note that solution C will give concentrations one-tenth of those in Table 2 of the potassium method. If experience shows that measurements are possible at lower levels, use standard solution D also.

(iii) Ionic strength adjustment solutions containing sodium should not be used.

A better choice would be caesium chloride solutions (0.25 mol l^{-1} for the high-level calibration and a ten-times dilution of that for low levels). Dissolve 42.1 g of caesium chloride in water and make up to 100 ml to prepare the more concentrated solution.

As the lithium electrode has a fairly high selectivity for hydrogen ions, an ionic strength adjustor with some buffering capacity may be needed. The caesium chloride solutions above could be replaced by caesium hydroxide solutions, or a Tris buffer might be tried. Dissolve 12.114 g of *tris*-(hydroxymethyl)aminomethane and 5.1 ml of concentrated hydrochloric acid in water and make up to 100 ml. More experience with the electrode is required before an ionic strength adjuster can be reliably recommended, however.

Bibliography

Eisenman, G., 1965, 'The electrochemistry of cation-sensitive glass electrodes', *Adv. Anal. Chem. Instrum.*, **4**, 213.

Friedman, S. M., 1967, 'H$^+$ and cation analysis of biological fluids in the intact animal', in *Glass Electrodes for Hydrogen and Other Cations*, G. Eisenman, (ed.), Arnold, London/Marcel Dekker, New York, 1967, Ch. 16.

Güggi, M., U. Fiedler, E. Pretsch, and W. Simon, 1975, 'A lithium ion-selective electrode based on a neutral carrier', *Anal. Lett.*, **8**, 857.

Determination of Sodium with a Sodium-responsive Glass Electrode

Almost all manufacturers include a sodium-responsive glass electrode among their products, the principal ones being the Beckman 39278, Corning 47621000, Electrofact OG512 and OG572, Electronic Instruments Ltd. GEA 33, Leeds & Northrup 117201, 117188 and 117409, Metrohm EA109Na, Orion 94-11, Philips GI5Na, Radiometer G502 and Schott 9601. For measuring sodium concentrations greater than 10 ppm there is little to choose between the various electrodes, but as the concentration decreases, the selectivity of the electrode must be carefully considered if bias is to be avoided. A neutral carrier membrane electrode is now available (Philips IS 561Na) and this is less subject to interference hy hydrogen ions than the glass electrode, but its selectivity with respect to other ions is generally poorer and it has a higher limit of detection. Not enough results have been obtained with the neutral carrier electrode for an analytical procedure to be recommended but it should be possible to treat it much as the lithium neutral carrier electrode ($q.v.$).

APPARATUS

Sodium glass electrode; reference electrode; pH meter; flow cell; pump(s)

Sodium measurements are very prone to contamination and the standard and sample solutions must be kept free from contact with glass or the hands of the analyst. Even a brief residence in a glass pipette can result in significant contamination (Nesbett and Ames, 1963). Sodium salts are so common in the laboratory that any apparatus must be particularly thoroughly washed before coming into contact with the sample. The only practical means of trace analysis is to place the electrodes in a flow cell through which the sample is pumped with the minimum risk of contamination. Complete systems for industrial on-line analysis are made by Electronic Instruments Ltd., Orion, Leeds & Northrup and Electrofact. Similar but simpler apparatus for use in the laboratory is made by Beckman and Electronic Instruments Ltd. (series 89). All parts of the apparatus that make contact with the sample should be made of plastic or stainless steel and no glass parts should be used, except for the electrodes themselves.

The flow cell should have separate chambers for the sodium and reference electrodes, joined in such a way that diffusion of salt from the reference electrode in the downstream chamber back to the glass electrode is kept to a minimum. If this is the case, the usual types of reference electrodes with concentrated potassium chloride electrolytes can be used.

The performance required of the pump(s) depends largely on the design of the flow cell. With the Electronic Instruments Ltd. cell, two streams of about 3–6 ml min^{-1} are needed, which can be achieved with simple peristaltic pumps. Other flow cells can require flow rates up to 100 ml min^{-1}.

REAGENTS

It is very difficult to prepare water containing less than 0.5–2 ppb of sodium and therefore to prepare accurate standard sodium solutions below about 100 ppb. The implications for calibration are considered below.

Water

For preparing standard solutions and final washing of apparatus, use distilled water that has been passed through a mixed-bed deionization unit. It is preferable to circulate water in a plastic reservoir through the mixed bed for some time before withdrawing a sufficient quantity for one batch of standards into a well washed plastic aspirator. Large amounts of water containing less than 2 ppb sodium can be obtained by this technique.

Standard sodium solution A (23 ppm)

Dry sodium chloride (analytical reagent grade) at 250° to 350°C for 1–2 hours. Dissolve 0.117 g in water and dilute with water to 2 litres in a calibrated flask. Store in a polyethylene bottle. This solution is stable for at least 6 months.

    1 ml ≡ 23 µg sodium

Standard sodium solution B (230 ppb)

Weigh 4.95 kg of water in a pre-weighed 5-litre polyethylene aspirator. Add 50.0 ml of standard sodium solution A, and mix. This solution is stable for at least 8 weeks.

    1 ml ≡ 0.23 µg sodium

Alkaline additive

Hydrogen ions interfere with the sodium electrode unless there is a 10^3–10^4-fold excess of sodium ions in solution. The pH of the sample is therefore raised to a constant high level (pH 10–11) by addition of a base. The most common additive is ammonia, but *tris*-(hydroxymethyl)aminomethane, cyclohexylamine, dimethylamine, diethylamine and other primary and secondary amines are used. If the base is added directly as a liquid or a solution, the level of sodium contamination will be increased and the range of measurement restricted. Ammonia and the volatile amines can be added as vapours by bubbling air first through the base in liquid or solution form and then through the sample solution. For ammonia and the lighter amines, a flow rate for the air approximately equal to that of the sample is sufficient to give the required pH, but with the heavier amines, either the air flow rate must be increased or the temperature of the amine reservoir raised above amibient. For these reasons, diethylamine and dimethylamine are the most convenient amines and diethylamine is probably preferable in terms of safety (toxicity and flashpoint). Dimethylamine has the advantage that it can be obtained as cylinders of pressurized gas and dosed directly into the sample through a pressure-regulating valve, thus eliminating the air-pump from the system. A further consideration is that the protonated form of the amine may itself interfere with the measurement. Electrodes have selectivities of $10^2 - 10^4$ for sodium over ammonium ions and at sodium concentrations below 200 to 2 ppb, depending on the selectivity, ammonium ions from the ammonia added to maintain pH 11 in the sample will interfere. The selectivities of the substituted ammonium ions are not well defined, but they are such that interference by these species is much smaller than that by the ammonium ion itself. The problem of the alkaline additive has been discussed by Galetti and Spear (1970) and Eckfeldt and Proctor (1971).

For most work, the cheapest and most convenient additive is ammonia, but for the lowest sodium concentrations (< 3 ppb), diethylamine or dimethylamine are better. Place the ammonia solution (sp. gr. 0.88) or the liquid amine or a 25% aqueous solution of the amine in a drechsel bottle and connect the dosing pump to the outlet from the air space above the liquid. The inlet tube to the bottle is fitted with a porous frit submersed in the liquid.

SAMPLE COLLECTION

Use a clean screw-capped polyethylene aspirator and collect at least 250 ml of sample. Great care should be taken to avoid contamination during sampling. Immediately before analysis, wash the inside of the tap of the aspirator by running about 50 ml of the sample through it, rinse the outside of the tap with deionized water and connect the sample pump tube to the tap. Samples should be stable for many days.

CONDITIONING AND STORAGE OF ELECTRODES

The glass electrode should be placed in deionized water for 48 h before use for the first time. Between analyses, leave the electrode in the flow cell with the tip immersed in either deionized water or a dilute standard solution, with the usual base added, rather than in the last sample solution, particularly if the sample contains substances likely to interfere with electrode response or deposit on any part of the flow system. Never allow the electrode to dry out. Better accuracy is obtained if the equipment is in continuous use; if the apparatus is used regularly it is therefore advantageous to keep the apparatus running continuously on deionized water whenever samples or standards are not being analysed.

CONCENTRATION RANGE AND UNITS

Results have been reported in the range 1 ppb–12% sodium, although the same calibration technique cannot be used over the entire range. Samples containing more than 250 ppm are probably best diluted before analysis. Some electrodes may not be useful in the lower ranges, e.g. Goodfellow, Midgley and Webber (1976) showed that Beckman 39278 and Electrofact OG572 electrodes were suitable for use above 10–20 ppb, but that EIL GEA 33 and Leeds & Northrup 117201 electrodes could be used below 1 ppb.

10^{-3} mol $l^{-1} \equiv 23$ ppm sodium; 4.348×10^{-5} mol $l^{-1} \equiv 1$ ppm sodium

ANALYTICAL PROCEDURE

Step 1 If the reference electrode has been removed from the flow cell, wash the tip with deionized water before inserting it in the flow cell.
Step 2 Connect an aspirator containing standard sodium solution B to the appropriate pump tube, open the tap of the aspirator and start pumping the solution and the ammonia/amine–air mixture through the cell.
Step 3 When the potential is steady, note the reading E_1.
Step 4 With the pump still operating, disconnect the pump tube from the tap of the aspirator and transfer it to the tap of the aspirator containing the sample. Open the tap of the sample aspirator.
Step 5 When the potential is steady, note the reading E_2.
Step 6 Calculate the difference $\Delta = E_2 - E_1$.
Step 7 Either read off the concentration from the calibration graph or (for concentrations above 200 ppb only) calculate it from the equation

$c = 230 \times$ antilog (Δ/k) ppb

where k is the slope of the calibration graph (see below).
Step 8 If more samples are to be analysed, repeat Steps 4–7 each time. The frequency with which standardisation (Steps 2–3) is required depends on the desired precision of the analysis, but it is recommended that it be done at least once a day.

Checking for Bias

A second standard sodium solution should be analysed with each batch of samples. The concentration should be chosen to be as near the lowest sample concentration as is convenient. Any change in the calibration slope will appear as a bias.

Preparation of the Calibration Graph

It is necessary to consider whether the calibration is likely to be biased in the range of sodium concentration expected and then to adopt the appropriate procedure.

(a) High-level calibration (200 ppb and greater)

It should be possible to obtain water of sufficient purity to permit an accurate calibration and almost all electrodes are sufficiently sodium-selective for interference from the alkaline additive to be negligible.

Prepare a daily calibration graph by carrying out Steps 2 and 3 of the analytical procedure with standard solution B and Steps 4–6 with a second, more concentrated (s_2 ppb), standard solution. Then either calculate the calibration slope $k = (E_2 - E_1)/(\log s_2 - \log 230) \cong 58$ mV per tenfold increase in concentration or plot the points (log 230, 0) and (log s_2, $E_2 - E_1$) and join them by a straight line.

If a permanent calibration graph is to be used (see p. 117 for conditions), carry out Steps 2 and 3 of the analytical procedure with standard solution B and Steps 4–6 with at least four other standard solutions of higher concentration. Plot the e.m.f. differences obtained from Step 6 against logarithms of the concentrations.

(b) Low-level calibration (1–200 ppb) without amine interference

If the presence of sodium in the water used to prepare the standards is the only source of error, calibration (a) should be used and extrapolated to the lower concentrations. As a rule, ammonia is the only amine to produce a significant interference, the extent of which may be approximately calculated from

$$c' = K_{NaNH_4} \times \text{antilog}\,(pH - 14)\ \text{mol}\,l^{-1}\ \text{apparent sodium}$$

where K_{NaNH_4} is the selectivity coefficient from Table 1.

An alternative procedure is to determine the sodium content of the water by flame photometry (Webber and Wilson, 1969a) or atomic absorption spectrometry and apply a correction to the nominal concentrations of low-level standards prepared by dilution of solutions A or B. Carry out Steps 2 and 3 of the analytical procedure with standard solution B and Steps 4–6 with at least four dilute standards. A plot of the e.m.f. against the logarithm of the corrected concentration should be a linear extension of the high-level calibration.

(c) Low-level calibration with amine interference

If it is shown that sodium and potassium in the water used to prepare the standard solutions is negligible, yet the calibration against the nominal standard concentrations prepared by dilution of standard solutions A and B is curved, the cause almost certainly lies in the alkaline additive. If the amine is added directly in liquid or solution form, it will almost certainly carry sodium with it, thus causing the curvature; changing to vapour addition of a volatile amine should cure this. Another

Table 1 Selectivity coefficients, K_{NaM}, for sodium-responsive electrodes

Electrodes	H^+	Li^+	K^+	NH_4^+	Ca^{2+}
Beckman 39278	35–115	5	$2 \times 10^{-4}/ 2 \times 10^{-2}$	1.5×10^{-4}	
Corning 4762 1000		4×10^{-3}	10^{-3}	3×10^{-4}	
Electrofact OG512	4	10^{-2}	3×10^{-2}	10^{-2}	
EIL GEA 33	17	0.1–0.2	6×10^{-4}	5×10^{-5}	
Leeds & Northrup 117201	10	10^{-1}	5×10^{-3}	10^{-3}	
Metrohm EA109Na	200	5×10^{-3}	5×10^{-4}	5×10^{-5}	
Orion 94–11	100	2×10^{-3}	9×10^{-3}	6×10^{-5}	
Philips G15Na	100		$< 10^{-3}$		
Radiometer G502Na	2–10	10^{-2}	$10^{-2} - 2 \times 10^{-4}$	3×10^{-5}	
Schott 9601			5×10^{-3}		
Philips IS 561 Na	0.5	4×10^{-2}	5×10^{-1}	2×10^{-1}	2×10^{-3}

possible cause is interference from the protonated form of the amine; usually this is caused by ammonium ions (the extent may be calculated as above) and can be eliminated by changing the additive to diethylamine. As both the interference and the sodium impurity associated with the alkaline additive should be constant, there will be no bias in the calibration graph, which is prepared as in (b) above. There will, however, be some loss of precision and it is not valid to extrapolate the high-level calibration.

(d) Low-level calibration with sodium impurity and amine interference

In principle, it is possible to measure the sodium impurity in the water as in (b), correct the nominal concentrations accordingly and use these values to obtain a curved calibration as in (c). This should not be resorted to unless there is no alternative — purify the water as far as possible by using a circulating system as described above, use a volatile alkylamine as the additive and if necessary change the type of electrode used before accepting such a calibration.

SOURCES OF ERROR

Effect of temperature

Changes in temperature affect the calibration slope of the electrode, causing an increasing error as the difference between the sample and standard solution concentrations increases. This is a particular problem if the extrapolated calibration (c) is used and greater than usual care is needed to see that the temperatures of sample and standard solutions are within 1 °C of each other and of the temperature at which the calibration graph was prepared. Bergner (1976) showed that sudden temperature changes affected various makes of electrodes in different ways: the responses of Radiometer G502Na and Philips G15Na electrodes were less temperature dependent in 1380 ppm sodium solution than those of the Orion 94—11A or Beckman 39278. The Beckman electrode also showed hysteresis and a factor in the Orion electrode's performance might have been changes in the interference by ammonium ions at different temperatures.

Effect of other substances

The selectivity of almost all sodium electrodes follows the order

$$Ag^+ \gg H^+ \gg Na^+ > Li^+ > K^+ \approx NH_4^+ > Cs^+ \approx Rb^+ \approx Ca^{2+} > Mg^{2+}$$

i.e. the electrodes have a greater selectivity for silver and hydrogen ions than for sodium. Silver is rarely present in natural or industrial waters and the hydrogen ion interference is overcome by the addition of a suitable base. As a general rule, the pH should be at least three units higher than the value of $-\log(c_{Na})$, where c_{Na} is the molar concentration of sodium. Potassium interference may be important in biological samples but the other alkali metals rarely need be considered. Ammonium ion is probably the most troublesome, because ammonia is added to suppress the hydrogen ion interference. The use of diethylamine instead of ammonia largely avoids the problem and, at the relatively high pH achieved with diethylamine additions, only about 2% of any ammonia in the sample is present as ammonium ion. The seriousness of the interference expected with different electrodes can be estimated from Table 1. The selectivity constants K_{NaM} refer to the equation

$$E = E^\circ + k \log (c_{Na} + K_{NaM} \cdot c_M^{1/m}),$$

where c_M is the molar concentration of interfering ion and m is the charge. The tabulated values were obtained from many sources and by different methods in different conditions; they can only be used to indicate the order of interference to be expected. The most thorough comparative study of interferences in sodium-responsive electrodes has been carried out by Wilson, Haikala and Kivalo (1975).

Bergner (1976) found that at concentrations above 1 mol l^{-1} the responses of sodium-responsive electrodes were affected by the nature of the anion present, so that the sodium concentration was apparently higher in solutions of sodium hydroxide than in solutions of sodium chloride of the same concentration. Carbonate ion depressed the apparent concentration even further. At such high concentrations ion-pairing will be significant, even for sodium ions (Midgley, 1975), and the samples should be diluted before analysis. Bergner recommended a twentyfold dilution with 1 mol l^{-1} ammonium carbonate which also served as a buffer, and at such high sodium concentrations, interference by ammonium ion and contamination by sodium in the ammonium carbonate should be negligible.

Octadecylamine causes the response of the electrode to become sluggish, with gross errors in the indicated sodium concentration. Washing with ethanol restores the performance.

PRECISION

Regrettably, little has ben published on the precision of analysis with the sodium electrode. Van den Winkel et al. (1972) found a relative standard deviation of 2% in determination of synthetic standard solutions containing 1.38–1380 ppm sodium with a Beckman 39278 electrode. The Electronic Instruments Ltd. GEA 33 electrode gave relative standard deviations of 5% on standard solutions between 25 ppb and 25 ppm sodium (Hawthorn and Ray, 1968). At lower sodium concentrations (2–15 ppb) the relative standard deviation is of the order of 25% (Webber and Wilson, 1969), which has also been found for the analysis of boiler feed water (Diggens, Parker and Webber, 1972). Reynolds (1971) reported relative standard deviations of 3% with an Orion 94–11 electrode used in the analysis of Alpine waters containing 0.07–1.38 ppm sodium. Moore (1967) found relative standard deviations of 0.2–0.6% when analysing biological samples such as urine, serum and cerebrospinal fluid containing 1750–2500 ppm sodium. The NAS 11–18 electrode used is equivalent to the Corning 47621000 or Orion 94–11. Bergner (1976) analysed synthetic solutions containing 23 000–115 000 ppm sodium with a Radiometer G502 electrode; the relative standard deviations were in the range 0.5–0.7%.

ACCURACY

By comparison with flame photometry, the sodium electrode gives an answer about 1 ppb high, over the range 2–20 ppb (Webber and Wilson, 1969). This has been confirmed in the analysis of boiler feed water by Diggens et al. (1972). In boiler water containing about 70 ppm sodium, the electrode read 2.2% higher than the flame photometer (Lower and Eckfeldt, 1969). In mineral waters, the electrode was on average low by 0.4% in the range 2.5–30 ppm (Van den Winkel et al., 1972). Reynolds (1971) found that in snow-fed Alpine waters containing about 0.1 ppm sodium, the flame photometer was on average 7% higher than the electrode, but in glacial waters flame photometry gave readings 30–200% higher due to interference by suspended matter. In urine samples (Jacobson, 1966) results between 2300 and 4600 ppm were an average of 3% lower than those indicated by flame photometry.

RESPONSE TIME

In solutions containing 1–20 ppb sodium, Webber and Wilson (1969) found that 30 min were required to obtain a near-equilibrium response. Wilson et al. (1975) followed the step change from 2.3 to 23 ppm with various electrodes. At 25 °C with EIL GEA 33 and Beckman 39278 electrodes the response was 95% complete after 10–15 min, while the Orion 94-11 took 30 min and the Radiometer G502Na about 50 min. At 10 °C these times were extended by factors of 5–7.

COMPARISON WITH OTHER METHODS

Webber and Wilson (1969) compared flame photometry and the sodium glass electrode as means of determining sodium in power station waters in laboratory conditions. At concentrations of about 1 ppb the electrode was more precise, but took longer to achieve a result. It was less dependent on experimental conditions and needed less attention from the analyst. The cost of the electrode system was also much lower. Since that time, the electrode technique has been improved at low concentrations by the use of volatile amines instead of ammonia as the alkaline additive. The advent of atomic absorption and flameless atomic absorption spectrophotometry has improved the precision of the rival techniques. At 1 ppb, a precision of 0.03 ppb can be expected instead of the 0.5 ppb of flame photometry. For continuous on-stream monitoring, there is no significant competition with the ion-selective electrode method. Table 1 shows that the selectivity of glass electrodes for sodium over other common cations is much better than that of the neutral carrier electrode (Philips IS 561Na) except in the case of hydrogen ions, which can normally be controlled by the addition of buffer. As the glass electrodes also have lower limits of detection than the neutral carrier electrode, they would generally be preferred for analytical work unless the addition of a buffer to control the pH is not possible.

OTHER APPLICATIONS

Sodium-selective electrodes have been widely used in clinical, biochemical and soil analyses (Eisenman, 1967), in the analysis of concentrated brines (Truesdell, Jones and VanDenburgh., 1965) and in thermodynamic studies (Dunsmore and Midgley, 1971).

TRACING FAULTS

In addition to the general problems of ion-selective electrode measurements (p. 127), the electrode may show no response to changes in sodium concentration because the glass membrane is broken or cracked, which may be checked by visual inspection. Another possibility is that the sample is not being pumped through the cell because of pump failure or a blocked sample line. The electrode may have a low sensitivity to sodium ions because the pH of the sample is too low. Measure the pH with a glass pH electrode and if it is not sufficiently alkaline (pH 10–11), check that the ammonia/amine dosing pump is working properly, that the tubing carrying the vapour is intact and that the reservoir containing the amine has not been exhausted. If the reservoir contains an aqueous solution of the amine, the amine will be preferentially removed by the stream of air, leading to a gradual decrease in the concentration of amine added to the sample and hence to a lower pH; the reservoir should be replenished regularly with fresh solution. If the sample has a naturally high acidity, the concentration of amine added must be increased by changing from an aqueous amine solution to the pure amine, by increasing the rate of air-flow through the reservoir or, ultimately, by changing from vapour addition to direct addition. In the last case the sodium blank would be expected to increase and a more curved calibration would be obtained.

If the electrode is allowed to dry out, virtually all aspects of its performance may deteriorate. The restoration of dried-out electrodes is problematic and it is usually more convenient to replace them with new ones. A properly designed flow cell does not allow the electrode to be left dry when the flow stops, but there is a danger of evaporation if the flow cell is not used regularly.

Bibliography

Bergner, K., 1976, 'Comparisons of some sodium-selective electrodes in concentrated solutions for use in automatic monitoring systems', *Anal. Chim. Acta*, **87**, 1.

Diggens, A. A., K. Parker, and H. M. Webber, 1972, 'The continuous determination of sodium in high purity water by using a sodium monitor incorporating a sodium-responsive glass electrode', *Analyst*, **97**, 198.

Dunsmore, H. S., and D. Midgley, 1971, 'Sodium glass electrode studies of sodium tartrate complexes', *J. Chem. Soc. (A)*, 3238.

Eckfeldt, E. L., and W. E. Proctor, 1971, 'Low level sodium ion measurement with the glass electrode', *Anal. Chem.*, **43**, 332.

Eisenman, G. (ed.), 1967, *Glass Electrodes for Hydrogen and Other Cations*, Edward Arnold, London/Marcel Dekker, New York.

Galetti, B. J., and J. F. Spear, 1970, 'Water quality measurements by sodium ion analysis for boiler feedwater applications', *Proc. Amer. Power Conf.*, **32**, 817.

Goodfellow, G. I., D. Midgley, and H. M. Webber, 1976, 'Factors affecting the limit of detection of sodium-responsive glass electrodes', *Analyst*, **101**, 848.

Hawthorn, D., and N. J. Ray, 1968, 'Determination of low levels of sodium in water by using a sodium-ion responsive glass electrode', *Analyst*, **93**, 158.

Jacobson, H., 1966, 'Direct determination of sodium and potassium in the presence of ammonium with glass electrodes', *Anal. Chem.*, **38**, 1951.

Lower, W. A., and E. L. Eckfeldt, 1969, 'Sodium ion monitoring', *Ind. Water Eng.*, **6**, 27.

Midgley, D., 1975, 'Alkali metal complexes in aqueous solution', *Chem. Soc. Rev.*, **4**, 549.

Moore, E. W., 1967, *'Hydrogen and cation analysis in biological fluids in vitro'*, in *Glass Electrodes for Hydrogen and Other Cations*, G. Eisenman (ed.), Edward Arnold, London/Marcel Dekker, New York, Ch. 15.

Nesbett, F. B., and A. Ames, 1963, 'Increase in sodium contamination from glass as a function of time after cleaning', *Anal. Biochem.*, **5**, 452.

Reynolds, R. C., 1971, 'Analysis of Alpine waters by ion electrode methods', *Water Resources Res.*, **7**, 1333.

Truesdell, A. H., B. F. Jones, and A. S. VanDenburgh, 1965, 'Glass electrode determination of sodium in closed basin waters', *Geochim Cosmochim. Acta*, **29**, 725.

Van den Winkel, P., J. Mertens, G. De Baenst, and D. L. Massart, 1972, 'Automatic potentiometric analysis of sodium in river and mineral waters', *Anal. Lett.*, **5**, 567.

Webber, H. M., and A. L. Wilson, 1969, 'The determination of sodium in high purity water with sodium-responsive glass electrodes', *Analyst*, **94**, 209.

Webber, H. M., and A. L. Wilson, 1969a, 'The flame-photometric determination of sodium in high purity water', *Analyst*, **94**, 569.

Wilson, M. F., E. Haikala, and P. Kivalo, 1975, 'An evalutaion of some sodium ion-selective glass electrodes in aqueous solution. Parts I and II', *Anal. Chim. Acta*, **74**, 395, 411.

Determination of Potassium

The three different kinds of potassium-selective electrode have quite different properties, which must be considered if the best choice is to be made for a particular application. Potassium-selective glass electrodes are durable and have stable standard potentials and sensitivities, but poor selectivity for potassium over the commonly occurring sodium and ammonium ions. Among such electrodes are the Electronic Instruments Ltd. GKN 33, the Corning 47622000, the Beckman cation glass electrode 39137 and the Electrofact OG612 and OG672 electrodes. Liquid ion-exchange electrodes of the neutral carrier type have relatively high selectivities for potassium but are generally less stable in all aspects of their performance than the glass membrane type and therefore require more attention. The neutral carrier electrodes are usually based on valinomycin, but the solvent and membrane support matrix vary between the different makes, which include the Orion 93-19, Philips IS 561-K, Radiometer F2312K, Metrohm EA301K, EDT Supplies EE-K, Activion 003 15 014, Simac K/1C and Beckman 39622. A third type, the Corning 47613200, had a liquid ion-exchange membrane containing potassium tetra(chlorophenyl)borate but is no longer available. This type of electrode is intermediate in selectivity between the glass and neutral carrier types, but is probably no more stable than the latter.

Some of the properties of the different electrodes are listed in Table 1. If selectivity over sodium is important, e.g. in most natural waters and biological fluids, the neutral carrier types are clearly the best. This is true also in regard to hydrogen and ammonium ion interferences, but these can often be overcome by chemical treatment of the sample. As far as cost and convenience of handling are concerned, the glass membrane electrodes are much superior. If potassium is present in a greater than tenfold excess over other alkali metal ions it should be possible to use a glass electrode without introducing a significant bias in the results, but such circumstances are rather rare.

USE OF POTASSIUM-SELECTIVE NEUTRAL CARRIER MEMBRANE ELECTRODES

These electrodes are far less prone to interference by sodium and hydrogen ions than the glass membrane types and are therefore more suitable for potassium determinations in most types of samples. In other aspects of their performance they require more careful attention than the glass electrodes.

Apparatus

Potassium electrode; reference electrode; magnetic stirrer; pH meter; plastic beakers

The properties of the various electrodes are summarized in Table 1. It should be noted that there are considerable differences between the results reported by different workers, depending on the conditions of the experiment and possibly on the age of the electrode. The Orion electrode, however, appears to be more prone to hydrogen ion interference. The Orion electrode is supplied with two sensing modules, each of which should last for about six months. The modules are screwed onto the body of the electrode before use. The Philips, Simac, EDT and Activion electrodes are supplied with plastic membranes containing the neutral carrier. Each membrane has a life of 4–6 months and has to be mounted in the electrode before use. The Beckman electrode has solid membranes already containing the neutral carrier. The relative merits of the three types of assembly have not yet been proved in practice.

Although saturated calomel electrodes with low rates of electrolyte outflow have been used as reference electrodes, this is not to be recommended. Either use a double junction reference electrode with a sodium chloride solution in the outer compartment or modify a calomel or silver–silver chloride reference electrode by using a sodium chloride solution instead of the usual potassium chloride solution. The concentration of sodium chloride to be used depends on the nature of the sample (see below).

Reagents

Water

Water of sufficient purity can be prepared by passing mains water through a mixed-bed deionization unit. Prepare a sufficient quantity of water for one batch of standard solutions and store it in a well washed plastic aspirator.

Standard potassium solution A (1000 ppm)

Dry potassium chloride (analytical reagent grade) at 110 °C for 1–2 h. Dissolve 1.907 g in deionized water and dilute with more water to 1 litre in a calibrated flask.

1 ml ≡ 1.0 mg potassium

Standard potassium solution B (100 ppm)

Pipette 50 ml of solution A into a 500-ml calibrated flask and make up to the mark with deionized water.

Table 1 Properties of potassium-selective electrodes

Parameter	Glass electrodes			Liquid ion-exchange		Neutral carrier electrodes			
	EIL GKN 33	Beckman 39137	Corning 47622000	Corning 476132000	Philips IS 561K	Beckman 39622	Orion 93-19	Activion 003 15 014	Radiometer F2312K
Linear response range (mol l^{-1})	$10^0 - 10^{-6}$		$10^0 - 10^{-5}$	$10^0 - 10^{-4}$	$10^{-1} - 10^{-5}/10^{-6}$	$10^{-1} - 10^{-5}$	$10^{-1} - 10^{-5}/10^{-6}$	$10^0 - 10^{-4}$	$10^0 - 10^{-5}/10^{-6}$
Temperature range (°C)				15–50	0–50		0–50		
Life (weeks)	> 26		> 26		18–26		26	8/12–26	
*Selectivity, K_{KM}, to:									
Cs$^+$	0.03			20	0.4	0.5	1.0	0.4	0.5
Rb$^+$	0.5			10	2–3	2	3–25	2	2.8
Na$^+$	0.13	0.05		0.01	3×10^{-6}	5×10^{-5}	2×10^{-4}	2.6×10^{-3}	7×10^{-5}
Li$^+$	0.06			0.004		3×10^{-4}	10^{-4}	2×10^{-3}	4×10^{-5}
NH$_4^+$	~1			0.025	0.01	0.02	0.03	0.3	0.02
H$^+$	50–100		3		10^{-5}	2×10^{-4}	0.01		
Mg^{2+}	0.03			0.003	6×10^{-6}	2×10^{-5}	$10^{-3} - 2 \times 10^{-4}$	2×10^{-3}	
Ca^{2+}	0.03–0.1			0.005	3×10^{-5}	2×10^{-5}	$2 \times 10^{-3} - 2 \times 10^{-4}$	2.5×10^{-3}	

*$E = E^\circ + k \log(c_{K^+} + K_{KM} \cdot c_M^{1/m})$, where m is the charge on the ion M.

$1\ ml \equiv 100\ \mu g$ potassium

Other potassium solutions are prepared by dilution of standard solutions A and B.

Standard potassium solution C (10 ppm)

Pipette 50 ml of standard solution B into a 500-ml calibrated flask and make up to the mark with deionized water.

$1\ ml \equiv 10\ \mu g$ potassium

Standard potassium solution D (1 ppm)

Pipette 50 ml of standard solution C into a 500-ml calibrated flask and make up to the mark with deionized water.

$1\ ml \equiv 1\ \mu g$ potassium

High-level ionic strength adjustor

Dissolve 14.6 g of sodium chloride (analytical reagent grade) in deionized water and make up to the mark in a 100-ml calibrated flask.

Low-level ionic strength adjustor

Either dissolve 1.5 g of sodium chloride (analytical reagent grade) in deionized water and make up to the mark in a 100-ml calibrated flask or pipette 10 ml of the high-level ionic strength adjustor into a 100-ml calibrated flask and make up to the mark with deionized water.

Reference electrode filling solutions

Pipette 1 ml of either the high- or low-level ionic strength adjustor, as appropriate, into a 100 ml calibrated flask and make up to the mark with deionized water.

Sample collection

Use a clean screw-capped polyethylene bottle and collect at least 250 ml of sample.

Conditioning and Storage of Electrodes

It is normally recommended by manufacturers that electrodes should be removed from the solution and stored in air after use. The electrodes can be used immediately on assembly or after storage, but there is some evidence that better performance can be obtained if the electrode is conditioned for about 30 min in a solution similar to those being analysed. The electrodes should not be left for long periods in solutions in which there is a serious interference effect. For long periods of storage, unscrew the sensing module of the Orion electrode and keep it in its glass vial. Oxidation of plastic membranes may cause them to harden and the electrode to lose its response: the shelf life of dry membranes is over six months.

Concentration Range and Units

The response is Nernstian above 10 ppm and possibly above 1 ppm, but the curved part of the calibration can be used down to 0.1 ppm or even lower. The exact range depends on the individual electrode and may change with its age.

$$10^{-3} \text{ mol l}^{-1} \equiv 39.1 \text{ ppm}; \quad 1 \text{ ppm} \equiv 2.558 \times 10^{-5} \text{ mol l}^{-1}$$

Analytical Procedure

The procedure to be used in the range 10–1 ppm depends on the linear range of the calibration graph, which should reach 10 ppm in all cases, but may extend to 1 ppm, depending on the individual electrode.

Step 1 Assemble the potassium-selective electrode according to the manufacturer's instructions.

Step 2 Fill the reference electrode with the appropriate filling solution, depending on whether measurements are to be made in the high- or low-level range. If the electrode has been used before with a different filling solution, rinse the interior with three portions of deionized water and one of the filling solution before finally filling it. If the electrode is of the double-junction type, the filling in the outer compartment should be completely replaced at least once a week. If the electrode is of the single-junction type, allow the internal element to equilibrate overnight with the reference solution before it is used for the first time. (With calomel electrodes an even longer equilibration time may be needed.)

High-level procedure (1000–10 or 1 ppm)

Step 3 Choose a standard solution of a concentration s_1 greater than the expected sample concentration. One of the standard solutions A, B and C is recommended, otherwise prepare by dilution of standard A or B as appropriate.

Step 4 Pipette 50 ml of the solution into a 100-ml plastic beaker containing a plastic-coated magnetic stirrer bar. Add 1 ml of high-level ionic strength adjustor.

Step 5 Rinse the electrodes with deionized water, dry them with a paper tissue and immerse the tips below the surface of the solution in the beaker. Check that there are no air bubbles at the surface of the membrane.

Step 6 Start the stirrer and adjust the control to give brisk agitation without creating a vortex. Stir all solutions at the same rate. Note the position of the electrode holder on the stand so that it can be returned to the same place every time.

Step 7 When a steady potential is reached, note the reading E_1.

Step 8 Repeat Steps 4 and 5 with a second standard solution of concentration, s_2, where $s_2 \leqslant 0.5\, s_1$.

Step 9 When a steady potential has been reached, note the reading E_2.

Step 10 *Either* Calculate the calibration slope, per tenfold increase in concentration.

$$k = \frac{E_1 - E_2}{\log s_1 - \log s_2} \text{ mV}$$

The value of k may be less than theoretical, but it should be greater than 50 mV per decade,

or

Prepare a temporary calibration graph by drawing a straight line through the points $(\log s_1, 0)$ and $(\log s_2, E_2 - E_1)$.

Step 11 Using the sample solution, repeat Steps 4 and 5.

Step 12 When a steady potential is reached, note the reading E_x.

Step 13 Calculate the e.m.f. difference $\Delta = E_x - E_1$.

Step 14 Either read the concentration corresponding to Δ off the calibration graph, or calculate it from

$$c_x = s_1 \times \text{antilog}\,(\Delta/k)$$

Step 15 For each sample solution repeat Steps 8, 9, 13 and 14. Restandardize with the first standard solution (Steps 4, 5 and 7) after every 10 samples or every 90 min, whichever is the sooner. Check the calibration slope (Steps 10–12) at least twice daily.

Worked example The e.m.f. observed with a 100 ppm standard solution is 5 mV and with a 20 ppm standard is 34 mV. The slope is therefore

$$k = \frac{5 - (-34)}{\log(100) - \log(20)} = \frac{39}{0.699} = 55.8 \text{ mV per decade}$$

The e.m.f. observed with the sample solution is -12 mV and the sample concentration is therefore

$$c_x = 100 \times \text{antilog}\left(\frac{-12 - 5}{55.8}\right) = 49.6 \text{ ppm}$$

Low-level procedure

Step 3A Pipette 50 ml of the sample into a 100-ml plastic beaker containing a plastic-coated magnetic stirrer bar. Add 1 ml of low-level ionic strength adjustor, either by pipette or with a convenient automatic dispenser.

Step 4A Rinse the electrodes with deionized water, dry them with a paper tissue and immerse the tips below the surface of the solution in the beaker. Check that there are no air bubbles at the surface of the membrane.

Step 5A Start the stirrer and adjust the control to give brisk agitation without creating a vortex. Stir all the solutions at this rate. Note the position of the electrode holder on the stand so that it can be returned to the same place every time.

Step 6A When a steady potential is reached, note the reading E_x.

Step 7A Read off the approximate concentration c_x corresponding to the potential E_x from the calibration graph.

Step 8A Prepare by dilution of standard solutions B or C four further standard solutions $s_1 - s_4$ of decreasing concentration in the range approximately $2c_x - 0.5c_x$.

Step 9A With each of the standard solutions $s_1 - s_4$ repeat Steps 3A and 4A and when a steady reading is obtained note the potentials $E_1 - E_4$ accordingly.

Step 10A Calculate the differences $\Delta_2, \Delta_3, \Delta_4$ from $\Delta_2 = E_1 - E_2$ etc.

Step 11A Plot the values 0, $\Delta_2, \Delta_3, \Delta_4$, against the logarithms of the corresponding concentrations $s_1 - s_4$.

Step 12A Calculate $\Delta_x = E_1 - E_x$.

Step 13A The concentration of the sample is that corresponding to Δ_x on the plot obtained in Step 11A.

Step 14A For each sample repeat Steps 3A, 4A, 6A, 12A and 13A. If E_x should lie outside the range E_1 to E_4 prepare further standards to extend the range as required. Restandardize with standard solution s_1 after every sixth sample or every hour, whichever is the sooner.

Preparation and use of calibration graphs

The calibration graphs described here are primarily for determining the characteristics of the electrode and not for use in routine analyses. A fresh calibration graph should be obtained every time the membrane is changed and preferably also after a prolonged period of storage.

High-level calibration With standard solution B, carry out Steps 4—7 of the high-level procedure and with standard solutions C and D and at least two other standards of intermediate concentration, repeat Steps 11—13. Plot the e.m.f. differences, Δ_x, against the corresponding concentrations and from the graph find the lower limit of the high-level procedure, i.e. the concentration at which the plot ceases to be linear. It may be necessary to take more standard solutions to define this point more clearly.

The use of a permanent calibration graph (p. 117) is not generally recommended, as the calibration slopes of neutral carrier membrane electrodes may decrease by up to 1 mV per week. The inclusion of a standard solution in each batch of samples would be essential to check that the permanent calibration graph was still valid.

Low-level calibration This serves only to indicate the first approximate concentration, c_x, in Step 7A of the low-level procedure and need not be determined

Table 2 Solutions for low-level calibration with standard solution B[a]

Size of portion (ml)	Cumulative addition (ml)	Final concentration (ppm)
0.1	0.1	0.20
0.2	0.3	0.60
0.2	0.5	0.99
0.5	1.0	1.96
2.0	3.0	5.66
2.0	5.0	9.09

[a] Use similar portions of standard solution C if still lower concentrations are required. The corresponding final concentrations are one-tenth those in the table

with the usual precision. This calibration should not be used as a permanent calibration graph. Carry out Steps 3A–5A with a portion of deionized water and add to the solution in the beaker successive portions (0.1, 0.2, 0.2 and 0.5 ml) of standard solution B by means of a graduated 1 ml pipette. Add by pipette two further portions of 2 ml each. After each addition, allow the potential to reach a steady value and note the reading. Plot the readings against the logarithms of the corresponding concentrations, given in Table 2.

Sources of Error

Effect of temperature

Changes in temperature affect both the slope of the calibration and the value of the standard potential. The standard and sample solutions should not differ in temperature by more than 1 °C.

Effect of other substances

Table 1 shows that the most serious interferences are by caesium and rubidium ions, which are very rare. Very large excesses of the other alkali metal ions and the alkaline earth ions will also cause interference. Ammonium ions will cause a positive bias of 1–5% if present at an approximately equal concentration to the potassium ions; the level of interference may be reduced by raising the pH of the sample to about 10.5, when only 10% of the total ammonia is present as ammonium ion. If ammonium ion interference is likely to occur, add 15 ml of 0.1 mol l^{-1} sodium hydroxide solution to the ionic strength adjustor before it is made up to the mark. Samples containing significant quantities of acids would require a higher concentration of sodium hydroxide, which can only be found by experiment.

Lal and Christian (1970) reported that the calibration slope was lower in solutions of potassium iodide than in other potassium salt solutions. In addition, the standard potential was different in potassium iodide, chloride and hydroxide solutions. For their electrode, Radiometer recommend the following maximum values of the product $c_K \cdot c_X$ of the molar concentrations of potassium and interfering anion (X) if an interference greater than about 5% is to be avoided: ClO_4^-, 10^{-7}; I^-, 10^{-5}; ClO_3^-, 2×10^{-4}; NO_3^-, 10^{-2}. This aspect of interference has not been thoroughly investigated, but the problem should be reduced by using the ionic strength adjustor to swamp effects from anions other than chloride.

Precision

The precision claimed by the manufacturers is equivalent to a standard deviation of about 0.5 mV or a relative standard deviation of about 2% in the concentration. Pioda et al. (1970) found a relative standard deviation of 1.7% for serum samples containing 150–300 ppm potassium, while Reynolds (1971) reported 3% for samples of Alpine waters containing 0.12–4 ppm potassium.

Accuracy

Frant and Ross (1970) and Pioda et al. (1970) both found no significant difference between analyses by electrode and by flame photometry in the concentration range 150–200 ppm. Reynolds (1971) reported that atomic absorption gave results about 5% higher than the electrode at concentrations below 1 ppm.

Response Time

A steady potential is normally attained within one minute in stirred solutions, but may be 5–10 min in unstirred solutions, especially at low concentrations. These times may be extended if an interfering ion is present in sufficient concentration.

Comparison with Other Methods

Pioda et al. (1970) found that atomic absorption spectroscopy and flame photometry gave, respectively, larger and slightly smaller standard deviations than the electrode in the analysis of serum. The agreement between the methods was very close. Reynolds (1971) found poorer agreement between results obtained by the electrode and atomic absorption methods with samples of Alpine waters containing approximately 1 ppm potassium and the latter method gave much higher results in waters containing particulate matter. The electrode is better than the spectroscopic methods for such turbid samples and is much less expensive in all cases. The glass electrode is much more prone to error in solutions containing sodium or ammonium ions, and even when allowance is made for these interferences, the precision is poorer than that for the liquid membrane electrode. The glass electrode is more robust and would be preferred to the liquid membrane type if interferences were not a problem. Mohan and Rechnitz (1970) used both glass and liquid membrane electrodes in a study of potassium complexing with adenosine triphosphate and obtained almost identical results with the two electrodes.

Tracing Faults

In addition to the common faults in electrode systems (p. 127), the following are particularly relevant:

Loss of sensitivity

If the calibration slope falls below 50 mV per tenfold change in concentration, the membrane is probably nearing the end of its useful life. Before changing it, however, check that the standard solutions and the ionic strength adjustor have been correctly prepared. Lack of sensitivity in sample solutions but not in standard solutions is probably due to interferences from other ions; in the case of hydrogen- and ammonium-ion interference it may be possible to overcome the problem as described in the section on sources of error.

Drift

Leakage of electrolyte from a reference electrode filled with a solution containing potassium ions will cause a drift towards an increasing e.m.f. Use the filling solution recommended.

USE OF POTASSIUM-SELECTIVE GLASS ELECTRODES

The durability of glass electrodes and the stability of their standard potentials make them more suitable for continuous analysis than the liquid membrane type. Their poor selectivities make them prone to a bias from other alkali metal ions or ammonium ions, but in the presence of a constant background of such interfering ions, the glass electrode may still be useful for monitoring changes in the potassium concentration. Friedman (1967) has described techniques for continuously monitoring both sodium and potassium concentrations with glass electrodes of appropriate selectivities and correcting for the sodium interference with the potassium electrode. In many respects the method is identical to that for determining sodium by glass electrode and detailed discussion of such material will not be repeated.

Apparatus

Potassium electrode; reference electrode; flow cell; pH meter; pump(s)

Apart from the potassium electrode itself, the apparatus is the same as that used for sodium determinations. Glass apparatus should not be used if the potassium concentration is less than 40 ppm, because of the possibility of contamination by sodium ions.

Reagents

Water

It should be possible to produce water containing a negligible amount of potassium by passing distilled water through a mixed-bed deionization unit. Sodium- and ammonium-ion impurities are likely to be much higher and care should be taken to avoid contamination by sodium from hands or glassware. The water should not be stored where contamination by ammonia vapour is possible. Prepare a sufficient quantity of water for one batch of standard solutions and store it in a well washed plastic aspirator.

Standard potassium solution A (1000 ppm)

Dry potassium chloride (analytical reagent grade) at 110 °C for 1–2 h. Dissolve 1.907 g in deionized water and dilute with deionized water to 1 litre in a calibrated flask.

1 ml ≡ 1.0 mg potassium

Standard solution B (10 ppm)

Weigh 4.95 kg of deionized water in a pre-weighed 5 litre polyethylene aspirator. Add by pipette 50 ml of standard solution A, and mix.

1 ml ≡ 10 mg potassium

Alkaline additive

Hydrogen ions interfere with the potassium electrode unless there is a thousandfold excess of potassium ions in solution. The pH of the solution is therefore raised to a constant high level (10–11) by addition of a base. Half-fill a drechsel bottle with diethylamine or diisopropylamine and connect the dosing pump to the outlet from the air-space above the liquid. The inlet tube to the bottle is fitted with a porous frit submersed in the liquid. A flow rate for the air approximately equal to that of the sample should be sufficient to give the required pH. Note that ammonia must not be used to adjust the pH of the sample.

Sample Collection

Use a clean screw-capped polyethylene aspirator and collect at least 250 ml of sample. Great care should be taken to avoid contamination during sampling. Immediately before analysis, wash the inside of the tap of the aspirator by running about 50 ml of the sample through it, rinse the outside of the tap with deionized water and connect the sample pump to the tap. Samples should be stable for many days.

Conditioning and Storage of Electrodes

The glass electrode should be placed in deionized water for 48 h before use for the first time. Between analyses, leave the electrode in the flow cell with the tip immersed in either deionized water or a dilute standard solution, with the usual base added, rather than in the last sample solution, particularly if the sample contains substances likely to interfere with the electrode response or deposit on any part of the flow system. Never allow the electrode to dry out. Better accuracy is obtained if the equipment is in continuous use; if the apparatus is used regularly it is therefore advantageous to keep the apparatus running continuously on deionized water whenever samples or standards are not being analysed. Eisenman (1965) found that potassium-selective electrodes improved in stability, response time and selectivity for at least two weeks after initial use. Even after a month the selectivity can still improve considerably (Moore, 1967). Electrodes have been known to have useful lives of up to 5 years but this will depend on individual circumstances.

Concentration Range and Units

Results have been reported in the range 0.1–1000 ppm.

$$10^{-3} \text{ mol } l^{-1} \equiv 39.1 \text{ ppm}; \quad 1 \text{ ppm} \equiv 2.558 \times 10^{-5} \text{ mol } l^{-1}$$

Analytical Procedure

Step 1 If the reference electrode has been removed from the flow cell, wash the tip with deionized water before inserting it in the flow cell.

Step 2 Connect an aspirator containing standard solution B to the appropriate pump tube, open the tap of the aspirator and start pumping the solution and the air/amine mixture through the cell.

Step 3 When the potential is steady, note the reading E_1.

Step 4 With the pump still operating, disconnect the pump tube from the tap of the aspirator and transfer it to the tap of the aspirator containing the sample. Open the tap of the sample aspirator.

Step 5 When the potential is steady, note the reading E_x.

Step 6 Calculate the difference $\Delta = E_x - E_1$.

Step 7 Either read off the concentration from the calibration graph or calculate it from the equation

$$c = 10 \times \text{antilog} (\Delta/k) \text{ ppm}$$

where k is the calibration slope.

Step 8 If more samples are to be analysed, repeat Steps 4–7 each time. The frequency with which standardization (Steps 2–3) is required depends on the desired precision of the analysis, but it is recommended that it be done at least twice a day.

Checking for Bias

A second standard potassium solution should be analysed with each batch of samples. The concentration should be chosen to be as near the lowest sample concentration as is convenient. Any change in the calibration slope will appear as a bias. If a bias is found, immediately repeat Steps 2–3 and calculate the slope from

$$k = \frac{E_1 - E_s}{1 - \log(c_s)}$$

where c_s is the concentration of the second standard (in ppm) and E_s is the potential observed with that solution. If k differs from the value given by the calibration graph, use it to calculate the concentrations of the sample solution.

Note: The analysis should be carried out at the temperature (± 1 °C) at which the calibration graph was prepared. If the slope has not changed, the standard potential is drifting and standard solution B must be run more frequently.

Preparation of the Calibration Graph

By dilution of standard solution A prepare at least four solutions spanning the range of concentration to be measured. Repeat Steps 2 and 3 of the analytical procedure and with the other standard solutions carry out Steps 4–6. The calibration should be linear with a slope of 55–62 mV per tenfold increase in concentration at room temperature.

The problems of calibrating sodium-responsive glass electrodes will be relevant here also, although interference by the amine additive may be more important in the case of potassium. For a general discussion on the use of a permanent calibration graph, see p. 117.

Sources of Error

Effect of temperature

Changes in temperature affect the slope of the calibration graph, causing an increasing error as the difference between the sample and standard concentrations increases, and also affect the standard potential.

Effect of other substances

The selectivity of potassium-sensitive glass electrodes follows the general order

$$H^+ \gg K^+ \simeq NH_4 > Rb^+ > Na^+ \simeq Cs^+ > Li^+ > Mg^{2+} \simeq Ca^{2+}$$

i.e. the electrodes have a greater selectivity for hydrogen ions than for potassium ions and it may be necessary to adjust the pH of a sample by adding a suitable base. The pH should be 3 units greater than $-\log [K^+]$, where $[K^+]$ is the molar concentration of potassium ions. The approximately equal selectivity for ammonium ions and potassium ions means that ammonia cannot be used to control the pH as with the sodium electrode. Ammonium interference can be minimized by working at as high a pH as possible, so that ammonium ions are converted into ammonia. If necessary, the ammonia may be removed before measurement by bubbling air or nitrogen through the alkaline sample. Sodium ion is commonly present at higher concentrations than the potassium ion and may interfere; either determine the sodium concentration independently with a sodium-selective electrode or by flame photometry or atomic absorption and apply a correction to the apparent potassium concentration, using the appropriate selectivity coefficient, or use the liquid membrane potassium electrode. If a selectivity coefficient is used, it should be determined empirically (p. 74), using sodium and potassium concentrations representative of the range of concentrations expected in the sample solutions. Use of the values in Table 1 is not recommended for accurate work.

Precision

No results seem to have been published for the precision of measurement in dilute solutions. EIL (1963) claim a reproducibility of about 2.5% for their GKN 33 electrode at concentrations of 40–40 000 ppm in the absence of sodium ions and less than 5% in the presence of an excess of sodium ions. Better precision can be obtained by working over a smaller range. Portnoy and Gurdjian (1966) found a relative standard deviation of 9% at a potassium concentration of about 90 ppm in cerebrospinal fluid with an NAS 27-4 electrode, believed to be similar, if not identical, to the Corning 47622000. At 160 ppm in whole blood and blood plasma, different workers found relative standard deviations of 9–14% (Moore, 1967).

Accuracy

Moore (1967) has reported that measurements by flame photometry and by electrode in biological samples containing about 160 ppm of potassium agree within 10%. The standard deviations of the results by the two methods are approximately the same. The precision of the electrode method should be better in the absence of significant concentrations of sodium ion. Jacobson (1966) found that the electrode gave values 1% lower in samples of human urine and 6% lower in dog urine (400–3500 ppm).

Response Time

Using an electrode in a flow cell as described in the analytical procedure may give response times, including mixing times, of about 5 min for a change from 4000 to 400 ppm solutions and twice as long for the change from 40 to 4 ppm; the time to reach a steady potential after dipping the electrode in a fresh solution should only be 1—2 min. Rechnitz (1967) found that less than 3 sec was needed for the Beckman 39137 electrode to reach equilibrium at 4000 ppm. Response times will be increased in the presence of a sufficient quantity of sodium ion to interfere.

Comparison with Other Methods

The possibility of bias between analyses by the glass electrode and flame photometric methods has been shown (Moore, 1967; Jacobson, 1966) even when the electrode results are corrected for sodium interference. There are reports of both larger and smaller standard deviations by the electrode method. The electrode is much cheaper than flame photometry and more suitable for continuous monitoring and for on-site analysis, but is generally less satisfactory. The liquid membrane electrode has most of the advantages of the glass electrode and much better selectivity and would be preferred for most analyses. The glass electrode is tougher, more stable and has a longer working life. Thermodynamic studies with both types of electrode (Mohan and Rechnitz, 1970) gave almost identical results.

Tracing Faults

For any faults likely to be encountered, see the general discussion on p. 127 and the appropriate section in the method for determining sodium.

Bibliography

Eisenman, G., 1965, 'The electrochemistry of cation-sensitive glass electrodes', *Adv. Anal. Chem. Instrum.*, **4**, 213.
Frant, M. S., and J. W. Ross, 1970, 'Potassium ion specific electrode with high selectivity for potassium over sodium', *Science*, **167**, 987.
Friedman, S. M., 1967, 'H^+ and cation analysis of biological fluids in the intact animal', in *Glass Electrodes for Hydrogen and Other Cations*, G. Eisenman, (ed.), Arnold, London/Marcel Dekker, New York, Ch. 16.
Jacobson, H., 1966, 'Direct determination of sodium and potassium in the presence of ammonium with glass electrodes', *Anal. Chem.*, **38**, 1951.
Lal, S., and G. D. Christian, 1970, 'Response characteristics of the Orion potassium ion selective electrode', *Anal. Lett.*, **3**, 11.
Mohan, M. S., and G. A. Rechnitz, 1970, 'Ion-electrode study of alkali metal adenosine triphosphate complexes', *J. Amer. Chem. Soc.*, **92**, 5839.
Moore, E. W., 1967, 'Hydrogen and cation analysis in biological fluids in vitro', in *Glass Electrodes for Hydrogen and Other Cations*, G. Eisenman (ed.), Arnold, London/Marcel Dekker, New York, Ch. 15.
Pioda, L. A., W. Simon, H.-R. Bosshard, and H. Ch. Curtius, 1970, 'Determination of potassium ion concentration in serum using a highly selective liquid-membrane electrode', *Clin. Chim. Acta*, **29**, 289.
Portnoy, H. D., and E. G. Gurdjian, 1965, 'Glass electrode measurement of cerebrospinal fluid sodium and potassium', *Clin. Chim. Acta*, **12**, 429.
Rechnitz, G. A., 1967, 'Cation-sensitive glass electrodes in analytical chemistry', in *Glass Electrodes for Hydrogen and Other Cations*, G. Eisenman (ed.), Arnold, London/Marcel Dekker, New York, Ch. 12.
Reynolds, R. C., 1971, 'Analysis of Alpine waters by ion electrode methods', *Water Resources Res.*, **7**, 1333.

Determination of Calcium Using Calcium-responsive Liquid Ion-exchange Electrodes

Most of the known exchangers in commercial liquid ion-exchange electrodes are calcium salts of high molecular weight organophosphoric acids and electrodes which use this type of exchanger are the Orion 93-20, EDT Supplies EE-Ca, Radiometer F2112Ca, Activion 003 15 012, Metrohm EA301Ca and the Simac Ca/1C. The Philips plastic membrane electrode, IS 561Ca, has a neutral carrier exchanger and its properties are somewhat different from those of the previous group of electrodes.

APPARATUS

Calcium electrode; reference electrode; pH meter; magnetic stirrer

For concentrations in the non-Nernstian range (10—ca. 1 ppm for the electrodes having calcium organophosphate exchangers), it has been found that a more reproducible response is obtained in flowing solution and therefore a flow cell, such as that used in the EIL series 89 Laboratory Monitors, is required. Alternatively, flow cells can be constructed from plexiglass (perspex) and used in conjunction with a pump (or pumps) to give flow rates of approximately 4—10 ml min^{-1} and 0.4—1 ml min^{-1} for the sample and buffer, respectively.

REAGENTS

Water

Distilled water which has been passed through a mixed-bed deionization unit such that its final specific conductivity is less than 0.2 μS cm^{-1} at room temperature is suitable for the preparation of the standard and buffer solutions.

Standard calcium solution A (1000 ppm)

Dry a sample of calcium carbonate (analytical reagent grade) in an oven at 110 °C for 3—4 h. Weigh out exactly 2.499 g and add to approximately 150 ml of water in a beaker. To this solution, add exactly 50.0 ml of 1 mol l^{-1} hydrochloric acid, stirring until the reaction is complete. Transfer the contents of the beaker to a 1-litre calibrated flask and make up to the mark with water. Store in a polyethylene bottle. Standard solutions made in this manner have been shown to be within 0.2% of the theoretical value (Hadjiioannou and Papastathopoulos, 1970).

1 ml solution A ≡ 1.0 mg calcium

Sodium acetate buffer solution

Dissolve 136.1 (±1) g of analytical reagent grade sodium acetate trihydrate in 500 ml water and transfer to a 1-litre calibrated flask. To this flask add 40 (±0.5) ml of acetylacetone and make up to the mark with water.

CONCENTRATION RANGE AND UNITS

The calcium electrodes generally have a Nernstian response in the range 400–10 ppm of calcium and therefore samples in this range are most conveniently analysed by Method A, below. The Nernstian range of the Philips electrode extends below 1 ppm and all the samples in the range 400–1 ppm can be analysed by this method. To avoid error, however, the linear response range of each electrode should first be checked by preparing a calibration graph (see below). If the calibration is non-linear in the expected concentration range of the samples, Method B should be used.

$$10^{-3} \text{ mol l}^{-1} \equiv 100.1 \text{ ppm CaO}_3 \equiv 40.08 \text{ ppm Ca}$$

$$9.99 \times 10^{-6} \text{ mol l}^{-1} \equiv 1 \text{ ppm CaCO}_3 \equiv 0.40 \text{ ppm Ca}$$

$$2.495 \times 10^{-5} \text{ mol l}^{-1} \equiv 2.498 \text{ ppm CaCO}_3 \equiv 1 \text{ ppm Ca}$$

ANALYTICAL PROCEDURE

Method A (400–10 ppm calcium)

Step 1A Place the calcium electrode in a standard solution (~40 ppm) for a 30 min conditioning period before use.

Step 2A Pipette 50 ml of standard solution, whose concentration s_1 is representative of the higher end of the expected sample concentration range, into a 100-ml beaker containing a magnetic stirrer bar and add 5 ml of buffer solution.

Step 3A Remove the electrodes from the solution in which they have been immersed, rinse them with deionized water and remove the surplus water with a tissue.

Step 4A Immerse the electrodes in the test solution and stir at a moderate rate. Both the depth of immersion and the stirring rate should be kept constant throughout the analysis.

Step 5A When a steady potential has been reached (usually within 3 min) note the reading E_1.

Step 6A Repeat Steps 2A–5A with a second standard solution of concentration s_2, representative of the lower end of the expected sample concentration range and not greater than $s_1/2$. Note the steady reading E_2.

Step 7A Calculate the calibration slope $\simeq 28$ mV per tenfold increase in concentration

$$k = \frac{E_1 - E_2}{\log s_1 - \log s_2}$$

or plot a temporary calibration graph by joining the points $(\log s_1, 0)$, $(\log s_2, E_2 - E_1)$.

Step 8A Pipette 50 ml of sample into a 100 ml beaker containing a stirrer bar and add 5 ml of buffer. The temperature of the sample should be within 1 °C of the temperature of the standard solutions used above.

Step 9A Repeat Steps 3A and 4A.

Step 10A When the potential is steady, note the reading E_x.

Step 11A Calculate the e.m.f. difference $\Delta = E_x - E_1$ and either calculate the calcium concentration, c, in the sample from

$$c = s_1 \times \text{antilog}\,(\Delta/k)$$

or read the concentration corresponding to Δ off the calibration graph

Step 12A Repeat Steps 8A–11A for each sample. It is recommended that the slope k be redetermined with every batch of about ten samples. If the electrode is in continuous use the standardization should be repeated three times a day.

Worked example

The value of k obtained from the response to 10 and 100 ppm standards was 29.7 mV per decade concentration change. The response to the 100 ppm standard was 9.7 mV and to the sample +2.0 mV. The concentration of the sample is given by,

$$c = 100 \times \text{antilog}\left(\frac{2 - 9.7}{29.7}\right)$$

$$= 100 \times \text{antilog}\,(-0.259)$$

$$= 55 \text{ ppm}$$

Method B (10–1 ppm calcium)

The following instructions are written for the use of a calcium electrode in a flow cell. If the latter is unavailable then Steps 2A–6A of Method A can be followed with the inclusion of two additional standard solutions whose concentrations lie between s_1 and s_2. Plot the calibration graph as described below and read the concentrations of the samples directly off this graph.

Step 1B As Step 1A.

Step 2B Rinse both the calcium and reference electrodes with deionized water and place them in the flow cell with the reference electrode downstream.

Step 3B Connect a 1 ppm standard solution and the buffer solution to the pump(s) via two separate lengths of narrow-bore polyvinylchloride or Teflon transmission tubing.

Step 4B Start the pump(s) and, when the flow has been sustained for a few minutes, check that no air bubbles are trapped at the membrane surface.

Step 5B When a steady potential has been recorded for a period of 2–4 min, note the reading E_1.

Step 6B Remove the transmission tube to a 2 ppm standard solution and when a steady e.m.f. has been reached note the reading E_2.

Step 7B Repeat Step 6B with 5 and 10 ppm standard solutions, noting the steady readings E_3 and E_4, respectively.

Step 8B Subtract the e.m.f. for each standard solution from E_4 and plot these differences Δ_1, Δ_2, Δ_3, and 0 versus the logarithms of the corresponding concentrations.

Step 9B Transfer the transmission tube to a sample and, when a steady e.m.f. has been reached, note the reading E_x. Note that the temperature of the samples should be within 1 °C of the temperature of the standard solutions.

Step 10B Calculate the e.m.f. difference $\Delta_x = (E_4 - E_x)$, and read the concentration of the sample corresponding to Δ_x off the calibration graph.

Step 11B Repeat Steps 9B and 10B for all the samples.

Step 12B Redetermine the value of E_4 for the 10 ppm standard after about eight samples have been analysed.

Preparation and Use of the Calibration Graph

High-level calibration (400–10 ppm) Prepare a calibration graph by carrying out Steps 2A–5A of the analytical procedure for at least four standard solutions covering the expected range of the samples. Plot the e.m.f. values against the logarithms of the concentrations as described on p. 118; although this graph is not used for the calculation of the results it is advisable to determine its shape following each fresh assembly of an electrode or period of prolonged storage, particularly if the samples lie near the limit of the electrode's linear range. The slope k obtained in Step 7A should be compared with that obtained from the corresponding portion of this graph and a decrease of about 5% in the value of k, at constant temperature, is an indication that the membrane requires replacement.

SOURCES OF ERROR

Effect of other substances

The most common source of interference in electrodes with organophosphate exchangers is hydrogen ion, which makes it necessary to buffer the sample to a pH between 7 and 11, especially when concentrations below 40 ppm are being determined. The solution of sodium acetate should maintain the pH of the sample within these limits but should the sample contain excessive acid, a more concentrated buffer is necessary. An increase in the concentration of sodium acetate from 0.1 to 0.2 mol l^{-1} could result in errors of the order of +3 mV at 4 ppm of calcium (Orion, 1966) and so it would be better to use potassium acetate solutions for the determination of low levels of calcium in samples of significant acidity since the interference from potassium is less than that from sodium (Moody, Oke and Thomas, 1970). The Philips neutral carrier electrode is much less prone to hydrogen ion interference and may be used down to pH 3, depending on the calcium concentration.

Divalent metal ions constitute the main interference in all the electrodes and the order of interference for electrodes having organophosphate exchangers is commonly $Zn^{2+} > Fe^{2+} > Pb^{2+} > Cu^{2+} > Ni^{2+} > Mg^{2+}$ but the reported magnitudes of selectivity coefficients are dependent on the make of the electrode and the manner in which the coefficients were calculated. An indication of the relative magnitude of these is given in Table 1. Acetylacetone is added with the buffer to mask magnesium (Hulanicki and Trojanowicz, 1974), which often coexists with calcium, but this may be omitted if a Philips neutral carrier electrode is used. Using an Orion 92-20 electrode, Hulanicki and Trojanowicz (1974) found that in the analysis of standard solutions containing 95 ppm calcium and amounts of magnesium varying from 24—96 ppm, the results had an average positive bias of 2%.

The level of free calcium ion in a sample can be reduced by complexation and two of the most common ions, bicarbonate and sulphate, have the ability to do

Table 1 Selectivity coefficients for some commercial calcium electrodes

Interfering ion	Selectivity coeffients*, K_{CaM} for:		
	Orion 29—20[a,b]	Philips IS 561Ca[c]	Activion 003 15 012[d]
Zn^{2+}	3.2 — 0.81	—	1.2
Fe^{2+}	0.8	—	0.8
Pb^{2+}	0.63	—	0.6
Cu^{2+}	0.27	—	—
Ni^{2+}	0.08	—	—
Sr^{2+}	0.017	10^{-2}	—
Mg^{2+}	0.055—0.14	4×10^{-5}	0.015
Ba^{2+}	0.033 — 0.010	4×10^{-4}	0.10
Na^+	1.74×10^{-4}	3×10^{-4}	0.003
K^+	6.6×10^{-5}	2×10^{-4}	2.2×10^{-5}

* $E = E° + k \log (c_{Ca} + K_{CaM} c_M^{2/z_M})$ where $k = \dfrac{RT}{2F} \ln (10)$ and z_M is the charge on M.

[a] Orion (1966).
[b] Moody, Oke and Thomas (1970).
[c] Philips (1976).
[d] Activion (1976).

this. The effect of bicarbonate can be most troublesome, as it is absorbed into a sample from the carbon dioxide of the air during the course of the analysis, and therefore it is advisable to measure the electrode response immediately after adding the buffer. In samples known to contain high levels of bicarbonate (>300 ppm) the error will be greater than 10% and a potentiometric titration method (Hadjiioannou and Papastathopoulos, 1970) should be considered. This also applies to samples containing levels of sulphate ion greater than about 5×10^{-4} mol l^{-1} (Orion, 1975).

PRECISION

Růžička and Tjell (1969), using a flow-cell technique, reported relative standard deviations for a single determination of less than 2% for calcium concentrations in the range 12–48 ppm.

ACCURACY

Excellent agreement was obtained between the calcium contents of certain soil extracts as measured with a calcium electrode and by atomic absorption (Woolson, Axley and Kearney, 1970), though a slight negative bias was found in the electrode values. In a study of the calcium content of tap-water, Moody *et al.* (1970) compared the results obtained with calcium electrodes with those obtained by an EDTA indicator titration. There was better than 98% agreement between the results obtained by these techniques in a series of samples analysed daily for ten separate days. Růžička and Tjell (1969) reported a small positive error for the analysis of calcium in solutions which also contained sodium, potassium and magnesium.

RESPONSE TIME

The response times of these electrodes tend to be longer than those obtained with glass and most solid-state electrodes. The response characteristics of an early Beckman electrode were investigated by Rechnitz and Hseu (1969) and responses were recorded inside one minute for small concentration changes of pure calcium solution about the 40 ppm level. Philips (1976) claim a response time of less than 30 sec when their neutral carrier electrode is transferred from a 10^{-3} to a 10^{-2} mol l^{-1} calcium solution. In general, response times as long as 5 to 10 min must be expected at calcium concentrations below 10 ppm and at these levels the efficiency of the stirring is most important.

LIFETIME OF ELECTRODES

The lifetimes of these electrodes are considerably shorter than those of most glass and solid-state electrodes. Operational lifetimes quoted by the manufacturers vary from 1 month to 6 months per membrane or module.

OTHER APPLICATIONS

The calcium electrode can be used to detect the end-point in compleximetric titrations of calcium and several methods have been published describing this technique (Hadjiioannou and Papastathopoulos, 1970; Tackett, 1969). Whitfield and coworkers (1969 and 1969a) have discussed the form of these potentiometric titration curves and their dependance on the selectivity properties of the calcium electrode. The use of calcium electrodes in biological studies has been fully described by Moore (1968).

COMPARISON WITH OTHER METHODS

The direct potentiometric technique has been shown to be applicable to a wide variety of samples and, unlike the standard EDTA titration, can be applied in coloured or turbid samples. An electrode measurement is much more rapid than a gravimetric determination in which calcium is precipitated as calcium oxalate, though the accuracy obtained by the gravimetric procedure after removal of any interferences is expected to be much better. The electrode has been used extensively in serum samples where its unique ability to measure activity, rather than concentration, is of paramount interest. Calcium can be determined by atomic absorption spectrophotometry over a wide concentration range with an accuracy and precision that cannot be matched by the electrode. This method is free from most interferences though particulate matter must be removed before analysis.

TRACING FAULTS

In addition to the general faults discussed on p. 127, the following should be noted.

Drift

In all of these electrodes a drift in E° is expected and this can be in the order of 5 mV d^{-1}. Drift that occurs in sample solutions and not in pure calcium standard solutions is associated with interference caused by other cations entering the organic phase and changing the E° value. Less extreme interference may cause a slower response. In cases of interference, the electrode can be restored to normal by standing it overnight in approximately 400 ppm calcium solution.

At low calcium levels (<10 ppm) it is sometimes difficult to obtain a steady reading. If the solution is contained in a beaker, an increase of stirring rate may help but the use of a flow cell is advantageous because a dynamic equilibrium is more easily maintained and reproduced.

Bibliography

Activion, 1976, *Ion Selective Electrode Technical Notes*, Activion Glass Ltd., Halstead, Essex.

Hadjiioannou, T. A., and D. S. Papastathopoulos, 1970, 'EDTA titration of calcium and magnesium with a calcium-selective electrode', *Talanta,* **17**, 399.

Hulanicki, A., and W. Trojanowicz, 1974, 'Direct potentiometric determination of calcium in waters with a constant complexation buffer', *Anal. Chim. Acta,* **68**, 155.

Moody, G. J., R. B. Oke, and J. D. R. Thomas, 1970, 'A calcium-sensitive electrode based on a liquid ion exchanger in a poly(vinylchloride) matrix', *Analyst,* **95**, 910.

Moore, E. W., 1969, 'Studies with ion-exchange calcium electrodes in biological fluids: Some applications in biomedical research and clinical medicine', in *Ion selective Electrodes,* R. A. Durst (ed.), National Bureau of Standards Special Publication 314, US Department of Commerce, Washington D.C.

Orion, 1966, *Calcium Ion Electrode Model 92-20, Instruction Manual,* Orion Research Inc., Cambridge, Mass.

Orion, 1975, *Calcium Ion Electrode Model 93-20, Instruction Manual,* Orion Research Inc., Cambridge, Mass.

Philips, 1976, *Ion-selective Plastic Membrane Electrodes IS-561 Series,* N. V. Philips, Eindhoven.

Rechnitz, G. A., and T. M. Hseu, 1969, 'Analytical and biochemical measurements with a new solid-membrane calcium-selective electrode', *Anal. Chem.,* **41**, 111.

Růžička, J., and J. C. Tjell, 1969, 'Ion-selective electrodes in continuous-flow analysis', *Anal. Chim. Acta,* **47**, 475.

Tackett, S. L., 1969 'Automatic titration of calcium with EDTA using a calcium-selective electrode', *Anal. Chem.,* **41**, 1703.

Whitfield, M., and J. V. Leyendekkers, 1969, 'Liquid ion-exchange electrodes as end-point detectors in compleximetric titrations. Determination of calcium and magnesium in the presence of sodium. Part I. Theoretical considerations', *Anal. Chim. Acta,* **45**, 383.

Whitfield, M., J. V. Leyendekkers, and J. D. Kerr, 1969a, 'Liquid ion-exchange electrodes as end-point detectors in compleximetric titrations. Part II. Determination of calcium and magnesium in the presence of sodium', *Anal. Chim. Acta,* **45**, 399.

Woolson, E. A., J. H. Axley, and P. C. Kearney, 1970, 'Soil calcium determination using a calcium-specific ion electrode', *Soil Sci.,* **109**, 279.

Determination of Water Hardness with Liquid Ion-exchange Electrodes

The hardness of water was a term originally adopted for the capacity of water for destroying the lather of soap and was originally determined by a titration with a standard soap solution. The most common cause of the destruction of lather is the presence of calcium and magnesium salts in the water and over the years the term 'water hardness' has been generally accepted (ASTM, 1973; BSI, 1962) as the total concentration of calcium and magnesium ions present in the sample and is usually expressed as a concentration of calcium carbonate. Hardness in water has been subdivided into two broad contributing sources: temporary and permanent hardness. *Temporary hardness* (alternatively called alkaline hardness) is that part of the total hardness which disappears on boiling and consists mainly of the bicarbonates and, to a lesser extent, the carbonates of calcium and magnesium. The hardness which remains after boiling, *permanent hardness* (alternatively called non-alkaline hardness), consists of calcium and magnesium salts such as sulphates and chlorides. Recently (Sekerka and Lechner, 1975), the term 'carbonate hardness' has been used instead of temporary hardness and the difference between total hardness and carbonate hardness has been assigned the term 'non-carbonate hardness'.

Standard methods for the determination of hardness involve either a lengthy gravimetric procedure (ASTM, 1973a) or a volumetric titration using EDTA and a colorimetric indicator (ASTM, 1973; BSI, 1962). Potentiometric methods have not yet been accepted as standards but their convenience makes them valuable, especially where a rapid (though less precise) determination is required. In some cases where only a coloured or turbid sample is available, electrodes provide the best means of detecting the end-point in an EDTA hardness titration.

Electrodes which are sold for the determination of water hardness are the Simac W/IC, Activion 003 15 0015 and the Orion 93-32. Any of these electrodes can be used in the following analytical methods.

Table 1 The classification of degree water hardness according to the scale used by the US Geological Survey

Classification of Hardness	Hardness (ppm calcium carbonate)
Soft	0–60
Moderately hard	61–120
Hard	121–180
Very hard	> 181

DETERMINATION OF WATER HARDNESS BY POTENTIOMETRIC TITRATION WITH EDTA

Apparatus

Divalent electrode; reference electrode; pH meter; magnetic stirrer; 10-ml capacity burette with subdivisions of 0.02 ml.

Reagents

Water

Distilled water which has been passed through a mixed-bed deionization unit such that its final specific conductivity is less than 0.2 $\mu S\ cm^{-1}$ at room temperature is suitable for the preparation of reagent solutions.

EDTA solution A

Dissolve 83.05 g of analytical reagent or purified grade of ethylenediaminetetraacetic acid disodium salt in about 600 ml of water. Transfer to a 1.litre calibrated flask and make up to the mark with water.

1 ml ≡ 10 mg calcium ≡ 25 mg calcium carbonate

EDTA solution B

Pipette 50.0 ml of EDTA solution A into a 500-ml calibrated flask and make up to the mark with water.

1 ml ≡ 1 mg calcium ≡ 2.5 mg calcium carbonate

EDTA solution C

Pipette 100.0 ml of EDTA solution A into a 500-ml calibrated flask and make up to the mark with water.

1 ml ≡ 2 mg calcium ≡ 5.0 mg calcium carbonate

Buffer solution

Dissolve 37.5 (±0.5) g of reagent grade glycine in 700 ml of water in a beaker and adjust the pH of the solution to 9.7 with 2 mol l^{-1} sodium hydroxide solution. Transfer the contents of the beaker to a 1-litre calibrated flask and make up to the mark with water.

Conditioning and Storage

Liquid ion-exchange electrodes can be stored in a small volume of deionized water if they are used on a day-to-day basis. If they are used irregularly, it is advisable to store them dry; this may require the removal of any internal aqueous reference solution or, in the case of the Orion electrode, returning the sensing module to its storage capsule. On returning electrodes to use after a period of storage it has been found that a short conditioning period in a solution of the ion to which they respond improves their performance.

Sample Collection

Samples can be collected in clean polyethylene bottles of sufficient size to allow for duplicate analyses. If the samples contain suspended solids they should be allowed to settle or if this fails the samples should be filtered before analysis. Techniques for dealing with turbid samples containing large amounts of suspended solids have been considered by Whitfield (1971).

Concentration Range and Units

This analysis is intended for samples whose hardness lies in the range 40–400 ppm calcium or 100–1000 ppm calcium carbonate and the procedure is divided into two ranges, 40–200 and 200–400 ppm calcium.

$$10^{-3} \text{ mol l}^{-1} \equiv 100.1 \text{ ppm CaCO}_3 \equiv 40.08 \text{ ppm Ca}$$
$$9.99 \times 10^{-6} \text{ mol l}^{-1} \equiv 1.0 \text{ ppm CaCO}_3 \equiv 0.40 \text{ ppm Ca}$$

Analytical Procedure

Concentration range 40–200 ppm calcium

Step 1 If the divalent electrode has been stored dry, place it in a 40 ppm calcium solution for a conditioning period of 30 min before use.

Step 2 Fill a 10-ml burette with EDTA solution B.

Step 3 Pipette 50 ml of sample into a 150-ml beaker containing a magnetic stirring bar and add 4 ml of buffer solution.

Step 4 Remove the divalent electrode from the solution in which it is immersed and place it in a holder together with a reference electrode. Rinse both electrodes with deionized water then remove surplus water carefully with a tissue.

Step 5 Immerse the electrodes in the test solution. Position the burette such that its tip is approximately 20 mm above the surface of the sample.

Step 6 Stir the sample at a moderate constant rate and check that no air bubbles are trapped at the membrane surface of the divalent electrode.

Step 7 Set the pH meter to read on the millivolt scale and note the initial reading of the burette V_1.

Step 8 Add EDTA solution in increments of 0.3–0.4 ml initially if the level of hardness is unknown, noting the millivolt reading approximately 10 sec after each addition. Plot millivolts versus burette reading. Note that the potential becomes more negative as the titration proceeds.

Step 9 When the rate of change of potential with volume added increases, indicating that the end-point is being approached, decrease the volume of the increments to 0.1–0.05 ml and allow 30 sec to elapse after each addition before noting the millivolt reading.

Step 10 Continue to titrate until the rate of change of potential with volume decreases, indicating that the end-point has been passed. A slight up-turn in the plot may be observed.

Step 11 From the plot, the end-point is taken as the point of maximum slope. Note the corresponding reading V_2 of the burette from the plot.

Step 12 Calculate the water hardness as ppm calcium from

$$\text{ppm Ca} = 20 (V_2 - V_1)$$

or calculate the water hardness as ppm calcium carbonate from

$$\text{ppm CaCO}_3 = 50 (V_2 - V_1)$$

Concentration range 200–400 ppm calcium

Step 1A As Step 1 above.

Step 2A Fill the burette with EDTA solution C.

Step 3A Repeat the procedure described in Steps 3–11 above, noting the corresponding burette readings V_1' and V_2'.

Step 4A Calculate the water hardness as ppm calcium from

$$\text{ppm Ca} = 40 (V_2' - V_1')$$

or calculate the water hardness as ppm calcium carbonate from

$$\text{ppm CaCO}_3 = 100 (V_2' - V_1')$$

Sources of Error

In the course of the titration, care must be taken to allow sufficient time for the electrode to reach equilibrium following addition of the titrant in the region of the end-point. At this stage, the response of the electrode becomes slower as the divalent ionic content tends to zero and the sodium ion interference (which is present from the disodium EDTA) becomes significant. A 30 sec waiting period has been recommended (Whitfield, Leyendekkers and Kerr, 1969) and this is the period given in the analytical procedure.

If the divalent electrode is used only for the water hardness titrations, it is important to check the sensitivity of the electrode before the titration by measuring its response in 10 and 100 ppm standard calcium solutions. Should the difference in potential between these two solutions fall below 25 mV at room temperature or should the electrode signal drift over a period of five minutes, then replacement of the membrane is probably necessary. A high background concentration of univalent ions can result in a reduction in the size of the jump in potential at the end-point, as these ions can interfere at the electrode.

Precision

Relative standard deviations of less than 0.5% were obtained by Hadjiioannou and Papastathopoulos (1970) using an Orion calcium electrode (Model 92-20) in semi-automatic titrations of solutions of calcium chloride in the concentration range 40—320 ppm calcium. Less precision would be obtained from a manual analysis, where a relative standard deviation of 1—2% may be expected.

Accuracy

Average errors of 0.3% were obtained by Hadjiioannou and Papastathopoulos (1970) for semi-automatic titrations of water hardness in samples containing both calcium and magnesium. The hardness range was 748—90 ppm calcium carbonate. After studying calcium titrations in the presence of relatively high concentrations of sodium ion, Whitfield et al. (1969) suggested that a titrimetric procedure, when applied to natural waters containing low levels of sodium, should have an accuracy of 1—2%.

Comparison with Other Methods

The titrimetric procedure for the determination of water hardness as the sum of the calcium and magnesium contents of the sample is considerably more robust than any direct potentiometric technique since it does not suffer from interference from bicarbonate and sulphate ions and is less susceptible to temperature effects. It has the advantage over conventional EDTA titrations using a visual indicator for end-point detection that it can be used in turbid or coloured samples. If a rapid determination of water hardness is required and no apparatus is available for an automatic titration, then the standard or known addition procedure, which follows, is probably more suitable than a manual titration.

DETERMINATION OF WATER HARDNESS BY KNOWN ADDITION POTENTIOMETRY

Apparatus

Divalent electrode; reference electrode; pH meter; magnetic stirrer.

Reagents

Water

Distilled water which has been passed through a mixed-bed deionization unit such that its conductivity at the outlet is less than $0.2\ \mu S\ cm^{-1}$ at room temperature is suitable for the preparation of solutions.

Calcium solution A

Dry analytical reagent grade or purified calcium carbonate in an oven for 3—4 h at 120 °C. Weigh out exactly 10.009 g of this salt and add to a beaker containing about 100 ml of water. Add cautiously, with stirring, exactly 200 ml of 1 N hydrochloric acid and, when the reaction is complete, transfer, with washing, the entire contents of the beaker of a 1-litre calibrated flask. Make up to the mark with water and mix.

1 ml ≡ 4 mg calcium ≡ 10 mg calcium carbonate

Alternatively, use a specially prepared standard calcium solution such as the Orion 0.1M calcium solution (932006) which has exactly the same concentration as solution A.

Calcium solution B

Pipette 50 ml of solution A into a 250 ml calibrated flask and make up to the mark with water.

1 ml ≡ 0.8 mg calcium ≡ 2.0 mg calcium carbonate

Calcium solution C

Pipette 50 ml of solution A into a 1-litre calibrated flask and make up to the mark with water.

1 ml ≡ 0.2 mg calcium ≡ 0.5 mg calcium carbonate

Sodium hydrogen carbonate solution A

Dissolve 21.02 g of analytical reagent grade sodium hydrogen carbonate in about 300 ml of water in a beaker, and transfer to a 500-ml calibrated flask and make up to the mark with water.

Sodium hydrogen carbonate solution B

Pipette 25 ml of solution A into a 500-ml calibrated flask and make up to the mark with water.

Concentration Range and Units

The analytical procedure is intended for the analysis of samples whose hardness is in the range 5–200 ppm calcium carbonate. but greater accuracy is achieved if the method is split into two ranges, 5–100 and 100–200 ppm.

$$10^{-3} \text{ mol l}^{-1} \equiv 100.1 \text{ ppm CaCO}_3 (\equiv 40.08 \text{ ppm Ca} \equiv 24.3 \text{ ppm Mg})$$
$$9.99 \times 10^{-6} \text{ mol l}^{-1} \equiv 1.0 \text{ ppm CaCO}_3 \quad (\equiv 0.40 \text{ ppm Ca} \equiv 0.243 \text{ ppm Mg})$$

Analytical Procedure

Step 1 If the divalent electrode has been stored dry, immerse it in a 40 ppm calcium solution for 30 min before use.

Step 2 Pipette 50 ml of the sample and 5 ml of the sodium hydrogen carbonate solution B into a 200 ml beaker containing a magnetic stirrer bar. Note: see Step 6.

Step 3 Remove the divalent electrode from the solution in which it is immersed and place it in a holder together with a reference electrode. Rinse both electrodes with deionized water, then remove the surplus water with a tissue.

Step 4 Immerse the electrodes in the sample and stir at a moderate rate without forming a vortex.

Step 5 When a steady reading has been obtained, note the potential E_1.

Step 6 Use E_1 to obtain an estimate, c_e, of the total hardness from a calibration graph for the electrode. This estimate is required to calculate the volume of the standard addition. Should the estimate indicate a hardness greater than 100 ppm, repeat the analysis but at Step 2 add 5 ml of sodium hydrogen carbonate solution solution A.

Table 2   Concentrations produced by known additions

Volume of calcium solution added, V_A (ml)	Concentrations, c_a ppm, of calcium carbonate in 50 ml sample produced by addition of calcium solution:		
	A	B	C
0.5	100	20	5
1.0	200	40	10
1.5	300	60	15
2.0	400	80	20

Step 7 Obtain from Table 2 the volume of calcium solution A, B or C to be added, choosing the volume of the appropriate solution that gives an addition, c_a, closest to the estimated concentration, c_e, from Step 6.

Step 8 Add by pipette the volume, V_A, of the appropriate standard calcium solution indicated by Table 2 and note the steady reading E_2.

Step 9 Add exactly 50 ml of deionized water and 5 ml of the same sodium hydrogen carbonate solution as in Step 2. It is important that the temperature of these additions be within 1 °C of the sample temperature in Step 1.

Step 10 Note the steady reading E_3.

Step 11 Calculate the calibration slope, $k = (E_2 - E_3)/(0.301) \simeq 27$ mV per tenfold increase in concentration.

Step 12 Calculate the water hardness, c, from

$$c = \frac{c_a}{\left[\left(\frac{V_A}{55} + 1\right)\left(\text{antilog}\frac{E_2 - E_1}{k}\right) - 1\right]} \text{ ppm calcium carbonate}$$

or

use the simpler formula which neglects volume corrections for the known addition V_A,

$$c = \frac{c_a}{\left(\text{antilog}\frac{E_2 - E_1}{k}\right) - 1} \text{ ppm calcium carbonate}$$

Step 13 Repeat steps 2–12 for all samples.

Worked example

The electrodes when immersed in a sample solution gave a value of $E_1 = +25.4$ mV, which gave an estimated water hardness of 27 ppm calcium carbonate from a calibration graph. The nearest concentration from Table 2 (25 ppm) corresponded to an addition of 0.5 ml of solution B. After the addition the steady reading was $E_2 = +33.6$ mV and following the subsequent 1:1 dilution the steady reading was $E_3 = +25.2$ mV.

$$k = \frac{(E_2 - E_3)}{0.301} = \frac{8.4}{0.301} = 27.9 \text{ mV per decade}$$

$$c = \frac{c_a}{\text{antilog}\frac{E_2 - E_1}{k} - 1} = \frac{25}{\text{antilog}\frac{8.2}{27.9} - 1}$$

$$= 25.8 \text{ ppm calcium carbonate}$$

Table 3 Addition of standard solution for calibration range 5–100 ppm calcium carbonate

Volume of calcium solution B added (ml)	Concentration (ppm calcium carbonate)
0.2	4.99
0.5	12.43
1.0	24.75
2.0	49.02
3.0	72.82
4.0	96.15

Preparation of the calibration graph

Range 5–100 ppm calcium carbonate A calibration curve suitable for the estimation of the hardness of the sample, c_e, can be prepared in the following manner: Add 10 ml of sodium hydrogen carbonate solution B to exactly 100 ml of deionized water contained in a 250-ml beaker. Stir the solution and with the electrodes immersed note the steady readings after each of the additions indicated in the Table 3.

Construct the calibration graph by plotting the e.m.f. after each addition versus the logarithm of the concentration indicated in Table 3. This calibration graph should be determined at least once a week.

Range 100–200 ppm calcium carbonate Add 10 ml of sodium hydrogen carbonate solution A to exactly 100 ml of deionized water contained in a 250-ml beaker. Note the steady readings after each of the additions indicated in Table 4.

Construct the calibration graph by plotting the e.m.f. after each addition versus the logarithm of the concentration indicated in Table 4. This calibration graph should be determined at least once a week.

Table 4 Addition of standard solution for calibration range 100–200 ppm calcium carbonate

Volume of calcium solution A added (ml)	Concentration (ppm calcium carbonate)
1.0	99.00
1.2	118.57
1.4	138.06
1.6	157.48
1.8	176.81
2.0	196.07

Sources of Error

The accuracy of the known addition technique depends on the constancy of activity coefficients, degree of complexation and calibration slope in the measurements of the potentials E_1, E_2, and E_3. For samples containing mainly hardness salts at concentrations corresponding to the lower ends of the analytical range in the method, the sodium bicarbonate will make the major contribution to the ionic strength and overall errors from variations in activity coefficients will be less than 5%. In samples of a similar nature having concentrations corresponding to the top of the analytical ranges, larger variations in activity coefficients will occur since the addition of calcium to the sample significantly alters the ionic strength. Samples which contain salts, other than those of calcium and magnesium, at molar concentrations approximately the same as the hardness salts, suffer more from errors in the variation of activity coefficients during the dilution stage (E_3) and in this type of sample the calibration slope would be best obtained from a calibration graph. Unfortunately, excess electrolyte cannot be added to dominate completely the ionic strength as the hardness electrode suffers from considerable interference from monovalent cations. The concentration of calcium carbonate obtained from the known addition procedure will include a contribution from any interfering cation present in the sample together with those added as reagents. The sodium bicarbonate buffers added will produce a positive bias of about 0.4 and 9 ppm in the lower and upper concentration ranges of the method, respectively.

The errors caused by variable degrees of complexation are dependent on the strength(s) of the complex(es) and the relative excess of the ligand(s) over the total calcium and magnesium present in the sample. Strong complexing agents of the multidentate type are rarely found in natural waters and if they are present their concentrations are very low and generally do not constitute an interference. Sekerka and Lechner (1975) reported that neither sulphate nor phosphate ions affected the potential of the electrodes if their molar concentrations were equal to, or less than, that of the total of calcium and magnesium ions present in the sample. The sodium hydrogen carbonate is added to the sample to provide an excess of the most commonly found complexing species and to control the pH (Reynolds, 1971).

This method is not suitable for the analysis of certain types of industrial water, e.g., samples from boilers that have been treated with phosphate.

Precision

A relative standard deviation of 2% was obtained in the concentration range 2—100 ppm calcium carbonate by an automated procedure using an Orion 92-32 electrode (Sekerka and Lechner, 1974). Less precision is likely for a manual technique, but by comparison with other similar procedures a relative standard deviation of 5% is to be expected.

Accuracy

Sekerka and Lechner (1974) compared the results obtained by the electrode with those obtained by an EDTA titration for a hardness range of 7.4—168 ppm calcium carbonate. A small negative bias, not greater than 4%, was found in the analysis of these natural waters.

The results obtained by this method will always be subject to the bias (discussed in *Sources of Error*) from the bicarbonate buffer, although this can be removed where necessary by applying a correction obtained from the analysis of a standard calcium solution of approximately the same concentration as that found in the sample.

Bibliography

ASTM, 1973, Part 23, D1126-67, *Standard Methods of Test for Hardness in Water*, American Society for Testing and Materials, Philadelphia.

ASTM, 1973a, Part 23, D511-52, *Standard Method of Test for Calcium and Magnesium Ion in Water*, American Society for Testing and Materials, Philadelphia.

BSI, 1962, BS 1427:1962, *Routine Control Methods of Testing Water Used in Industry*, British Standards Institute, London.

Hadjiioannou, T. P., and D. S. Papastathopoulos, 1970, 'EDTA titration of calcium and magnesium with a calcium-selective electrode', *Talanta,* **17**, 399.

Reynolds, R. C., 1971, 'Analysis of Alpine waters by ion electrode methods', *Water Resources Res., 7*, 1333.

Sekerka, I., and J. F. Lechner, 1974, 'Automated simultaneous determination of water hardness, specific conductance and pH', *Anal. Lett., 7*, 399.

Sekerka, I., and J. F. Lechner, 1975, 'Simultaneous determination of total, non-carbonate and carbonate water hardness by direct potentiometry', *Talanta,* **22**, 459.

Whitfield, M., J. V. Leyendekkers, and J. D. Kerr, 1969. 'Liquid ion-exchange electrodes as end-point detectors in compleximetric titration. Part II. Determination of calcium and magnesium in the presence of sodium', *Anal. Chim. Acta,* **45**, 399.

Whitfield, M., 1971, *Ion Selective Electrodes for the Analysis of Natural Waters*, Australian Marine Sciences Association, Sydney, N.S.W.

Determination of Silver

Suitable electrodes are silver metal electrodes and those sulphide-selective membrane electrodes with silver sulphide membranes, e.g. Orion 94-16, Beckman 39610, Leeds & Northrup 117408, Metrohm EA306S/Ag, Philips IS 550-S, Radiometer F1212S and F1712S, Activion 003 15 007, Coleman 3-805, and Tacussel PAG 2. The silver metal electrode may be in the form of wire, foil, a solid billet or as silver-plated platinum. The Beckman 39654 is a combination electrode of the Ag_2S membrane type. The membrane electrodes suffer less interference from strongly oxidizing solutions, but no other advantages have been clearly demonstrated and they are many times more costly than a silver metal electrode. Halide-selective membrane electrodes based on the corresponding silver halide can also be used, in principle, to determine silver, but they have higher limits of detection than silver metal or silver sulphide electrodes and are more prone to attack by complexing agents. For these reasons, their use is not normally recommended.

APPARATUS

Electrode; mercury–mercurous sulphate or double-junction reference electrode; pH meter; magnetic stirrer; PTFE, polyethylene or Vycor beakers

In practice, double-junction reference electrodes have been used rather than the mercury–mercurous sulphate type. In either case, a sleeve-type junction is preferable to one with a ceramic frit.

REAGENTS

Water

Deionized water with a specific conductivity of less than $0.2 \,\mu S \, cm^{-1}$ at 25 °C is suitable for preparing reagent and standard solutions. Particular care should be taken to keep the water (and any solutions) away from hydrogen sulphide, and concentration solutions of ammonia or hydrochloric acid, as absorption of any of these compounds will cause errors.

Stock silver solution (1000 ppm)

Dissolve 1.574 g of silver nitrate (analytical reagent grade, dried at 150 °C for 24 h) in water and make up to the mark in a 1-litre calibrated flask. Store in a brown glass bottle or in a dark place.

1 ml ≡ 1 mg silver

Standard silver solutions

Prepare by dilution of the stock solution. Store in the dark. Solutions in the concentration range 1–10 ppm should not be kept for more than a week and more dilute solutions should be prepared daily.

Potassium nitrate solution (1 mol l^{-1})

Dissolve 101.11 g of potassium nitrate (analytical reagent grade) in water, transfer to a 1-litre calibrated flask and make up to the mark with water.

Filling solution for double-junction reference electrodes

Add 10 ml of 1 mol l^{-1} potassium nitrate solution to 100 ml deionized water, and mix.

SAMPLE COLLECTION

Collect samples in clean polyethylene bottles and store in the dark. The sample should be analysed as soon as possible and and no later than six hours after being taken. Great care should be taken to avoid contaminating the sampling bottle and its cap with chloride ions from the skin.

CONDITIONING AND STORAGE OF ELECTRODES

At the end of each batch of analyses, rinse the sensing electrode with deionized water and dry it with a soft tissue. Store membrane electrodes with a protective cap over the sensing surface. Rinse the electrode occasionally with dilute ammonia solution, particularly if it has been immersed in a solution contaminated with chloride ions.

Change the solution in the outer chamber of a double-junction reference electrode at least once a week.

CONCENTRATION RANGE AND UNITS

Silver-selective electrodes have Nernstian calibrations at concentrations down to 10 ppb in dilute solutions of silver nitrate, but the range extends to much lower concentrations of free silver ion when moderate to high total concentrations (>100 ppb) of silver are in equilibrium with relatively large concentrations of anions form complexes or sparingly soluble silver salts, e.g. thiosulphate, bromide, sulphide. Veselý, Jensen and Nicolaisen (1972) found that the Nernstian calibration extended to 10^{-23} mol l^{-1} (10^{-15} ppb) for electrodes in equilibrium with sulphide solutions. The electrodes can therefore be used to determine silver in solutions of strong complexing agents, even though the concentration of free silver ion is much less than 10 ppb; a known addition method is used rather than the conventional type of calibration.

$$10^{-3} \text{ mol l}^{-1} \equiv 107.9 \text{ ppm}$$
$$9.268 \times 10^{-6} \text{ mol l}^{-1} \equiv 1 \text{ ppm}$$

ANALYTICAL PROCEDURES

Method A

Concentrations above 1 ppm in the absence of complexing agents

Step 1A Pipette 50 ml of a standard solution of concentration s_1 into a 100-ml beaker containing a PTFE-coated stirrer bar. Add 5 ml of 1 mol l^{-1} potassium nitrate solution. The concentration, s_1, should be close to, but greater than, the highest expected sample concentration.

Step 2A Rinse the electrodes with a portion of the standard solution and immerse them in the solution in the beaker.

Step 3A Start the stirrer, adjust the control so that the solution is stirred without a vortex being formed, and note the setting. Mark the position of the electrode holder on the stand.

Step 4A When a steady reading has been observed, note the e.m.f., E_1.

Step 5A Repeat Steps 1A and 2A with a second standard solution of concentration, s_2, at the lower end of the expected concentration range of the samples and not more than half the concentration s_1.

Step 6A Keeping the conditions the same as in Step 3A, when a steady reading has been obtained, note the potential E_2.

Step 7A *Either* calculate the calibration slope, per tenfold increase in concentration

$$k = \frac{E_1 - E_2}{\log(s_1) - \log(s_2)} \simeq 59 \text{ mV}$$

or

prepare a temporary calibration graph by drawing a straight line through the points ($\log s_1$, 0) and ($\log s_2$, $E_1 - E_2$).

Step 8A Repeat Steps 1A and 2A with the sample solution.

Step 9A Keeping the conditions the same as in Step 3A note the steady potential E_x.

Step 10A Calculate the potential difference, $\Delta = E_1 - E_x$.

Step 11A *Either* calculate the concentration of the sample from
$c = s_1/\text{antilog}\,(\Delta/k)$
or
read it off the calibration graph.

Step 12A For each sample repeat Steps 8A–11A. The value of E_1 should be determined at least twice in a full working day (Steps 1A–4A). The calibration slope should be checked at the start of each day (Steps 5–7).

Worked example

The e.m.f. values observed with a 100 ppm standard, a 10 ppm standard and a sample solution are 350.2 mV, 293.0 mV and 331.6 mV, respectively. The calibration slope $k = (350.2 - 293.0)/(\log 100 - \log 10) = 57.2$ mV per tenfold change in concentration.

$$c = 100/\text{antilog}\left[\frac{350.2 - 331.6}{57.2}\right] = 47.3 \text{ ppm}$$

Preparation and use of a permanent calibration graph

If the conditions discussed on p. 117 are maintained, Steps 5A–7A may be omitted from the routine analytical procedure and a permanent calibration graph prepared instead. Carry out Steps 1A–4A of the analytical procedure, then with at least four other standard solutions of lower concentration repeat Steps 8A–10A. Plot the values of Δ on the y-axis against the logarithms of the corresponding concentrations on the x-axis, including the first standard solution as the point ($\log s_1$, 0).

Method B

Determination of silver in presence of complexing agents and in the range 10 ppb–1 ppm in the absence of complexing agents. See Chapter 6 for a fuller appreciation of the factors involved in the known addition method.

Step 1B Pipette 50 ml of sample into a 100-ml beaker containing a plastic-coated stirrer bar and add 5 ml of 1 mol l^{-1} potassium nitrate solution.

Step 2B If it is known by experience to be necessary, add 1 ml of a concentrated solution of a complexing agent (Note 1, below).

Step 3B Rinse the electrodes with deionized water and immerse them in the solution in the beaker.

Step 4B Start the stirrer and adjust the control so that the solution is stirred without a vortex being formed.

Step 5B When a steady reading has been observed, note the e.m.f., E_1.

Step 6B Use Note 1 to decide if complexing agent should be added: if so, repeat Steps 2B and 5B.

Step 7B Add a known volume, v ml, of standard silver solution of concentration

s ppm; v should be no greater than 1 ml and s should be chosen so that $v \cdot s \simeq 50c'$, where c' is the expected concentration of the sample (Note 2).

Step 8B When a steady reading has been obtained, note the e.m.f., E_2.

Step 9B Calculate the factor $Q = $ antilog $\{(E_2 - E_1)/k\}$, where $k \simeq 59$ mV per tenfold increase in concentration is the calibration slope or find Q using Appendix 4.

Step 10B Calculate the concentration (Note 3).

$$c = v \cdot s/50(Q - 1) \text{ ppm}$$

Step 11B If $E_2 - E_1 > 35$ mV or <10 mV and the best accuracy is desired, use the value of c calculated in Step 10B to choose a better value of v or s and repeat the procedure.

Note 1 The method is inaccurate if the fraction of silver present as the free ion is significantly changed when the standard solution is added in Step 7B. The action to be taken in Steps 2B or 6B depends on the concentration of the complexing agent in the sample, L mol l^{-1}, and the stability constant, β, values of which may be obtained from Ringbom (1963) and Martell and Sillén (1964 and 1971):

(a) $\beta L \leq 0.01$ – No addition is necessary; this includes the case in which no complexing agent is present.

(b) $\beta c^* \leq 10^{-3}$ – No addition is necessary if $50 L \geq v \cdot s/100$, otherwise add complexing agent until the second condition is satisfied. Note that s must be expressed in ppm and that c^* is the molar concentration of free silver ion (see below).

(c) $\beta c^* > 10^{-3}$ – Add complexing agent until both the conditions in (b) are satisfied.

The free silver ion concentration, c^*, may be estimated by finding the concentration corresponding to E_1 on a calibration graph prepared in the absence of complexing agents, e.g. either the permanent or temporary calibration graphs for Method A. These graphs may be extrapolated if necessary.

Note 2 An estimate of c' may be obtained from the concentration corresponding to E_1 on a calibration graph prepared with solutions containing approximately the same concentrations of complexing agent as in the samples.

Note 3 The calculation in Step 10B is not corrected for the dilution of the sample by the added standard solution. If v is small, the error is negligible for most purposes, as may be seen below from the expression with a volume correction

$$c = \frac{v \cdot s}{50[Q(1 + v/V_2) - 1]} \text{ ppm}$$

where $V_2 = 56$ if complexing agent is added in Steps 2B or 6B and 55 otherwise.

Preparation and use of calibration graphs for known addition procedure

Calibration graphs can serve three purposes in known addition procedures and two distinct types of graph are needed to cover all of them. The most important use is the determination of the calibration slope, k. This may be done exactly as in Method A for either a permanent or temporary calibration graph. In strongly complexing media, however, those graphs may not extend to sufficiently low concentrations of free silver ion and the calibration procedure given below should be preferred, as it proves that the electrode is functioning correctly in the conditions of the actual analysis.

The second use is to estimate the free silver ion concentration, c^* mol l^{-1}, for Step 6B/Note 1. Data from either calibration from Method A may be used for this purpose also, although it may be necessary to extrapolate them to much lower concentrations. As the concentrations are now required in molar units, it may be convenient to redraw the graphs as plots of e.m.f. against the logarithm of the molar concentration.

Finally, the calibration given below is used for finding c' in Step 7B/Note 2.

To prepare a calibration graph that compensates for the effect of complex formation carry out Step 1B with 50 ml of deionized water or a solution containing as nearly as possible the same concentrations of complexing agents as are expected to be present in the samples. Carry out Step 2B if necessary. Add a small volume, $v_1 \leqslant 0.2$ ml, of standard silver solution of concentration s to the solution in the beaker (a micrometer syringe is a convenient means of making the addition). Carry out Steps 3B—5B. Add four further increments of standard silver solution, noting in each case the total volume added, v_i, and the steady potential, E_i. The total volume added should not exceed 1 ml. Calculate the concentration $c_i = v_i \cdot s/(50 + v_i)$ corresponding to each addition and plot the potentials, E_i, on the y-axis against the logarithms of the concentrations on the x-axis. The concentration of the standard solution should be chosen so that the final concentrations, c_i, span the expected concentration range of the samples.

The second and third applications of the calibration graphs are for making approximations only, and the graphs do not have to be redefined very frequently nor is such strict temperature control necessary as is usual. In defining the calibration slope, however, the same precautions are necessary as in direct potentiometry.

Worked examples

(a) No complexing agent present. The e.m.f. values before and after the addition of 1 ml of 5 ppm standard silver solution to 50 ml of sample were 170.0 and 192.5 mV respectively. The calibration slope was 58.0 mV per decade. The concentration of the sample was therefore

$$c = \frac{1 \times 5}{50 \times [\text{antilog}\,(22.5/58) - 1]} = 0.069 \text{ ppm} = 69.3 \text{ ppb}$$

(b) In the presence of an excess of a complexing agent with a stability constant of 10^6 for silver, a silver electrode with a calibration slope of 58 mV per decade has a potential of 50 mV before 0.5 ml of 1 g l^{-1} silver standard are added and 70 mV afterwards. The volume of sample taken was 50 ml and 5 ml of sodium nitrate solution were added. Therefore

$$c = \frac{0.5 \times 1000}{50 \times [\text{antilog}\,(20/58) - 1]} = 8.25 \text{ ppm}$$

Note that the e.m.f. indicated only about 1 ppb of silver present as the uncomplexed ion, as indicated by a calibration in which no complexing agent was added.

SOURCES OF ERROR

Effect of Temperature

Changes in temperature affect the slope of the calibration graph and the standard potentials of the electrodes. The standard and sample solutions should not differ in temperature by more than 1 °C.

Effect of Other Substances

Mercury ions must be absent from the solution. Durst and Taylor (1967) found that ferric iron interfered with a silver metal electrode if the molar ratio of ferric to silver ions exceeded 0.1. Otherwise, heavy metal cations are not expected to interfere unless present in a very great excess ($>10^6$ fold), as has been confirmed by Veselý et al. (1972) for membrane electrodes and Durst and Taylor (1967) for metal electrodes. Alkaline earth cations and alkali metal cations are even less likely to interfere.

Anions that form sparingly soluble silver salts will cause a negative bias if introduced accidentally into the sample solution even in trace amounts, e.g. chloride, bromide, iodide, sulphide, and thiocyanate. Sulphate and hydroxide can interfere only at high concentrations. Silver forms complexes very readily with complexing agents and if these are present, only the free silver can be determined by direct potentiometry. The total silver in the presence of an excess of complexing agent can be determined by known addition potentiometry. If only a small concentration of ligand is present, more must be added before the known addition technique is used (p. 98). Exposure to high (100 ppm) concentrations of chloride ion makes membrane electrodes lose their sensitivity to silver (Müller, West and Müller, 1969), but the response can be restored by washing for a few seconds in 0.005 mol l^{-1} ammonia solution and then in deionized water, followed by immersion in 0.05 mol l^{-1} sulphide solution. The surface of some membrane electrodes may be restored by polishing.

PRECISION

Müller et al. (1969), who calibrated an Orion electrode very carefully, obtained a maximum error for eighteen determinations of silver concentration in each of six synthetic standard solutions (13.5–80 ppb) of 0.2%. They also showed that the e.m.f. of the electrode was very reproducible over a three-month period. Durst and Taylor (1967) obtained relative standard deviations of 1–2.5% for determinations in the range 1.3–130 ppm by null-point potentiometry with silver metal electrodes. At lower concentrations the relative standard deviations were in the range 5–10%. Tamele, Irvine and Ryland (1960) reported relative standard deviations of about 4% for the determination of 10–100 ppm silver by direct potentiometry with a silver metal electrode, but measurements at lower concentrations were much less precise. The relative standard deviation for the determination of 4 g l^{-1} silver in a cyanide solution by a known addition procedure was 5.5% (Lapatnick, 1974).

ACCURACY

Müller *et al.* (1969) obtained a probable error for a single measurement with an Orion electrode in the range 13.5–80 ppb of 0.15–0.2%. Errors of less than 3.5% were reported by Durst and Taylor (1967) for solutions containing 1.3–130 ppm silver analysed by null-point potentiometry using silver metal electrodes. The mean errors for the determination of silver in a cyanide plating solution were found by Lapatnick (1974) to be −1.7% in the range $12–120 \text{ g l}^{-1}$ and 8.3% in the range $2–12 \text{ g l}^{-1}$ for a known addition method.

RESPONSE TIME

Response times of 1–2 min have been reported for membrane electrodes at concentrations above 1 ppm, extending to several minutes at lower concentrations (Veselý *et al.*, 1972; Crombie, Moody and Thomas, 1975). Similar times have been found for silver metal electrodes, but Tamele *et al.* (1960) found that silver metal electrodes took 15 min to reach equilibrium in 10–100 ppm solutions and several hours at lower concentrations.

OTHER APPLICATIONS

Silver-selective membrane electrodes are also used for the determination of sulphide. Silver electrodes are used for potentiometric titrations of silver, chloride, bromide, iodide, cyanide, thiocyanate, thiols and possibly several other anions. Membrane electrodes can be used for the determination of cyanide by direct potentiometry if argentocyanide ion is added to the solution.

TRACING FAULTS

The general points discussed on p. 127 should be considered, but the following have particular relevance for silver electrodes:

Low sensitivity and slow response

If the surface of a silver sulphide membrane electrode becomes tarnished the electrode may suffer from a reduction of the calibration slope and long response times, especially at low concentrations; renew the surface of the membrane according to the manufacturer's instructions.

High sensitivity

The proportion of silver lost by adsorption on the walls of containers increases as the conentration decreases, with the effect that the calibration slope is apparently greater than theoretical.

Bibliography

Crombie, D. J., G. J. Moody, and J. D. R. Thomas, 1975, 'Observations on the calibration of solid-state silver sulphide membrane ion-selective electrodes', *Anal. Chim. Acta,* **80**, 1.

Durst, R. A., and J. K. Taylor, 1967, 'Modified linear null-point potentiometry', *Anal. Chem.,* **39**, 1374.

Lapatnick, L. N., 1974, 'Application of ion-selective electrodes to the analysis of silver-plating baths', *Anal. Chim. Acta,* **72**, 430.

Martell, A. E., and L. G. Sillén, 1964, *Stability Constants,* Special Publication No. 17, The Chemical Society, London.

Martell, A. E., and L. G. Sillën, 1971, *Stability Constants, Supplement No. 1,* Special Publication No. 25, The Chemical Society, London.

Müller, D. C., P. W. West, and R. H. Müller, 1969, 'Determination of silver ion in parts per billion range with a selective ion electrode', *Anal. Chem.,* **41**, 2038.

Ringbom, A., 1963, *Complexation in Analytical Chemistry,* Interscience, New York and London.

Tamele, M. W., V. C. Irvine, and L. B. Ryland, 1960, 'Potentiometric determination of sulphide ions and the behaviour of silver electrodes at extreme dilution', *Anal. Chem.,* **32**, 1002.

Veselý, J., O. J. Jensen, and B. Nicolaisen, 1972, 'Ion-selective electrodes based on silver sulphide', *Anal. Chim. Acta.,* **62**, 1.

Determination of Copper

Several different types of copper electrode are available. All have solid-state membranes, the commonest being composed of a mixture of silver sulphide and cupric sulphide (Orion 94-29A, Radiometer F3002, Leeds & Northrup 117403). The Radiometer F1112Cu has a single crystal of $Cu_{1.8}Se$ for a membrane. Other electrodes include the Beckman 39612, Metrohm EA306Cu, Philips IS 550-Cu, Simac Cu/1C, Tacussel PCU2, Activion 003 15 011, Coleman 3-804 and EDT Supplies EE-Cu. The Beckman 39655 is a combination ion-selective and reference electrode. All the electrodes are used in essentially similar ways but for most of the above types insufficient data has been published for their characteristics to be assessed with confidence.

APPARATUS

Copper electrode; reference electrode; pH meter; magnetic stirrer

For most purposes any of the common reference electrodes may be used.

REAGENTS

Water

Distilled water passed through a mixed-bed deionization column should have a copper content of about 0.1 ppb, which is adequate for all measurements with electrodes. Raw water treated by a mixed-bed deionization unit can have a copper content of 5 ppb, which is adequate for many purposes.

Standard copper solution A (1000 ppm)

Dissolve 3.930 g of cupric sulphate pentahydrate (analytical reagent grade) in deionized water and make up to the mark in a 1-litre calibrated flask.

$$1 \text{ ml} \equiv 1 \text{ mg copper}$$

Standard copper solution B (100 ppm)

Pipette 50 ml of standard solution A into a 500-ml calibrated flask and make up to the mark with deionized water.

$1\ ml \equiv 100\ \mu g$ copper

Standard copper solution C (10 ppm)

Pipette 50 ml of standard solution B into a 500-ml calibrated flask and make up to the mark with deionized water. Store in a polyethylene bottle.

$1\ ml \equiv 10\ \mu g$ copper

Other standard solutions should be prepared as required by dilution of the above standards. Solutions with a concentration below 10 ppm should be prepared daily and stored in polyethylene bottles.

5N sodium hydroxide

Dissolve 200 g of sodium hydroxide pellets in deionized water and make up to the mark in a 1-litre calibrated flask with water. Store in a polyethylene bottle.

0.1N sodium hydroxide

Either pipette 10 ml of 5N sodium hydroxide solution into a 500-ml calibrated flask and make up to the mark with deionized water, or prepare the solution from an ampoule of concentrated volumetric solution.

5N nitric acid

Carefully add, with stirring, 317 ml of concentrated nitric acid (analytical reagent grade) to about 400 ml of deionized water. Transfer the solution to a 1-litre calibrated flask and make up to the mark with deionized water.

Ionic strength adjustment solution

Dissolve 21.2 g of sodium nitrate (analytical reagent grade) in water and make up to 100 ml in a calibrated flask. This solution contains $2.5\ mol\ l^{-1}\ NaNO_3$.

Bromocresol purple indicator

Dissolve 0.05 g of bromocresol purple in 25 ml deionized water + 0.095 ml 0.1N sodium hydroxide solution. If the solid indicator is slow to dissolve, it is permissible to add a slight excess of sodium hydroxide solution, but the indicator solution should not in that case be used for a conventional volumetric titration. Transfer the indicator solution to a 50-ml calibrated flask and make up to the mark with deionized water.

CONDITIONING AND STORAGE

The various types of electrode differ considerably in this respect. The Orion, Beckman and Radiometer F1112Cu electrode need no conditioning and after use should be rinsed with deionized water, dried with a paper tissue and stored dry in air, preferably with a protective cap over the membrane. After a time, the electrodes may become tarnished and they should be polished to restore the surface. Orion electrodes lose their metallic sheen and become matt black; they can be restored with polishing paper supplied by the Orion company. Radiometer F1112Cu electrodes turn from a metallic blue to a green—bronze as they age; they can be polished with Radiometer V604 polishing paste. Radiometer F3002 electrodes are first impregnated with the sensitizing mixture according to the manufacturer's instructions and then conditioned by soaking them for one or preferably two days in 0.1 mol l^{-1} disodium EDTA solution. The electrode should be stored in the disodium EDTA solution when not in use, although storage in deionized water has no rapid deleterious effect; the electrode should not be stored dry as this affects its sensitivity.

A single application of the sensitizing mixture on the Radiometer F3002 electrode has been known to last for months; the electrode itself should last for years. The other electrodes should all have lifetimes measured in years. Persistent use in solutions that cause tarnishing of the membrane will shorten the lifetime, as the electrodes will need to be polished or re-sensitized more often.

SAMPLE COLLECTION

Collect samples in polyethylene bottles. Before use, clean the bottles by filling them with dilute hydrochloric acid (1 + 1) and setting them aside for 2—3 days; invert them for a period during this time to remove any contamination from the necks and stoppers. Wash well with water. Sampling lines should not be made of copper or brass. The samples should be analysed without delay, as copper may be adsorbed on the walls of the container. If it is not possible, on a routine basis, to analyse the samples immediately, the following trial should be carried out: collect a representative batch of samples and analyse them by Method A, below, (i) immediately (ii) after 6 h (iii) after 24 h (iv) every 24 h until the longest storage time has been passed. If the indicated copper concentration does not change (within experimental error) over the likely period of storing the sample, sample as above and use Method A, otherwise sample as follows and use Method B.

Before use, fill each sample bottle with a suitable volume (e.g. 500 or 1000 ml) of deionized water. Mark the level of the water in each bottle. When sampling, fill the bottle to the mark and then add 5N nitric acid in the proportion of 10 ml for each litre of sample. Shake the bottle and store it.

CONCENTRATION RANGE AND UNITS

The methods should work over the range 650—0.05 ppm for most electrodes. The Radiometer F1112Cu has a linear range down to 0.5 ppm, but below that the calibration would be curved. Higher concentrations may be determined, but the concentration of the ionic strength adjustor in Method A should be increased by a factor of ten for every tenfold increase in maximum copper concentration. Method B is less likely to be needed at high concentrations, unless the sample has a high acidity.

10^{-3} mol l^{-1} ≡ 63.54 ppm

1 ppm ≡ 1.574×10^{-5} mol l^{-1}

ANALYTICAL PROCEDURE

Method A

Step 1A Pipette 50 ml of standard solution into a 100-ml plastic beaker containing a plastic-coated stirrer bar. Add 1 ml of ionic strength adjustor. The standard solution should have a concentration s_1 close to, but greater than, the highest expected sample concentration.

Step 2A Rinse the electrodes with deionized water, dry them with a paper tissue and immerse them in the solution.

Step 3A Start the stirrer, adjust the control so that the solution is stirred briskly without a vortex being formed and note the setting. Mark the position of the electrode holder on the stand.

Step 4A When a steady reading has been obtained, note the potential E_1.

Step 5A Repeat Steps 1A and 2A with a second standard solution of concentration s_2 not more than half the concentration s_1.

Step 6A Keeping the conditions the same as in Step 3A, allow the electrodes to reach a steady reading and note the potential E_2.

Step 7A Linear calibration

Either calculate the calibration slope per tenfold increase in concentration

$$k = \frac{E_1 - E_2}{\log s_1 - \log s_2} \simeq 28 \text{ mV}$$

or

prepare a temporary calibration graph by drawing a straight line through the points $(\log s_1, 0)$ and $(\log s_2, E_1 - E_2)$.

Non-linear calibration

Read the apparent concentration, c_2, off the calibration graph. If c_2 and s_2 do not agree within a prescribed limit, say 5% repeat Steps 5A and 6A and if c_2 and s_2 are still unacceptibly different, recalibrate the electrode.

Step 8A Repeat Steps 1A and 2A with the sample solution.

Step 9A Keeping the conditions the same as in Step 3A, when a steady reading has been attained, note the potential E_x.

Step 10A Calculate the potential difference, $\Delta = E_1 - E_x$.

Step 11A Either calculate the concentration of the sample from

$$c = s_1/\text{antilog}\,(\Delta/k)\text{ ppm}$$

or read it off the calibration graph.

Step 12A For each sample repeat Steps 8A–11A. The value of E_1 should be determined at least twice on a full working day (Steps 1A–4A). The calibration slope should be checked at the start of each day (Steps 5A–7A).

Method B

Step 1B Pipette 100 ml of a standard solution of concentration s_1 into a plastic beaker containing a magnetic stirrer bar. Add 1 ml of 5N nitric acid. *Note:* Do not use hydrochloric acid. Add from a graduated pipette 0.9 ml of 5N sodium hydroxide solution.

Step 2B Either add 2 drops of bromocresol purple indicator or immerse in the solution a calibrated combination pH electrode that has been rinsed with deionized water and wiped dry with a paper tissue.

Step 3B Add from a burette 0.1N sodium hydroxide solution until either the indicator is pale blue (almost colourless in fluorescent light) or the pH electrode indicates pH 5.5–6.0. If the end-point is overshot, add 0.1N nitric acid dropwise until the correct colour or pH is attained. Note the total volume of solution added, v, whether as acid or base; this should be approximately 1 ml.

Step 4B Add $(3-v)$ ml of deionized water from a graduated pipette.

Step 5B Rinse the copper and reference electrodes with deionized water, wipe them with a paper tissue and immerse them in the solution.

Step 6B Adjust the stirrer control to give brisk stirring without causing a vortex. Note the position of the electrode holder on the stand.

Step 7B When a steady potential has been reached, note the reading E_1.

Step 8B Repeat Steps 1B–4B and 5B with a second standard solution of concentration s_2 such that s_2 is no more than half s_1 and spans the expected concentration range of the samples.

Step 9B When a steady potential has been reached, note the reading E_2.

Step 10B **Linear calibration**

Either calculate the calibration slope per tenfold increase in concentration

$$k = \frac{E_1 - E_2}{\log s_1 - \log s_2} \simeq 28 \text{ mV}$$

or

prepare a temporary calibration graph by drawing a straight line through the points $(\log s_1, 0)$ and $(\log s_2, E_1 - E_2)$.

Non-linear calibration
Read the apparent concentration, c_2, off the calibration graph. If c_2 and s_2 differ by more than a prescribed amount, say 5%, repeat Steps 8B and 9B and if c_2 and s_2 are still unacceptably different, recalibrate the electrodes.

Step 11B Pipette 100 ml of sample solution into a plastic beaker containing a magnetic stirrer bar. Add 0.9 ml of 5N sodium hydroxide solution.

Step 12B Repeat Steps 2B and 3B.

Step 13B Add $(2-v)$ ml of deionized water from a graduated pipette.

Step 14B Repeat Step 5B.

Step 15B When a steady potential has been reached, note the reading E_x.

Step 16B Calculate the potential difference $\Delta = E_1 - E_x$.

Step 17B Either read the concentration of the sample from the calibration graph or calculate it from

$$c = s_1 / \text{antilog}\,(\Delta/k) \text{ ppm}$$

Step 18B For each sample, repeat Steps 11B–17B. The value of E_1 should be determined at least twice in a full working day (Steps 1B–5B, 7B). The calibration slope should be checked at the start of each day (Steps 8B–10B).

Worked Example

The potentials recorded in 10 and 3 ppm standard solutions are 50 and 35 mV, respectively. The potential in the sample is 42 mV. The calibration slope is

$$k = (50 - 35)/[\log(10) - \log(3)] = 28.69 \text{ mV per decade.}$$

The concentration is

$$c = 10/\text{antilog}\left(\frac{50-42}{28.69}\right) = 5.26 \text{ ppm.}$$

Preparation and use of a permanent calibration graph

If conditions permit a permanent calibration graph to be used (see p. 117), Steps 5A–7A and 8B–10B may be omitted from the analytical procedure for concentrations in the linear part of the calibration only. The linear range can vary considerably between electrodes, even those of the same manufacturer, and the characteristics of the electrode must be determined by a proper calibration procedure before a routine two-point calibration is accepted.

By dilution of standard solutions A, B or C, prepare 500 or 1000 ml each of at least five standard solutions spanning the concentration range to be measured (the wider the range, the more solutions are required). With the most concentrated, repeat Steps 1A–4A or 1B–7B of the appropriate analytical procedure and with the others the Steps 8A–10A or 1B–5B, 15B, 16B.

SOURCES OF ERROR

Effect of Temperature

Changes in temperature affect the slope of the calibration graph and the standard electrode potential. The standard and sample solutions should not differ in temperature by more than 1 °C.

Effect of other substances

Mercury and silver ions must be absent from the solutions as they poison the surfaces of the membranes. Most other cations have to be present in a very large excess before they interfere (see Table 1); ferric ion can interfere at relatively low concentrations, but this can be eliminated as long as the pH of the solution is above 4. If possible, the electrodes should not be used in strongly acidic solutions, as very large positive bias may occur. The extent of the bias will depend on the age and condition of the electrode and it would be misleading to prescribe fixed limits for pH. The Orion and Beckman electrodes appear to be affected worse than the Radiometer F1112Cu or Radiometer F3002 electrodes (Hansen, Lamm and

Table 1 Cationic selectivity coefficients, $\log K_{CuM}$, for copper electrodes

Cation	Electrode type A[a]	B[b]
Cu^+	14	11
Ag^+	12	6
Hg^{2+}	16	4
Pb^{2+}	−9.5	−3.5
Cd^{2+}	−9	−4.4
Ni^{2+}		−4.3
Co^{2+}	−10.5	−4.8
Zn^{2+}	−10	−4.2
Mn^{2+}	−25	−4.6
Sr^{2+}	−10	−5.3
Fe^{3+}	−1	

[a]For no interference

$$\log [Cu^{2+}] - \frac{2}{m} \log [M^{m+}] > \log K_{CuM}$$

(concentrations in mol l^{-1}).
Calculated from solubility products for Ag$_2$S/CuS electrodes such as Orion 94−29A and Radiometer F3002.
[b]Quoted by manufacturers for Radiometer F1112Cu electrode as

$$E = E° + \frac{RT \ln(10)}{2F} \log(a_{Cu} + K_{CuM} \cdot a_M^{2/z_M})$$

Růžička, 1972; Midgley, 1976) and should not be used below pH 3. The Radiometer F1112Cu electrode can be used at concentrations down to about 5 ppm copper and the Radiometer F3002 down to about 0.5 ppm in 0.1N nitric acid solutions. Both may be used at lower concentrations if a curved calibration graph is acceptable. The Radiometer F1112Cu gives stable potentials in acidic solutions, but the other electrodes may not reach a steady e.m.f. if going from a less to a more concentrated solution and the potential may pass through a minimum for the opposite change. Clearly, it is best to use Method B whenever possible if the sample has a high acidity and it may be necessary to use more concentrated solutions than the 5N nitric acid and sodium hydroxide in order that the variations in ionic strength due to variable acidity in the sample remain negligible.

Anions, particularly halides, can affect the electrodes in various ways, depending on the composition of the membrane. Electrodes with silver salts in their composition, i.e. the Orion and Radiometer F3002 electrodes and possibly others also, are subject to interference by the conversion of the silver salt in the membrane into the appropriate halide. This process occurs only if the following relationship is satisfied

$$[Cu^{2+}][X^-]^2 > \frac{K^2_{AgX} K_{CuS}}{K_{Ag_2S}} \quad (1)$$

where K_{AgX}, K_{CuS} and K_{Ag_2S} are the solubility products of the various salts. The ion X^- may be chloride, bromide, iodide or thiocyanate and analogous conditions could be written if the membrane contained silver and copper selenides and tellurides instead of the sulphides. The limiting values of the left-hand side of relationship (1) are given in Table 2. The Radiometer F1112Cu electrode does not contain silver salts and relationship (1) does not apply, but the cuprous ions in the membrane can react with halide ions, which have maximum permissible concentrations of 175 ppm (chloride) and 80 ppm (bromide).

Chloride also causes tarnishing of the membrane in the Orion and Radiometer F3002 electrodes (Midgley, 1976a) even at lower concentrations than predicted by relationship (1). Other halides would be expected to produce a similar effect, but they have not been tested so far. Tarnishing causes a sluggish response and may also

Table 2 Maximum values[a] of $[Cu^{2+}][X^-]^2$ for CuS–Ag$_2$S membranes

Anion	mol^3 l^{-3}	(ppm)^3
Cl$^-$	6.3×10^{-8}	7.9×10^4
Br$^-$	5.0×10^{-13}	3.2×10^0
I$^-$	1.4×10^{-20}	2.2×10^{-7}
CNS$^-$	2.5×10^{-12}	8.5×10^0

[a] Although the silver halide will not be deposited until the tabulated value is exceeded, interference will occur at much lower values.

change the standard potential and reduce the linear range of the calibration. Very severe tarnishing results if the electrodes are immersed in sulphide solutions. The tarnish on an Orion electrode can be removed by polishing the membrane with the polishing strip sold by the manufacturers. The membrane of the Radiometer F3002 electrode is simply replaced once it is damaged in any way. The other electrodes do not appear to have been tested for this effect, but they can all be expected to be affected in some way by halides.

Copper forms many strong complexes with anionic and neutral ligands in solutions and as the electrode responds only to free, i.e. uncomplexed, copper ions, the presence of such ligands will affect the determination of the total copper present. Among the commonest ligands are the basic species hydroxide ions, carbonate ions and ammonia; these can be effectively removed by adding acid to the sample and using Method B. If organic chelating agents are present, the known addition method should be used.

In strongly reducing solutions the copper will be present as cuprous ion and the electrodes will not be suitable for measurement in such media, which are, however, fairly rare. Strongly oxidizing solutions may result in rapid tarnishing of the membrane and may also interfere directly with the measurement.

Effect of Light

The Orion electrode is affected by light and should be shielded from direct sunlight. The sensitivity to light varies between individual electrodes. Light has not been found to affect the Radiometer F3002 electrode.

Effect of Stirring

The potential of the Orion electrode has sometimes been found to change with the rate of stirring of the solution. The effect has not been reported consistently, but as a precaution the stirring conditions should be kept constant. The potential of the Radiometer F3002 electrode may be different in stirred and unstirred solutions, but the rate of stirring does not seem to be critical.

PRECISION

The precision of analysis with Orion and Radiometer F3002 electrodes in synthetic standard solutions has been determined by Midgley (1976) and is shown in Table 3. Smith and Manahan (1973) used an Orion electrode to analyse tap-water samples containing 3–50 ppb copper by a known addition method. The relative standard deviation for the recovery of a 9.0 ppb spike added to the samples was 7.3%. In synthetic standard solutions, the same workers obtained a relative standard deviation of 4.5% for copper concentrations 27–90 ppb and 13% at 9 ppb. Van der Meer, Den Boef and Van der Linden (1976) used an electrode essentially the same as the Radiometer F3002 to analyse standard solutions by compleximetric (EDTA) and addition titrations, using the Gran plots of the type E(a). 1 and type G (p. 90),

Table 3 Relative standard deviations for the analysis of synthetic standard solutions

Electrode	Copper concentration (ppm)	$S_w{}^a$ (%)	$S_t{}^b$ (%)
Orion 94-29A	6.300	1.6	2.4
	0.630	3.3	3.3
	0.063	7.5	10.2
Radiometer F3002	6.300	–	6.8
	0.630	–	5.1
	0.063	–	9.5

$^a S_w$ is the relative within-batch standard deviation for a single determination (5 degrees of freedom).
$^b S_t$ is the relative total standard deviation for a single determination. The between-batch standard deviation was non-significant for the Orion electrode and was not determined for the Radiometer F3002.

respectively. The relative standard deviations were less than 0.2% at copper concentrations of 6.6 and 66 ppm by the compleximetric titration and 2–3% at 0.66 and 66 ppm by the addition titration. At 66 ppb, the addition titration gave relative standard deviations of 3–8%, depending on the conditions.

ACCURACY

Smith and Manahan (1973), using an Orion 94-29A electrode, obtained recoveries of 98.9–101.1% in synthetic standard solutions in the range 9–900 ppb copper. The recovery of a 9 ppb spike added to tap water varied from 93.3 to 113.3%, with an average of 103%. Van der Meer et al. (1976) obtained results less than 1% low when 0.66–66 ppm standard copper solutions were analysed by compleximetric and addition titrations (further details under *Precision*).

RESPONSE TIME

The time taken for the potential to reach equilibrium after immersion of a rinsed electrode in a stirred portion of solution either ten times more dilute or ten times more concentrated than the last one is less than two minutes for the Orion 94-29A electrode at concentrations greater than or equal to about 5 ppm and five minutes at lower concentrations. The Radiometer F1112Cu electrode took less than five minutes at all concentrations between 63 and 0.063 ppm. The Radiometer F3002 electrode reached equilibrium in about one minute at concentrations above 0.6 ppm and no more than five minutes at lower concentrations. As the electrodes aged, their response times increased, but they could be reduced again by polishing or renewing the membrane.

COMPARISON WITH OTHER METHODS

Direct comparisons with other techniques do not appear to have been made, but for the determination of total copper the electrodes are probably inferior to the atomic absorption methods. Copper ions, however, readily form complexes with a wide variety of neutral and anionic species and the electrodes are useful when the free copper concentration is required (Stiff, 1971; Lamm, Hansen and Růžička, 1972).

OTHER APPLICATIONS

Copper electrodes make good indicators for potentiometric titrations (p. 226). Titrations of copper with complexing agents such as EDTA, EGTA and NTA can be used to determine either the copper concentration or the concentration of the chelating agent (Hansen et al., 1972; Rechnitz and Kenny, 1970; Veselý, 1971). The electrodes can also be used for chelometric indicator titrations of cadmium, lead, zinc, calcium, lanthanum, thorium, samarium, zirconium, mercury, iron and aluminium (Ross and Frant, 1969; Baumann and Wallace, 1969; Šůcha and Suchánek, 1970; van Oort, van den Bergen and Griepink, 1974; Hansen et al., 1972).

TRACING FAULTS

In addition to the faults common to most systems (p. 127), the following are especially relevant to copper-selective electrodes.

Low sensitivity

If this condition appears only with sample solutions and not with standard solutions, it is probably caused by the presence of an interfering substance in the sample. The most likely source of interference is the acidity of the sample; check the pH and use Method B if necessary.

If the sensitivity in standard solutions is reduced, the membrane has probably become tarnished. Polish or renew the membrane as recommended by the manufacturer.

Slow response times

These are likely to arise from the same causes as loss of sensitivity, but high concentrations of chloride or bromide ions may make the potential drift slowly to more positive values. On changing to a solution with a lower copper concentration the potential may go through a minimum before increasing almost linearly with time.

Sudden changes in potential

A very sharp change in potential of as much as several millivolts may be due to the effect of light or of irregular stirring. Keep the electrode out of direct sunlight, especially if its brightness is varying, and check that the rotation of the stirrer is not impeded.

Bibliography

Baumann, E. W., and R. M. Wallace, 1969, 'Cupric-selective electrode with copper(II)—EDTA for end point detection in chelometric titrations of metal ions', *Anal. Chem.*, **41**, 2072.

Hansen, E. H., C. G. Lamm, and J. Růžička, 1972, 'Selectrode — The universal ion-selective solid-state electrode. Part II. Comparison of copper(II) electrodes in metal buffers and compleximetric titrations', *Anal. Chim. Acta*, **59**, 403.

Lamm, C. G., E. H. Hansen and J. Růžička, 1972, '*In situ* use of the ion selective electrode, the Selectrode^{TM}, in studies of soil—plant relationships', *Anal. Lett.*, **5**, 451.

Midgley, D., 1976, 'Comparison of copper(II) ion-selective electrodes for measurements at micromolar concentrations', *Anal. Chim. Acta*, **87**, 7.

Midgley, D., 1976a, 'Halide and acid interferences with solid-state copper(II) ion-selective electrodes', *Anal. Chim. Acta*, **87**, 19.

Rechnitz, G. A., and N. C. Kenny, 1970, 'Determination of nitrilotriacetic acid (NTA) with ion-selective membrane electrodes', *Anal. Lett.*, **3**, 509.

Ross, J. W., and M. S. Frant, 1969, 'Chelometric indicator titrations with a solid-state cupric ion-selective electrode', *Anal. Chem.* **41**, 1900.

Smith, M. J., and S. E. Manahan, 1973, 'Copper determination in water by standard addition potentiometry', *Anal. Chem.*, **45**, 836.

Stiff, M. J., 1971. 'The chemical states of copper in polluted fresh water and a scheme of analysis to differentiate them', *Water Res.*, **5**, 585.

Šůcha, L., and M. Suchánek, 1970, 'Indirect complexometric determination of aluminium using a solid-membrane cupric ion-selective electrode', *Anal. Letters*, **3**, 613.

van Oort, W. J., V. W. J. van den Bergen, and B. Griepink, 1974, 'Chelometric titration of copper(II), zinc(II), cadium(II) and lead(II), using one simple ion selective electrode', *Z. anal. Chem.*, **269**, 184.

Van der Meer, J. M., G. den Boef, and W. E. Van der Linden, 1976, 'The micro determination of copper(II) with a solid-state copper-selective electrode', *Anal. Chim. Acta*, **85**, 317.

Veselý, J., 1971, 'Selektive Kupferelektrode', *Coll. Czech. Chem. Commun.*, **36**, 3364.

Compleximetric Titrations

The same apparatus and essentially the same technique can be used for a number of different purposes. Chelating agents can be determined by titration with a metal ion, using the appropriate metal ion-selective electrode as indicator. It is desirable to use as the titrant the metal ion that forms the strongest complex with the chelating agent and whose electrode has the lowest limit of detection, as these conditions determine the sharpness of the end-point and therefore the precision with which the analysis can be carried out. For most purposes, cupric ions and copper-selective electrodes are the best, but cadmium and, to a lesser extent, calcium electrodes may also be useful. Lead and cadmium complexes are generally of similar stability, but as lead is the more likely to form complexes with common inorganic species, e.g. sulphate and halides, cadmium is usually the better choice of the two. Interferences are clearly an important factor in deciding the titration conditions. The procedure is given as Method A.

Almost any metal for which there is an ion-selective electrode can be determined by titration with a chelating agent, but not at such low concentrations as can be achieved by direct potentiometry. Compleximetric titrations can, however, be used to determine metals for which no electrodes exist. The method to be used depends on the relative stabilities of the complexes formed with the chelating agent by the determinand and a suitable electroactive indicator ion.

Metals that form stronger complexes than the electroactive ion can be determined by back titration. A known concentration of chelating agent is added in excess to the sample and the concentration of the unreacted portion determined by titration as above. The metal ion concentration is found from the difference of the total and unreacted chelating agent concentrations. This technique, described in Method B, can sometimes be applied even if the determinand forms weaker complexes than the titrant, but the end-point will frequently be defined with less precision and accuracy than would be found with a strongly complexed determinand.

Metals that form weaker complexes than the electroactive ion can be determined in the presence of a known concentration of that ion by titration with a chelating agent. The electroactive ion is added to the sample either as a simple salt, in which case two steps will be observed in the titration curve, or as a solution containing

Table 1 Stability constants[a] of EDTA complexes

Metal	Fe^{3+}	In^{3+}	Th^{4+}	Cr^{2+}	Bi^{3+}	Sn^{2+}	Hg^{2+}	Ti^{3+}	
log K	29.1	25.0	23.2	23.0	22.8	22.1	21.8	21.3	
Metal	Cu^{2+}	Ni^{2+}	Pb^{2+}	Cd^{2+}	Zn^{2+}	Co^{2+}	Al^{3+}	Fe^{2+}	Mn^{2+}
log K	18.8	18.6	18.0	16.5	16.5	16.3	16.1	14.3	14.0
Metal	Ca^{2+}	Be^{2+}	Mg^{2+}	Sr^{2+}	Ba^{2+}	Ag^{+}			
log K	10.7	9.3	8.7	8.6	7.8	7.3			

[a] K = [ML]/[M][L] at 20–25 °C

Table 2 Abbreviations for chelating agents

EDTA	ethylenediaminetetraacetic acid
EGTA	ethyleneglycol bis(2-aminoethylether)tetraacetic acid
CDTA	1,2-diaminocyclohexane tetraacetic acid; also known as DCTA
TEPA	tetraethylenepentamine
NTA	nitrilotriacetic acid
Trien	triethylenetetramine
Chel DP	ethylenediaminedi(o-hydroxyphenylacetic acid)

exactly equal quantities of the electroactive ion and the chelating agent. In the latter case, only one step occurs in the titration curve. The procedure is given as Method C.

Whichever method of determining a non-electroactive metal is adopted, it is desirable that the stabilities of the complexes formed by the two metal ions should differ by as much as possible, although both should also be as strong as possible. Method C would generally best be carried out with a copper electrode, as copper tends to form stronger complexes than other electroactive ions, while the calcium electrode could almost never be used. The calcium electrode could be used for almost all back titrations (Method B), but the relative difficulty of handling liquid ion-exchange electrodes would promote the use of cadmium or copper electrodes if possible. The choice between using Method B and a cadmium electrode or Method C and a copper electrode may not be easy. Guidance as to the relative stabilities of complexes is given in Table 1 for EDTA complexes, and a similar order will be found for many other chelating agents, but quantitative information on each particular system should be sought from compilations of stability constants (Ringbom, 1963; Martell and Sillén, 1964 and 1971). The following methods suppose the use of the copper electrode, but adaptation to any other electrode should be straightforward.

Table 2 gives the abbreviations for the common chelating agents that have been used in potentiometric titrations.

APPARATUS

Copper-selective electrode; reference electrode; magnetic stirrer; pH meter; burette

Hansen, Lamm and Růžička (1972) used several types of electrode for EDTA–copper titrations and found that the size of the break at the end-point varied considerably. In the same titration, the breaks given by Radiometer F3002, Orion 94–29A and Beckman 39612 electrodes were about 120, 60 and 20 mV, respectively. The Radiometer F1112Cu electrode gave a satisfactorily large, but unspecified, break. Van der Meer, Den Boef and Van der Linden (1975) obtained a break of about 120 mV with an Orion electrode in an EDTA titration of comparable concentration and therefore there may be considerable variations between the performances of individual electrodes even when they are of the same make. The performance of some electrodes may be poor enough to affect the practicality of the titration.

For further details of the electrodes, see the section on the determination of copper (or those on cadmium or calcium if those electrodes are to be used).

REAGENTS

Water

Distilled water that has been passed through a mixed-bed deionization column will be sufficiently free of heavy metals for no bias to arise from this source.

Stock copper titrant solution ($0.1\ mol\ l^{-1}$)

Dissolve 24.968 g of copper sulphate pentahydrate or 24.160 g of copper nitrate trihydrate (analytical reagent grade) in water and make up to 1 litre in a calibrated flask.

Working copper titrant solutions

Prepare by dilution of the stock solution. The concentration of the titrant should be 10–100 times greater than that of the chelating agent in the sample.

Buffer solutions

Acetate buffer (pH 3.6–5.6). Dissolve 57 ml of glacial acetic acid (analytical reagent grade) in about 750 ml water, adjust the pH to the desired value by adding pellets of sodium hydroxide and make up to 1000 ml.

Triethanolamine buffer (pH 7–9). Dissolve 70 ml of triethanolamine in about 600 ml of water, adjust the pH to the desired value by adding nitric acid (12–90 ml of 5 mol l^{-1} acid or 4–30 ml of concentrated acid) and make the volume up to 1 litre.

Tris buffer (pH 7–9). Dissolve 60.5 g of *tris*-hydroxymethylaminomethane in about 600 ml of water, adjust the pH by adding nitric acid (6–90 ml of 5 mol l^{-1} acid or 2–30 ml of concentrated acid) and make up to 1 litre.

Ammonia buffer (pH 10). Dissolve 570 ml of concentrated ammonia (sp. gr. 0.88) and 70 g of ammonium chloride in water and make up to 1 litre.

METHOD A. DETERMINATION OF CHELATING AGENTS

Step 1 Pipette V_s ml of sample into a beaker containing a magnetic stirrer bar.

Step 2 Add V_b ml of buffer solution. V_b should be about one-tenth of V_s and may be added with a graduated pipette or from an automatic dispenser.

Step 3 Rinse the electrodes with water and immerse them in the solution. Start the magnetic stirrer.

Step 4 Add a portion of titrant solution (t mol l^{-1}) from a burette and when a steady potential has been reached, note the volume added, v_i, and the e.m.f., E_i.

Step 5 Repeat Step 4 as many times as is necessary to prepare an adequately defined titration curve. In the region of the end-point, indicated by the large change in e.m.f. on each addition, the volume added should not exceed about 1% of the expected value of the equivalent volume. Larger additions (5–10% of the expected titre) may be made away from the end-point.

Step 6 Plot E_i against v_i and find the volume, v_e, corresponding to the point of inflection of the titration curve (Note 1, below).

Step 7 Calculate the concentration of chelating agent (Note 2) from

$$c = t \cdot v_e / V_s \text{ mol l}^{-1}$$

Note 1 Other methods may be used to find v_e, e.g. the titration results may be plotted differentially (p. 89) or the appropriate Gran plots E(b).1 and E(b).2 may be used (p. 90). No results have been reported on the basis of plot E(b).1, but Van der Meer *et al.* (1975) have used a simplified form of E(b).2 with a correction for the effect of dilution in EDTA–copper titrations.

Note 2 Step 7 assumes that the complexing agent and copper form a 1:1 complex, which is true for most ligands, e.g. EDTA, NTA and similar compounds. If the complex has the formula $Cu_x L_y$, the concentration is calculated from

$$c = \frac{y \cdot t \cdot v_e}{x \cdot V_s} \text{ mol l}^{-1}$$

Sources of error

Effect of temperature

Large changes in temperature during the titration may distort the shape of the titration curve. Ensure that the titration vessel is insulated from the heat produced by the stirrer.

Effect of other substances

In the presence of ions that form stronger complexes with ligand than copper does, e.g. Fe^{3+} and Hg^{2+}, only the free ligand will be determined. If the ions are weakly complexing, e.g. alkali metal and alkaline earth ions, Mn^{2+}, the copper will displace them from their complexes and the end-point will correspond to the total ligand concentration, although it may be possible to observe a first, rather inaccurate, end-point corresponding to the free ligand only. In the intermediate case it may not be possible to obtain any sharp end-point, especially if a mixture of heavy metals is present. Hannema and den Boef (1970) considered the sharpness of end-points in the presence of metal ions and Van der Meer *et al.* (1975) have studied EDTA–copper titrations in the presence of Fe^{3+}, Zn^{2+}, Ni^{2+}, Cd^{2+} and Mn^{2+}.

Inorganic anions may alter the shape of the titration curve slightly, but should not change the position of the end-point. Anions such as carbonate and sulphide that can precipitate copper should be removed from the sample by acidifying it and passing nitrogen through the solution. In concentrated solutions of chloride or sulphide ion, the membrane will tarnish and the electrode will lose its sensitivity, which can be restored by suitable treatment (see *Determination of Copper*).

In a mixture of chelating agents it may be possible to obtain more than one end-point, corresponding to the individual components, or only the end-point for the total concentration of ligands. Whether one ligand will interfere with the determination of another depends on the stability constants of their copper complexes and their relative concentrations.

Effect of pH

The size of the break in potential at the end-point depends on the pH of the solution. When the pH is low, say less than 4, the degree of protonation of the complexing agent will be considerable and its ability to form a copper complex will be reduced, with the result that the end-point break is smaller. The weaker the acidic groups on the ligand, the higher the pH needs to be if a good end-point is to be obtained. In alkaline solution, copper tends to precipitate as the hydroxide, again reducing the end-point break. As a result of these two effects, the optimum pH tends to be in the range 5–8 and the chosen value should be maintained by adding a buffer to the solution. Suitable buffers for adding in the ratio of 1:10 by volume are given in the reagents section. Buffers containing carbonate and phosphate should not be used.

Precision

Van der Meer *et al.* (1975) titrated EDTA and Trien with copper at pH 4.75 and found precisions of 1–1.4% at 5×10^{-5} mol l^{-1} of ligand, 0.6% at about 10^{-4} mol l^{-1} and less than 0.1% at 10^{-3} mol l^{-1} for triplicate analyses.

Accuracy

The results of Van der Meer *et al.* (1975) for EDTA and Trien (see above) are 5–7% high at about 5×10^{-5} mol l^{-1}, 1.5–3% high at about 10^{-4} mol l^{-1} and within 0.01% at 10^{-3} mol l^{-1}.

Applications

Rechnitz and Kenny (1970) titrated $5 \times 10^{-5} - 10^{-3}$ mol l^{-1} NTA at pH 9.6. Van der Meer *et al.* (1975) titrated $5 \times 10^{-5} - 10^{-3}$ mol l^{-1} of EDTA and Trien at pH 4.75. Hansen *et al.* (1972) titrated 2.5×10^{-3} mol l^{-1} solutions of NTA at pH 4.75 and EDTA and Chel DP at pH 10.

So far, applications have been limited to the determination of very strongly complexing species, such as aminopolycarboxylates, but it should be feasible to determine some of the weaker chelating agents, e.g. citric and tartaric acids. As the complexes are weaker, the end-points will be less sharp and the limits of detection will be greater.

METHOD B. BACK TITRATION OF METAL IONS (NOTE 1)

Step 1 Carry out Step 1 of Method A.

Step 2 Pipette V_c ml of standard chelating agent solution of concentration c_T mol l^{-1} into the beaker. V_c and c_T should be chosen such that $c_T \cdot V_c / V_s$ is approximately equal to twice the expected concentration of metal ions in the sample (Note 2).

Step 3 Carry out Step 2 of Method A.

Step 4 Carry out Steps 2–6 of Method A.

Step 5 Calculate the concentration of metal ions in the sample (Note 3)

$$m = \frac{c_T \cdot V_c - t \cdot v_e}{V_s} \text{ mol l}^{-1}$$

Note 1 The best conditions for this type of titration occur when the determinand is more strongly complexed than the titrant: a single sharp end-point should be observed, corresponding to the chelating agent that has not reacted with the determinand. Even if the determinand forms weaker complexes than the titrant, it may be possible to use the procedure, but the end-points obtained will be less accurate and precise than if the determinand were more strongly complexed. If the stability constant of the determinand complex is smaller by a factor of 10^3 or less there may be a fairly large and steep increase in potential when the excess of chelating agent is used up, but the titration curve will be asymmetric. If the determinand complex is weaker by a factor of about 10^6, the jump in potential will be smaller and less steep, although the curve will be more symmetrical. As the titrant displaces the determinand from its complex, a second end-point will be observed, corresponding to the total chelating agent added. The weaker the determinand complex relative to the titrant complex, the sharper the second end-point becomes and the fainter the first end-point becomes, until for very weak determinand complexes only the end-point corresponding to the total chelating agent can be discerned.

Note 2 If the determinand and the chelating agent are slow to react, as is likely with aluminium and chromium complexes, boil for 5—10 min at this stage, but cool to room temperature before proceeding to Step 3.

Note 3 Step 5 assumes that both the determinand and the titrant form 1:1 complexes with the chelating agent. This is almost always the case, but if the determinand forms a complex with the formula $D_p L_q$ and the copper titrant forms the complex $Cu_x L_y$ the concentration is calculated from

$$m = \frac{p\left(c_T \cdot V_c - \frac{y}{x} t \cdot v_e\right)}{q \cdot V_s} \text{ mol l}^{-1}$$

Sources of Error

Effect of temperature

See Method A.

Effect of pH

See Method A.

Effect of other substances

Other metal ions that form stronger complexes than the titrant will not be distinguished from the determinand, the concentration of which will be overestimated. Metals forming much weaker complexes than the titrant will not interfere unless present at high concentrations, but metals forming only slightly weaker complexes may make it difficult to see an end-point at all. For these reasons, it is desirable that the titrant metal ion should form as strong a complex as possible, while still enabling the back titration to be done.

The presence in the sample of traces of chelating agents will give a low bias to the result, unless they react much less strongly with the titrant than the chelating agent added in Step 2. Inorganic anions are unlikely to interfere, but see Method A.

Precision, Accuracy, Applications

Table 3 summarizes some of the results obtained by the back-titration procedure. Both aluminium and chromium(III) solutions have to be boiled after the EDTA has been added and then cooled before being titrated. As iron(III) can be determined without a boiling step, and the sum of iron(III) and either chromium(III) or

Table 3 Back titration of metal ions

Determinand	Chelating agent	Titrant and electrode	Concentration range tested (ppm)	Error (%)	Precision[a] (%)	Conditions	Reference[b]
Aluminium	CDTA	copper	8–80	0.1–0.5	0.2	pH 5.5 (hexamethylene-tetramine + HCl)	1
Iron(III)	CDTA	copper	50			pH 5.5 (hexamethylene-tetramine + HCl)	1
Iron(III)	EDTA	copper	28			pH 4.7 (acetate)	2
Zinc	EDTA	copper	3/33/0.6	0.2/3.3/12.5	0.2/0.5/1.8	pH 4.7 (acetate)	2
Manganese(II)	EDTA	copper	22			pH 4.7 (acetate)	2
Nickel(II)	EDTA	copper	29			pH 4.7 (acetate)	2
Calcium	EGTA	copper	15	0.3	1.4	pH 11 (cyclohexylamine–cyclohexylammonium nitrate, 0.25 mol l^{-1})	3
Magnesium	EDTA	copper	10	0.6	0.6	pH 10 (ammonia–ammonium nitrate, 0.25 mol l^{-1})	3
Chromium(III)	EDTA	magnesium[c]		−3.3		pH 8–9	4
Lanthanum	EDTA	magnesium[c]		−4.7		pH 8–9	4

[a] Relative standard deviation.
[b] 1 Šůcha and Suchánek (1970); 2 Van der Meer et al. (1975); 3 Van der Meer et al. (1976); 4 Chang and Cheng (1975).
[c] Water hardness electrode.

aluminium with a boiling step, mixtures of iron(III) with either of the others can be resolved. Calcium and magnesium can be back titrated even though they form very much weaker EDTA complexes than copper, because the ammonia or cyclohexylamine in the buffer solution complex the copper ions selectively and strongly so that copper does not displace the alkaline earth ions from their complexes, as it would in an acetate buffer.

Zinc and manganese are not particularly suitable for back titration with copper because, once the excess of EDTA has been used up, the copper titrant displaces them from their complexes. As Van der Meer et al. (1975) have shown, however, the zinc and manganese EDTA complexes are sufficiently strong that the titration curve has an inflection corresponding to the free chelating agent, although this is poorly defined when compared to those obtained in the back titration of iron or nickel.

METHOD C. DETERMINATION OF METAL IONS BY TITRATION WITH A CHELATING AGENT (NOTE 1)

Step 1 Carry out Steps 1 and 2 of Method A.
Step 2 Pipette V_I ml of standard copper solution (c_I mol l^{-1}) into the beaker. V_I and c_I should be chosen so that the molar ratio of determinand to copper ions should be between 100:1 and 10:1.
Step 3 Carry out Steps 3–5 of Method A, using as titrant a solution of a suitable chelating agent.
Step 4 Plot E_i against v_i and find the volumes, v_{Cu} and v_e, corresponding to the first and second points of inflexion of the titration curve.
Step 5 Calculate the concentration of determinand from

$$m = \frac{t(v_e - v_{Cu})}{V_s} \text{ mol l}^{-1}$$

Note 1 Metals that form stronger complexes than copper can be determined by an almost identical procedure. Repeat Steps 1–4. Only one point of inflexion will be observed; call this v_e. The determinand concentration is given by $m = (t \cdot v_e - c_I \cdot V_I)/V_s$ mol l^{-1}. Alternatively, perform another titration with V_s ml of deionized water and call the point of inflexion v_{Cu}; m is then calculated as in Step 5.

Sources of Error

Effect of temperature

See Method A.

Effect of pH

See Method A.

Effect of other substances

The presence in the sample of stronger chelating agents than the titrant will cause the results to have a low bias, i.e. only the free metal would be determined and not the total. Very weak complexing agents should not interfere, unless present at high concentrations.

Interference by metal ions depends on the relative strengths of the complexes formed by the determinand and indicator ions.

(a) Determinand more strongly complexed than the indicator ion All metal ions present that form stronger complexes than the indicator ion will interfere, as will any indicator ion in the sample originally. Metal ions that are less strongly complexed than the indicator ion can interfere if present in great excess and may reduce the sharpness of the end-point, even if they do not change its position.

(b) Determinand less strongly complexed than the indicator ion Metal ions that form stronger complexes than the indicator ions will not interfere, although they will contribute to larger values of v_{Cu} than would be expected from the added indicator alone. Indicator ion in the original sample will not interfere. Metals forming much weaker complexes than the determinand will not interfere unless present in great excess. Metals forming complexes similar in strength to these of the determinand will interfere. If the interferent forms a complex intermediate in strength between those of the determinand and indicator ions, it may give rise to another end-point between these for the two other ions; in this case, take the second end-point as v_{Cu} and the third as v_e (Step 4 of Method C). The accuracy obtainable if this intermediate point of inflection occurs should be checked by titration of a synthetic solution containing the concentrations of determinand and interferent estimated from the titration of the sample.

Precision, accuracy, Applications

Table 4 summarizes many of the results obtained by titration; with one exception, a copper electrode was used. The accuracy of these titrations has not been adequately reported and the titration should be carried out with standard solutions before real samples are analysed, so that any bias is shown up. If necessary, titrations of standard solutions can be used to provide a correction factor for the bias.

In a variation of the procedure, van Oort et al. (1974) determined lead, zinc and cadmium by titration to a constant potential with a copper-selective electrode. Zinc was determined with fairly good accuracy (less than 1% error) and precision (less than 1% relative standard deviation), but the results for cadmium and, to a greater extent, lead were less accurate and precise. Olsen et al. (1976) titrated the following ions with EDTA and/or CDTA using a copper-selective electrode but without adding copper ions: Ni^{2+}, Ca^{2+}, Cd^{2+}, Zn^{2+}, Ba^{2+}, Hg^{2+}, Pb^{2+}, Mn^{2+}, Sr^{2+}, Mg^{2+}. The largest relative standard deviation reported was 0.6% and the mean relative standard

Table 4 Potentiometric indicator titration of metal ions

Determinand	Chelating agent	Indicator	Concentration range tested (ppm)	Error[a] (%)	Precision[b] (%)	Conditions	Reference[c]
Ca(II)	EGTA	Cu	20			NH$_3$ buffer	1
	EDTA	Cu	3960	−1.4	0.8	pH 10 (NaOH)	2
	EDTA	Cu	100			pH 10 (NH$_3$)	3
Ca(II) + Mg(II)	EDTA	Cu	20 + 12			NH$_3$ buffer	1
Zn(II)	EDTA	Cu	0.6–65			pH 4.7 (acetate)	4
	TEPA	Cu	65			NH$_3$ buffer	1
	EDTA	Cu	162			pH 10 (NH$_3$)	3
	EDTA	Cd	<260			pH 5.5 (acetate)	5
Hg(II)	EDTA	Cu	27 000	−1.3	0.2	pH 5 (acetate)	2
Fe(III)	EDTA	Cu	56			pH 4.7 (acetate)	4
	EDTA	Cu	5445	0.5	0.2	pH 5 (acetate)	2
Ni(II)	TEPA	Cu	59			NH$_3$ buffer	1
	EDTA	Cu	59			pH 4.7 (acetate)	4
Mn(II)	EDTA	Cu	55			pH 4.7 (acetate)	4
Th(IV)	EDTA	Cu	57 400	−0.1	0.1	pH 5 (acetate)	2
Zr(IV)	EDTA	Cu	9400	0.1	0.1	pH 5 (acetate)	2

[a]Compared with visual indicator titration.
[b]Relative standard deviation.
[c]1 Ross and Frant (1969); 2 Baumann and Wallace (1969); 3 Hansen et al. (1972); 4 Van der Meer et al. (1975a); 5 Taga et al. (1976).

deviation for all the ions was 0.25%. Less than 1% error was found in all the titrations, with no bias in a particular direction. Although the titration curves were generally asymmetric, the end-points were reasonably sharp. The solutions contained $1-5 \times 10^{-3}$ mol l^{-1} metal ions and were adjusted to pH 12 by adding ammonia to a final concentration of 3 mol l^{-1}.

Comparison with Other Methods

Compleximetric titrations using visual indicators suffer from many of the same problems as the potentiometric procedures. In particular, the interference in the determination of chelating agents and in the determination of metals by back titration will be of the same nature, but the choice of titrant in the visual procedures is not restricted to those metals for which there are ion-selective electrodes. The potentiometric titration curve, however, conveys a great deal of information to the experienced analyst and changes in the shape of the curve will often indicate the presence of an interferent and something of its nature.

The extensive studies of C N. Reilley and co-workers have shown that mercury indicator electrodes can be used for all types of compleximetric titrations, but halide ions interfere even in trace quantities. Davies (1964) has summarized the application of mercury electrodes in this field.

Tracing Faults

For problems relating to particular indicator electrodes, see the sections on their use in direct potentiometry. In addition there may be difficulties peculiar to the titration itself. Slow response times may be caused by slow formation of one of the chelates rather than by any property of the electrode; in either case, they can lead to a loss of sharpness of the endpoint in automatic titrations. When the chelating agent is present in excess, the concentration of free indicator ion may be so low that the limit of detection of the electrode is approached and symmetrical step changes in the titration curve cannot be obtained. This is especially a problem with liquid ion-exhange indicator electrodes, as these are subject at low concentrations of free indicator ion to direct interference at the electrode by alkali metal and other ions that do not normally interfere significantly by competing for the chelating agent.

Bibliography

Baumann, E. W., and R. M. Wallace, 1969, 'Cupric-selective electrode with copper(II)—EDTA for end point detection in chelometric titrations of metal ions', *Anal. Chem.*, **41**, 2072.

Chang, F. C., and K. L. Cheng, 1975, 'Determination of tri- and tetravalent ions with a divalent ion-selective electrode', *Anal. Chim. Acta*, **76**, 177.

Davis, D. G., 1964, 'Potentiometric titrations', in *Comprehensive Analytical Chemistry,* Vol IIA, C. L. Wilson and D. W. Wilson, (eds.), Elsevier, Amsterdam, London and New York, Ch. 3.

Hannema, U. and G. den Boef, 1970, 'Titration curves of compleximetric back-titrations', *Anal. Chim. Acta*, **49**, 43.
Hansen, E. H., C. G. Lamm, and J. Růžička, 1972, 'Selectrode – The universal ion-selective solid-state electrode', *Anal. Chim. Acta*, **59**, 403.
Martell, A. E., and L. G. Sillén, 1964, *Stability Constants*, Special Publication No. 17, The Chemical Society, London.
Martell, A. E., and L. G. Sillén, 1971, *Stability Constants, Supplement No. 1*, Special Publication No. 25, The Chemical Society, London.
Olsen, V. K., J. D. Carr, R. D. Hargens, and R. K. Force, 1976, 'Potentiometric response of silver(I) sulphide/copper(II) sulphide membranes to chelons and applications for end-point detection in chelometric titrations', *Anal. Chem.*, **48**, 1228.
Rechnitz, G. A., and N. C. Kenny, 1970, 'Determination of nitrilotriacetic acid (NTA) with ion-selective membrane electrodes', *Anal. Lett.*, **3**, 509.
Ringbom, A. 1963, *Complexation in Analytical Chemistry*, Interscience, New York and London.
Ross, J. W., and M. S. Frant, 1969, 'Chelometric indicator titrations with the solid-state cupric ion-selective electrode', *Anal. Chem.*, **41**, 1900.
Šůcha, L., and M. Suchánek, 1970, 'Indirect Complexometric determination of aluminium using a solid-membrane cupric ion-selective electrode', *Anal. Lett.*, **3**, 613.
Taga, M., M. Mizuguchi, H. Yoshida, and S. Hikima, 1976, 'Potentiometric EDTA titration of zinc with cadmium ion-selective electrode', *Jap. Analyst*, **25**, 362.
Van der Meer, J. M., G. den Boef, and W. E. Van der Linden, 1975, 'Solid-state ion-selective electrodes as end-point detectors in compleximetric titrations. Part II. Back-titrations in acidic media', *Anal. Chim. Acta*, **79**, 27.
Van der Meer, J. M., G. den Boef, and W. E. Van der Linden, 1975a, 'Solid-state ion-selective electrodes as end-point detectors in compleximetric titrations, Part I. The titration of mixtures of two metals', *Anal. Chim. Acta*, **76**, 261.
Van der Meer, J. M., G. den Boef, and W. E. Van der Linden, 1976, 'Solid-state ion-selective electrodes as end-point detectors in compleximetric titrations, Part III. Selection of some experimental conditions for back-titrations in alkaline medium', *Anal. Chim. Acta*, **85**, 309.
van Oort, W. J., V. W. J. van den Bergen, and B. Griepink, 1974, 'Automated titration of sub-microgram amounts of metal ions in aqueous solutions', *Z. Anal. Chem.*, **269**, 184.

Determination of Cadmium

Electrodes available include the Orion 94–48, Radiometer F3003, EDT Supplies EE-Cd, Leeds & Northrup 117401, Philips IS 550-Cd, Metrohm EA306Cd and Tacussel PCD 1. The first four types are known to have membranes containing cadmium sulphide and silver sulphide. In the absence of reported user experience, no method is given here and it is recommended that the procedure for using the copper electrode should be followed.

APPARATUS

Cadmium electrode; reference electrode; pH meter; magnetic stirrer

Calomel or silver–silver chloride reference electrodes may be used. Use plastic containers whenever possible, especially for solutions containing low concentrations of cadmium.

REAGENTS

Water

Pass distilled water through a mixed-bed deionization column. The specific conductivity of this water as it leaves the column should be less than $0.2 \, \mu S \, cm^{-1}$.

Standard cadmium solution A (1000 ppm)

Dissolve 2.744 g of cadmium nitrate tetrahydrate, $Cd(NO_3)_2 \cdot 4H_2O$, in deionized water and make up to the mark in a 1-litre calibrated flask.

   1 ml ≡ 1 mg cadmium

Standard cadmium solution B (100 ppm)

Pipette 50 ml of standard solution A into a 500-ml calibrated flask and make up to the mark with deionized water.

   1 ml ≡ 100 μg cadmium

Standard cadmium solution C (10 ppm)

Pipette 50 ml of standard solution B into a 500-ml calibrated flask and make up to the mark with deionized water. Store in a polyethylene bottle.

1 ml ≡ 10 μg cadmium

Other standard solutions should be prepared as required by dilution of standard solutions A, B or C. Those below 10 ppm should be prepared daily as required and stored in a polyethylene bottle.

5 N sodium hydroxide, 0.1 N nitric acid, bromocresol purple indicator; ionic strength adjustment solution

See *Determination of Copper*.

CONDITIONING AND STORAGE

The Orion electrode needs no conditioning and after use should be rinsed with deionized water, dried with a paper tissue and stored dry in air, preferably with a protective cap over the membrane. After a time the electrodes may become tarnished and they should be polished with emery cloth to restore the surface. Radiometer F3003 electrodes are first impregnated with the sensitizing mixture according to the manufacturer's instructions and then conditioned in 100-ppm cadmium solution for 1—2 days. The electrode should be stored in deionized water when not in use and should last for at least three weeks before re-impregnation is necessary.

SAMPLE COLLECTION

See *Determination of Copper*.

ANALYTICAL PROCEDURE

See *Determination of Copper*, including the section on preparing the calibration graph.

Concentration range and units

The methods should work over the range 100—0.1 ppm. More concentrated solutions should be diluted with deionized water before analysis.

10^{-3} mol l^{-1} ≡ 112.4 ppm; 1 ppm ≡ 8.897×10^{-6} mol l^{-1}

SOURCES OF ERROR

Effect of temperature

Changes in temperature affect the slope of the calibration graph and the standard potential. The standard and sample solutions should not differ in temperature by more than 1 °C.

Effect of other substances

Mercury, silver and copper ions must be absent from the solutions as they poison the membrane of the electrode. Lead ions will interfere if the molar ratio of cadmium is less than about 3 (ratio of about 1.5 by weight). Free ferric ion should not be allowed to exceed the level of cadmium, which is achieved by adjusting the pH of the sample to above pH 4. Other cations that form insoluble sulphides will interfere if present in sufficient quantity. The selectivity coefficients in Table 1 have been reported for Orion (Brand, Militello and Rechnitz, 1969) and Philips electrodes.

Sulphide ions may damage the membrane and bromide and iodide ions should generally be avoided. The concentration of chloride ion should not exceed 350 ppm.

Cadmium ions are precipitated by anions such as hydroxide, carbonate and phosphate and even the soluble cadmium will be partly complexed. Free cadmium can be determined in solutions containing complexing agents, but measurement of total cadmium may be difficult. Basic solutions should be acidified with nitric acid and then neutralized, as in Method B for the determination of copper.

Table 1 Selectivity coefficients for cations[a]

Cation	Fe^{2+}	Tl^+	Mn^{2+}	Al^{3+}	Ni^{2+}	Co^{2+}	Zn^{2+}	Ca^{2+}	Mg^{2+}
Log K_{CdM}	2.3	2.1	0.4	−0.9	−1.5	−1.7	−3.4	−3.7	−3.8

[a]For no interference $\log [Cd^{2+}] - \frac{2}{m} \log [M^{m+}] > \log K_{CdM}$

PRECISION AND ACCURACY

No results are available for actual analyses carried out with the electrode. Philips claim a reproducibility of ± 0.5 mV for their electrode (equivalent to ± 4.5% in concentration terms).

RESPONSE TIME

The electrode should reach a steady potential 2–5 min after immersion in a solution that is a factor of ten different in concentration from the previous one. If the membrane needs polishing or renewing, the response time will lengthen.

COMPARISON WITH OTHER METHODS

No direct comparisons have been made. The electrode is sufficiently subject to interferences for atomic absorption or colorimetric methods to be generally preferred for total cadmium determinations. The electrode may be useful, however, for continuous monitoring and especially for field tests. The electrode may be used to determine free cadmium in solutions containing complexing agents, but it will still be more subject to interference than the polarographic methods.

OTHER APPLICATIONS

The electrode has been used as in indicator for compleximetric titrations of cadmium with EDTA (Brand et al., 1969; Růžička and Hansen, 1973; Mascini and Liberti, 1973) and with NTA (Mascini and Liberti, 1973) and for precipitation titrations with 8-hydroxyquinoline (Mascini and Liberti, 1973). Růžička and Hansen (1973) also used it for the indirect compleximetric determination of calcium with EDTA.

TRACING FAULTS

See *Determination of Copper*.

Bibliography

Brand, M. J. D., J. J. Militello and G. A. Rechnitz, 1969, 'Potentiometric measurements with a new solid-state cadmium ion selective electrode', *Anal. Lett.*, **2**, 523.

Mascini, M., and A. Liberti, 1973, 'The preparation and analytical evaluation of a new heterogeneous membrane electrode for cadmium', *Anal. Chim. Acta*, **64**, 63.

Růžička, J., and E. H. Hansen, 1973, 'Selectrode^{TM} — the universal ion-selective electrode, Part IV. The solid-state cadmium (II) Selectrode in EDTA titrations and cadmium buffers', *Anal. Chim. Acta*, **63**, 115.

Determination of Lead

Electrodes available include the Orion 94–82, Radiometer F3004, Leeds & Northrup 117407, EDT Supplies EE-Pb, Tacussel PPB 1, Simac Pb/1C and the Activion 003 15 016. Most of them are solid-state electrodes, having membranes containing lead sulphide and silver sulphide, but the Activion electrode has a neutral carrier encapsulated in a PVC membrane. Lead electrodes have mainly been used for titrations and almost no results have been reported for the direct determination of lead by potentiometry. In the absence of user experience, no method is given here and it is recommended that the procedure for using the copper electrode should be followed.

APPARATUS

Lead electrode; double-junction reference electrode; pH meter; magnetic stirrer

The outer compartment of the reference electrode should be filled with 1 mol l^{-1} sodium nitrate solution to prevent the precipitation of lead chloride that would occur with single-junction reference electrodes. Use plastic containers whenever possible, especially for solutions containing low concentrations of lead. Any glass apparatus should be made of lead-free glass.

REAGENTS

Water

Pass distilled water through a mixed-bed deionization column.

Standard lead solution A (1000 ppm)

Dissolve 1.599 g of anhydrous lead nitrate (analytical reagent grade) in water and make up to the mark in a 1-litre calibrated flask.

$$1 \text{ ml} \equiv 1 \text{ mg lead}$$

Standard lead solution B (100 ppm)

Pipette 50 ml of standard solution A into a 500-ml calibrated flask and make up to the mark with deionized water.

$1 \text{ ml} \equiv 100 \ \mu g$ lead

Standard lead solution C (10 ppm)

Pipette 50 ml of standard solution B into a 500-ml calibrated flask and make up to the mark with deionized water. Store in a polyethylene bottle.

$1 \text{ ml} \equiv 10 \ \mu g$ lead

Other standard solutions should be prepared as required, by dilution of the above standards. Solutions with a concentration below 10 ppm should be prepared daily and stored in polyethylene bottles.

5N sodium hydroxide, 0.1N sodium hydroxide; 5N nitric acid; bromocresol purple indicator; ionic strength adjustment solution

see *Determination of Copper*.

CONDITIONING AND STORAGE

The Orion and Leeds & Northrup electrodes need no conditioning and after use should be rinsed with deionized water, dried with a paper tissue and stored dry in air, preferably with a protective cap over the membrane. After a time, the electrodes may become tarnished and they should be polished with emery cloth to restore the surface. Radiometer F3004 electrodes are first impregnated with the sensitizing mixture according to the manfacturer's instructions and then conditioned by soaking them for 48 h in 100 ppm lead solution. The electrode should be stored in deionized water when not in use, giving the electrode a lifetime of 3—4 weeks before re-impregnation is necessary (Hansen and Růžička, 1974). The electrode should not be stored dry in air as the membrane will deteriorate.

SAMPLE COLLECTION

See *Determination of Copper*.

ANALYTICAL PROCEDURE

See *Determination of Copper*, including the section on preparation of the calibration graph.

Concentration range and units

The methods should work over the range 200–0.2 ppm. More concentrated solutions should be diluted with deionized water before analysis.

10^{-3} mol l^{-1} ≡ 207.2 ppm

1 ppm ≡ 4.826×10^{-6} mol l^{-1}

SOURCES OF ERROR

Effect of temperature

Changes in temperature affect the slope of the calibration graph and the standard and sample solutions should not differ in temperature by more than 1 °C.

Effect of other substances

Mercury, silver and copper ions must be absent from solutions as they poison the surfaces of the membranes. Ferric ions can cause a redox interference if the molar ratio of lead ions to ferric ions is less than about 10, but ferric ion interference can be effectively eliminated by keeping the pH above 4. Cadmium ions should not be present in more than an approximately threefold molar excess (1.5 times excess by weight). Other cations that form insoluble sulphides will interfere if present at sufficiently high concentrations. Rechnitz and Kenny (1970) determined some selectivity coefficients for the Orion 94–82 electrode, but found that they varied considerably with concentration. In general, a thousandfold excess of transition metal ions can be tolerated, except for those mentioned above, and alkali metal ions and alkaline earth ions do not interfere. Prolonged exposure to solutions below pH 3 will damage the membrane and it is recommended that acidic solutions be neutralized before measurement.

Lead ions are complexed or precipitated by a very wide range of anions, including hydroxide, chloride, sulphate, phosphate and carbonate. While free lead may be determined in solutions containing complexing anions, analysis for total lead is difficult. Basic solutions should be acidified with nitric acid and then neutralized, as in Method B for the determination of copper.

PRECISION

Almost no information is available, but Rechnitz and Kenny (1970) found relative standard deviations for a single determination of lead in urine of 8.8 to 18.4% at lead concentrations of 76–90 ppb.

ACCURACY

Rechnitz and Kenny (1970) found that a standard colorimetric method indicated only 51–67% of the values obtained by the electrode for the lead content of urine (76–90 ppb).

RESPONSE TIME

The electrode should reach a steady potential 2–5 min after immersion in a solution that is a factor of ten different in concentration from the previous one. As the electrode ages, longer times may be required and the membrane should be polished or renewed.

COMPARISON WITH OTHER METHODS

The only direct comparison that has been made is that by Rechnitz and Kenny (1970), described in the section on accuracy. The electrode is sufficiently subject to interferences, however, for atomic absorption or colorimetric methods to be generally preferred for total lead determinations. The electrode may be useful for determining free lead in solutions of complexing agents, but it will still be subject to interference by copper ions which can be avoided by polarographic methods.

OTHER APPLICATIONS

The main use of the lead electrode is not for the direct determination of lead, but for the titrimetric determination of sulphate ($q.v.$), including the determination of the sulphur content of materials as sulphate after oxidation. Other ions that have been determined titrimetrically are phosphate (Selig, 1970), oxalate (Selig, 1970a), and tungstate, chromate, pyrophosphate and hexacyanoferrate(II) (Hansen and Růžička, 1974; Mascini and Liberti, 1972).

TRACING FAULTS

See *Determination of Copper*. Note that lead ions are precipitated by sulphate and high levels of chloride and that a stable liquid junction potential may be difficult to obtain with a single-junction reference electrode, resulting in a noisy and/or drifting signal. The recommended double-junction reference electrode should be used.

Bibliography

Hansen, E. H., and J. Růžička, 1974, 'Selectrode – The universal ion-selective electrode, Part VIII. The solid-state lead(II) Selectrode in lead(II) buffers and potentiometric titrations', *Anal. Chim. Acta*, **72**, 365.

Mascini, M., and A. Liberti, 1972, 'Preparation and analytical evaluation of a new lead(II) heterogeneous membrane electrode', *Anal. Chim. Acta*, **60**, 405.

Rechnitz, G. A., and N. C. Kenny, 1970, 'Potentiometric measurements in aqueous, non-aqueous and biological media', *Anal. Lett.*, **3**, 259.

Selig, W., 1970, 'Potentiometric microdetermination of phosphate with an ion-selective lead electrode', *Mikrochim. Acta*, 564.
Selig, W., 1970a, 'Semi-microdetermination of oxalate with a lead-specific electrode', *Microchem. J.,* **15**, 452.

Determination of Aluminium

As there are no aluminium-selective electrodes, a titrimetric procedure must be used. Among those proposed are precipitation as cryolite in ethanolic solution, using as an indicator either a fluoride-selective electrode (Baumann, 1970) or a platinum electrode in the presence of ferrous ions containing a trace of ferric ion (McCallum, 1956)

$$Al^{3+} + 6 NaF \rightarrow Na_3AlF_6 + 3 Na^+$$

With a fluoride-selective electrode, the increase in the fluoride concentration after the end-point is observed directly but in the other case the fluoride ions present in excess of the aluminium complex the ferric ions, changing the ratio of the concentrations of free ferric and ferrous ions and, therefore, producing a change in the redox potential.

A simpler procedure is achieved by titrating to a fixed potential of a fluoride-selective electrode in a buffered solution (Jaselkis and Bandemer, 1969; Homola and James, 1976). The fixed potential represents, at a constant pH, a constant level of free fluoride. If aluminium is present, it will form fluorocomplexes and more fluoride will need to be added in order to reach the fixed potential. The volume of fluoride solution added (corrected for a blank) is proportional to the concentration of aluminium in the solution. The fixed potential does not correspond to an inflexion in the titration curve, nor to the formation of a particular complex, as the species AlF_2^+, AlF_3, AlF_4^- and possibly AlF^{2+} will coexist in the conditions chosen. The fixed potential is chosen so that the ratio of bound fluoride to aluminium is between 2 and 3. The method is described in greater detail below.

Šůcha and Suchánek (1970) determined aluminium indirectly by compleximetric titration. A known amount of CDTA is added to the sample, which is then boiled for 2–3 min to promote the slow aluminium–CDTA complexation reaction. The solution is cooled, its pH adjusted to 5.5 with hexamethylenetetramine–HCl buffer and the excess of CDTA determined by titration with standard copper solution, using a copper-selective electrode as an indicator. The titration procedure is described in the section on compleximetric titrations (*q.v.*). Other ions that form stronger CDTA complexes than copper will interfere, but they may often be determined separately by omitting the boiling stage of the procedure, e.g. ferric iron.

TITRATION TO A FIXED POTENTIAL

Apparatus

Fluoride-selective electrode; reference electrode; pH meter; magnetic stirrer; plastic beakers

Fluoride-selective electrodes are listed in *Determination of Fluoride*. There may be an advantage in using a reference electrode with a body made of plastic rather than glass, e.g. the Orion 90—01 silver chloride electrode and the Beckman 40459 calomel electrode.

Reagents

Water

Use a mixed-bed deionization unit to produce water with a specific conductivity of less than 0.2 μS cm^{-1}.

Fluoride titrant solution, 0.1 mol l^{-1}

Dissolve 4.199 g of sodium fluoride, 3.035 g of sodium acetate anhydrate or 5.035 g of sodium acetate trihydrate and 82 ml of 1 mol l^{-1} acetic acid or 9.3 ml of glacial acetic acid (all analytical-reagent grade) in water and dilute to 1 litre in a calibrated flask. Store in a polyethylene bottle.

Standard aluminium solution A (100 ppm)

Dissolve 8.792 g of analytical reagent grade potassium alum (Al K(SO$_4$)$_2 \cdot$ 12 H$_2$O) in water and make up to 1 litre in a calibrated flask.

 1 ml $\equiv$ 500 μg aluminium

Prepare the aluminium standards required by dilution of this solution.

pH 4 buffer solution

Dissolve 62.3 g sodium acetate anhydrate or 103.3 g sodium acetate trihydrate and 212.4 ml glacial acetic acid in water and make up to 1 litre.

Calibration Procedure

Step 1 Fill a burette with the fluoride titrant solution.

Step 2 Pipette 100 ml of standard or blank solution and 5 ml of buffer solution into a plastic beaker containing a magnetic stirrer bar.

Step 3 Remove the electrodes from the solution in which they have been immersed, rinse them with deionized water, wipe them dry with a soft tissue and immerse them in the solution in the beaker.

Step 4 Start the magnetic stirrer and adjust the control to give a steady rate of stirring. It is a sensible precaution to note the stirrer setting and the depth of immersion of the electrodes and to keep the same conditions for all titrations.

Step 5 Add a suitable volume of titrant and allow the electrodes to reach a steady potential. Record the e.m.f. and the volume added. See Note 1.

Step 6 Repeat Step 5 as often as is needed to define the titration curve with adequate precision. A minimum of six additions is necessary but a greater number is desirable.

Step 7 Plot the e.m.f. values on the y-axis against the volume of titrant added on the x-axis.

Step 8 Repeat Steps 1—7 with at least five standard solutions and a blank solution (deionized water). Plot all the titration curves on the same graph.

Step 9 Select the e.m.f. corresponding to the best sensitivity; this is a compromise between obtaining the maximum spacing between curves and using the steepest parts of the curves, but the former is the more important factor (see Note 2). Draw a line through this e.m.f. parallel to the volume axis and note the volumes where it intersects each titration curve.

Step 10 Subtract the volume intercept of the blank solution from the intercepts for each standard solution and plot these differences against the concentrations of the standard solutions. The plot should be essentially linear. See Note 3.

Note 1 The volume of titrant required is about 12 ml for a 100 ppm solution and about 2 ml for the blank. Suitable additions for these solutions in Step 5 are, therefore 1 ml and 0.2 ml, respectively; always make additions that correspond exactly to the graduations on the burette.

Note 2 In the useful range of e.m.f. values the titration curves become more widely spaced and only slightly less steep as the fluoride concentration increases. Beyond this range the curves become markedly less steep and also more closely spaced. Homola and James (1976) titrated to a potential corresponding to about 10^{-3} mol l^{-1} uncomplexed fluoride at pH 4, i.e. that given by the addition of 1 ml of titrant to the blank solution. This gives a molar ratio of bound fluoride to aluminium of about 3 and occurs at about -10 mV for an Orion 94—09 fluoride electrode used with a calomel reference electrode. With other fluoride and reference electrodes, the e.m.f. could be quite different. Jaselkis and Bandemer (1969) titrated to +80 mV with a similar electrode pair at pH 3.8. corresponding to a ratio of bound fluoride to aluminium of 2.15 and an uncomplexed fluoride concentration of about 2×10^{-4} mol l^{-1}.

Note 3 A more precise calibration graph can be obtained by repeating the procedure, this time making closely spaced additions in the region of the selected e.m.f.

Concentration Range and Units

The method works in the range 6–100 ppm aluminium. At higher concentrations the slope of the titration curve decreases and, therefore the precision of the analysis deteriorates; concentrated samples should be diluted before analysis.

$$10^{-3} \text{ mol l}^{-1} \equiv 26.98 \text{ ppm aluminium}$$
$$3.706 \times 10^{-5} \text{ mol l}^{-1} \equiv 1 \text{ ppm aluminium}$$

Analytical Procedure

Step 1 Repeat Steps 1–4 of the calibration procedure with the sample solution.

Step 2 Slowly add titrant until the e.m.f. is within 20 mV of the target e.m.f. selected in the calibration procedure and then allow the electrodes to reach a steady potential. Record the e.m.f. and the volume added.

Step 3 Add a small volume of titrant and allow the electrodes to reach a steady potential. Record the e.m.f. and the total volume added. A suitable volume increment is an exact multiple of the smallest division on the burette equal to 3–5% of the volume added by Step 2.

Step 4 Repeat Step 3 until the e.m.f. is about 20 mV past the target potential.

Step 5 Plot the e.m.f. values against the volume added and find the volume corresponding to the target e.m.f.

Step 6 Correct the volume for the blank.

Step 7 Find the concentration of the sample from the corrected volume and the calibration graph.

Step 8 Repeat Steps 1–7 for each sample.

Step 9 It is recommended that a standard solution be included in each batch of samples to check the validity of the calibration graph.

Sources of Error

Effect of Temperature

Changes in temperature will not only affect the standard potential and calibration slope of the electrode, but also the stability constants of the fluoroaluminate complexes. It is desirable always to work with the samples at the temperature at which the calibration was prepared and to bring the standard and sample solutions to this temperature by immersing them in a water-bath thermostatically controlled at a temperature a few degrees above ambient.

Effect of pH

The pH should be in the range of 3.5–4.0. Variations in pH affect the accuracy of the method as part of the fluoride is complexed as HF and HF_2^- and the solutions must be buffered.

Effect of other substances

Metal ions that form strong fluoro-complexes will interfere, e.g. beryllium, ferric iron, rare earths, hafnium, zirconium. 50 ppm of ferric iron caused an interference effect of 0.5 ppm Al in a 15 ppm aluminium solution. 200 ppm of calcium produced an error of 3.6 ppm in the determination of 80 ppm aluminium. 500 ppm lead, 1500 ppm copper and 150 ppm magnesium had no significant effect (Jaselkis and Bandemer, 1969). Anions that do not form complexes or precipitates with aluminium do not interfere, e.g. sulphate, nitrate, perchlorate, chloride, bromide, iodide and acetate. Only 20 ppm of phosphate can be tolerated at pH 3.8, but up to 120 ppm at pH 3.0 (Jaselkis and Bandemer, 1969).

Precision

The relative standard deviation of seven determinations of 10 ppm aluminium was 1.1% (Homola and James, 1976). The relative standard deviation of Jaselkis and Bandemer's titrations varied from 2.5% at 6 ppm to 0.6% at 100 ppm.

Accuracy

An 88.35% aluminium alloy dissolved in acid gave a solution concentration of 5.43 ppm aluminium, which was analysed at 5.36 ppm (Jaselkis and Bandemer, 1969).

Other Applications

Metals can be analysed by first dissolving them in the miniumum of hydrochloric acid, diluting to a known volume with water and proceeding as above.

Comparison with Other Methods

Atomic absorption spectroscopy is the best method of determining aluminium, but methods involving simple apparatus will still find a use. The titration described above can be completed more quickly than those based on the precipitation of cryolite and the samples need less preparation. Baumann (1970) achieved a precision of 1—2% for the determination of 200—600 ppm aluminium by precipitation titration using a fluoride-selective electrode as indicator, no better than those of the simpler technique. Potentials tend to drift in the ethanolic solutions necessary for the precipitation. In the redox electrode indicator method (McCallum, 1956) the ferrous reagent solution has a life of only 4 h, air must be excluded from the titration vessel and 5 min are required for a steady potential to be reached after each addition. Interferences from calcium, beryllium and ferric iron are less serious, however.

Aluminium can only be determined with EDTA and similar compounds by back titration. Many people find the end-point hard to judge in the colorimetric indicator method and the potentiometric indicator method of Šůcha and Suchánek (1970) involves a time-consuming boiling step. Metallic interferences are more serious than in the fluoride titration, but fluoride ion itself does not interfere. The nature of the sample will, therefore, determine the method adopted.

Colorimetric methods such as the Eriochrome Cyanine R method suffer from having a colour-development time and are not suitable for turbid or coloured samples.

Bibliography

Baumann, E. W., 1970, 'Determination of aluminium by potentiometric titration with fluoride', *Anal. Chem.*, **42**, 110.

Homola, A., and R. O. James, 1976, 'Determination of aluminium in paper machine white water by potentiometric titration with fluoride ion', *Anal. Chem.*, **48**, 776.

Jaselkis, B., and M. K. Bandemer, 1969, 'Determination of micro and semimicro amounts of aluminium using fluoride activity electrode', *Anal. Chem.*, **41**, 855.

McCallum, J. R., 1956, 'The potentiometric method for the determination of aluminium on a semimicro scale', *Can. J. Chem.*, **34**, 915.

Šůcha, L., and M. Suchánek, 1970. 'Indirect complexometric determination of aluminium using a solid-membrane cupric ion-selective electrode', *Anal. Lett.*, **3**, 613.

Determination of Total and Free Carbon Dioxide with a Gas-sensing Membrane Electrode

The most convenient electrode for general purposes is the Radiometer E5036 used with the Radiometer D616 flow cell. Similar electrodes are made as components of blood-gas analysis apparatus by Corning, Instrumentation Laboratories and Radiometer (type E5037), and are not normally suitable for separate use.

APPARATUS

Electrode; flow cell; pH meter; peristaltic pump(s)

Note that the electrode is really a complete cell and needs no external reference electrode. In the Radiometer electrode, the signal from the internal reference electrode is led along the copper screening of the cable. If the electrode is used with a different manufacturer's pH meter, the plug will probably have to be changed, and it may be necessary to solder a lead to the screening to take the signal to the reference terminal of the meter. Rinse the inside of the plastic body with deionized water and assemble the electrode according to the manufacturer's instructions.

Either one twin-channel or two single-channel peristaltic pumps capable of flow rates in the range $1-5$ ml min^{-1} are needed. Preferably, one channel should pump at 5–10 times the rate of the other, but this is not necessary.

Reagents

Water

Water of low carbon dioxide content should be used for the preparation of the standard solutions. Water taken directly from the outlet of a mixed-bed deionizer is a convenient source.

Sulphuric Acid, 1N

Dissolve 26.5 ± 0.5 ml of concentrated acid (sp. gr. 1.84) in about 700 ml of deionized water and make up to 1 litre with water. The acid can also be obtained ready diluted. If carbon concentrations below 10 ppm are to be measured, the dilute acid should be purged with nitrogen before use.

Standard carbon dioxide solutions

Solution A (1000 ppm) Dissolve 1.910 ± 0.001 g of sodium hydrogen carbonate (analytical reagent grade) in water, make up to 1 litre with water in calibrated flask and mix. Store in a glass-stoppered glass bottle.

  1 ml ≡ 1000 µg carbon dioxide

Solution B (100 ppm) Pipette 50 ml of solution A into a 500-ml calibrated flask, fill to the mark with water and mix. Store in a glass-stoppered glass bottle.

  1 ml ≡ 100 µg carbon dioxide

Solution C (10 ppm) Pipette 10 ml of solution A into a 1-litre calibrated flask, fill to the mark with water and mix.

  1 ml ≡ 10 µg carbon dioxide

Sample Collection

Carbon dioxide may be absorbed from the atmosphere unless stringent precautions are taken. The sample should be collected in a glass bottle such as that illustrated in Figure 7.2 by the procedure described on p. 109. The sample should be analysed as soon as possible.

Concentration Range and Units

The electrode has been tested successfully over the range 0.1–100 ppm CO_2.

1×10^{-3} mol l^{-1} ≡ 44.0 ppm CO_2 ≡ 100.1 ppm $CaCO_3$

9.99×10^{-6} mol l^{-1} ≡ 0.44 ppm CO_2 ≡ 1 ppm $CaCO_3$

2.273×10^{-5} mol l^{-1} ≡ 1 ppm CO_2 ≡ 2.275 ppm $CaCO_3$

DETERMINATION OF TOTAL CARBON DIOXIDE

Analytical Procedure

Assemble the electrode and place it in the flow cell. The two peristaltic pump tubes are joined by a Y-piece which leads to the inlet to the cell. Connect the free end of one tube, which should preferably have a flow rate 5–10 times slower than the other, to the bottle of 1N sulphuric acid.

Step 1 Connect the sample line to the bottle of standard solution C and start pumping both standard and acid.

Step 2 When the electrode gives a steady potential, which for the Radiometer electrode should be in the range -70 to -100 mV, note the reading E_1.

Step 3 Fit a soda-lime guard tube to the outlet of the sample bottle and the sample line to the three-way tap.

Step 4 In a rapid succession, switch on the pump, turn the three-way tap to connect the pump with the bottle and release the screw clip on the outlet.

Step 5 When the response is steady, note the millivolt reading E_x.

Step 6 Calculate the difference in readings between the sample and the standard solution

$$\Delta = E_1 - E_x$$

Step 7 Using this difference, read off the concentration from the calibration graph, or, for concentrations above 0.5 ppm only, calculate the concentration from the equation

$$c = 10 \times \text{antilog}\,(\Delta/k) \text{ ppm}$$

where $k \simeq 56$ mV per tenfold increase in concentration is the slope of the linear part of the calibration graph.

Step 8 If more samples are to be analysed, repeat Steps 3–7 each time. The frequency with which restandardization is required (Steps 1 and 2) depends on the desired precision of the analysis, but it is recommended that it be done at least twice in a full working day.

Checking for bias

A second standard carbon dioxide solution should be analysed with each batch of determinations. The concentration should be chosen so that the two standard solutions roughly span the range of sample concentrations. Any change in slope will appear as a bias.

Preparation of the Calibration Graph

For measurements in the range above 5 ppm carbon dioxide, the calibration is straightforward, but at lower levels a more complicated procedure is unavoidable. See p. 117 for a discussion of the factors involved in using a permanent calibration graph; if water at a fixed temperature is pumped through the water-jacket of the Radiometer flow cell, the conditions for using a permanent graph are quite favourable.

High-level calibration (5–100 ppm) By dilution of standard solutions A or B prepare 500 or 1000 ml each of solutions containing 50, 20 and 5 ppm carbon dioxide. With solution C carry out Steps 1 and 2 of the analytical procedure and with solution B and the other solutions repeat Steps 3–6.

Calculate the value of Δ for each solution and plot these differences on the y-axis against the logarithm of the carbon dioxide concentration on the x-axis. Standard solution C should be included as the point (log 10, 0). The calibration graph should be linear and the slope, k, will be positive.

In this concentration range it is permissible to prepare a temporary calibration graph for each batch of samples by carrying out Steps 1 and 2 and then repeating them with a second standard solution of concentration s_2 noting the e.m.f. E_2. Either prepare the graph by drawing a straight line through the points (log 10, 0) and (log s_2, $E_1 - E_2$) or calculate the calibration slope per tenfold increase in concentration

$$k = \frac{E_1 - E_2}{\log 10 - \log s_2} \simeq 56 \text{ mV}$$

Low-level calibration (0.1–10 ppm) To avoid contamination from the carbon dioxide in air, low-level calibration is best carried out by using one standard solution (solution C) and diluting *in situ* by varying the relative flow rates of this solution and the sulphuric acid solution.

First, with solution C repeat Steps 1 and 2 of the analytical procedure, using the same combination of flow rates as for the analysis of samples. Call the pump rate of solution C, f_C and the pump rate of the acid f_A. Calculate the factor $P_C = f_C/(f_C + f_A)$. Choose four different combinations of pump rates for solution C and the acid, giving increasing dilutions of solution C to cover the range of concentration required. Repeat Steps 4–6 of the analytical procedure with each combination and calculate a factor P analogous to P_C. If it is planned to use a permanent calibration graph, repeat these determinations until the graph is defined with the desired precision. The values of f_C and f_A should be found empirically by weighing the quantity of water delivered in a given time by each pump tube after all air bubbles have been cleared from the lines. Pump tubes tend to change their delivery with time and a 'permanent' calibration graph may not be valid for much more than two weeks, even if all the other factors involved are favourable.

Calculate the value of Δ for each combination of pump tubes used and plot these differences against the logarithm of the concentration ($10\,P/P_C$) on the x-axis. The combination of pump tubes used in the analysis of samples gives the point (log 10, 0). The calibration should be linear down to a level of 0.5 ppm, but will be curved below that region.

Sources of Error

Variation in sensitivity

The sensitivity may drop if the glass electrode or the membrane is damaged.

Effect of Temperature

Changes in temperature affect the sensitivity of the electrode, but for most practical purposes this is negligible. More serious are the large millivolt shifts produced by ambient-temperature changes. The temperatures at which standard and sample solutions are measured should not differ by more that 1 °C. By connecting the water jacket round the flow cell to a thermostatically controlled water bath, this source of error can be eliminated.

Bias

Bias will be produced by carbon dioxide absorbed from, or lost to, the atmosphere unless due precautions are taken during sampling and analysis.

Effect of other Substances

The only expected interferents are volatile acidic species which, like carbon dioxide, can cross the membrane and change the internal pH. Volatile basic species such as ammonia could interfere, producing low results, but they are retained by the acidic medium. Ionic species do not themselves interfere, but may be converted into volatile forms by the acid, e.g. sulphite. The following have been tested and shown to produce no significant interference at carbon dioxide concentrations between 1 and 10 ppm; the levels quoted should not be taken as the permissible maxima.

Cations 1 ppm each of Ca^{2+} and Mg^{2+}; 1 ppm each of Cu^{2+}, Fe^{3+} and Ni^{2+}.
Anions 35 ppm Cl^-, 62 ppm NO_3^-, 97 ppm PO_4^{3-}.
Amines 1 ppm ammonia + 10 ppm morpholine + 4 ppm cyclohexylamine; octadecylamine (saturated).

The effect of likely acidic species is shown in Table 1.

Table 1   Interference effect of acidic species

Substance	Concentration (ppm)	Apparent concentration (ppm) of carbon dioxide due to interference at actual concentrations of:	
		1 ppm	10 ppm
Sulphite	100	0.75	4
	10	0	—
Acetate	100	—	0
	10	0	—

Precision

Precision tests have been carried out with standard solutions prepared by the low level calibration method: the results are given in Table 2.

Table 2 Precision of measurements with Radiometer E5036 electrode

CO_2 concentration (ppm)	23.2	6.13	5.00	2.32	1.98
Standard deviation (ppm)	2.0	1.2	0.3	0.13	0.3

The relative standard deviations at 23.2, 5.00 and 2.32 ppm are less that 10%, but those at 6.13 and 1.98 are up to four times larger. This difference shows the effect of a large dilution factor in preparing the standards. In the first group, the stock solutions (100 and 10 ppm) were diluted only by factors of 2 to about 4, whereas in the second the 100 ppm stock solution was diluted approximately 16 and 50 times.

Accuracy

Few comparative tests have been carried out, but analysis of feed water from a power station by the electrode and by using the low-temperature section of a Beckman, model 915, total carbon analyser gave good agreement at concentrations between 3 and 9 ppm (Midgley, 1975). Jensen, Van Gundy and Stolzy (1965) observed a linear correlation between results in gaseous samples with the electrode and by gas chromatography.

Response Time

At carbon dioxide concentrations above 1 ppm, the electrode required 3–7 min to reach equilibrium for a ten fold increase in concentration and about twice as long for the reverse change, including about 2 min wash-out time for the flow cell and its associated tubing. Below 1 ppm, the response time can extend to 30 min for a ten fold decrease in concentration.

Lifetime of a Single Assembly

An electrode operated continuously in a power station for 10 weeks without needing attention and with no loss of sensitivity or increase in response time.

Other Applications

Apart from the determination of total carbon dioxide, the electrode can be used to measure the free carbon dioxide (*q.v.*) If the pH of the original sample is known, the distribution of carbonate species in solution can be calculated. The use of the electrode for determining carbon dioxide in blood is well established (Severinghaus, 1968; Smith and Hahn, 1969). Carbon dioxide in sea water and deep-sea sediment (Moore, Roberson and Nygren, 1962) and in soil-root systems (Jensen, *et al.* 1965) has also been determined.

Comparison with Other Methods

The electrode method is the simplest means of determining carbon dioxide, the only one capable of discriminating between free and total carbon dioxide and the one most easily adapted to continuous monitoring. Methods that involve separation of the carbon dioxide from the water by gas stripping followed by detection by infra-red gas analyser, flame ionisation detector or gas chromatography, involve far more manipulation of the sample with the risk of contamination and depend heavily on the quality of the blank at low concentrations. Titration methods also involve separation of the gas and are rather time consuming.

DETERMINATION OF FREE CARBON DIOXIDE

Analytical Procedure

Assemble the electrode according to the manufacturer's instructions and place it in the flow cell.

- **Step 1** Connect the sample line, pumping at a rate f_C, to the bottle of standard solution C and the acid line, pumping at a rate f_A, to the bottle of sulphuric acid. Join the lines by a Y-piece leading to the inlet to the cell and start pumping.
- **Step 2** When the electrode gives a steady potential note the reading E_1.
- **Step 3** Disconnect the acid line and join the sample line directly to the inlet to the cell.
- **Step 4** Fit a soda-lime guard tube to the outlet of the sample bottle and the sample line to the three-way tap.
- **Step 5** In rapid succession, switch on the pump, turn the three-way tap to connect the pump with the bottle and release the screw clip on the outlet.
- **Step 6** When the response is steady, note the potential E_x.
- **Step 7** Calculate the difference in readings between the sample and the standard solution from

$$\Delta = E_1 - E_x$$

- **Step 8** Using this difference, read off the apparent concentration from the calibration graph and multiply it by the factor $(f_C + f_A)/f_C$ to get the true concentration.
- **Step 9** If more samples are to be analysed, repeat Steps 4—8 each time. The frequency with which restandardisation is required (Steps 1—3) depends on the desired precision of the analysis, but it is recommended that it be done at least twice in a full working day.

Preparation of the Calibration Graph

The calibration graphs (for high and low levels) prepared for the determination of total carbon dioxide are used, but remember that since no acid is added to the sample in this case, the concentration read off the graph must be multiplied by the factor $(f_A + f_C)/f_C$ to allow for the dilution of the standard solutions.

Concentration Range

0.1–100 ppm, as for total carbon dioxide.

Sources of Error

These are the same as for total carbon dioxide, with two variations: The temperature effects on the electrode are the same, but the temperature also affects the proportion of total carbon dioxide in the sample present as free carbon dioxide. The analyst must decide whether to bring all the samples to a standard temperature in a water-bath, which would be better for comparative and most control purposes, or to measure at the original temperature of the source of the sample. The interferences are essentially the same, but only the free acidic forms are significant, i.e. sulphur dioxide rather than sulphite and acetic acid rather than acetate. Since both are more strongly ionized in water than carbon dioxide, lower proportions of them are present in their free forms and the risk of interference is smaller.

Comparison with Other Methods

Other methods do not directly measure the free carbon dioxide concentration in water. The total concentration can, however, be used to calculate the free carbon dioxide concentration if the pH of the sample is known.

Tracing Faults

Complete loss of response

The most likely cause is the appearance of hole in the membrane. This allows sulphuric acid to leak through, giving a potential of about 200 mV. Dismantle the electrode, rinse it with deionized water and reassemble with a new membrane. If sulphuric acid is not being mixed with the sample the sensitivity may be partially or completely lost; check the pump and the bottle of acid.

Partial loss of sensitivity

If this occurs at both high and low concentrations of carbon dioxide the cause may be incorrect assembly of the electrode or lack of acidification of the sample. When the effect is seen only at low concentrations, it is usually due to contamination of the acid reagent or the water used for preparing standard solutions. Check that the glass electrode has been fully screwed into the plastic body. If the membrane is wrinkled, reassemble the electrode with a new one. If it is not possible to take the water for preparing standard solutions directly from a mixed-bed deionizer, remove carbon dioxide by flushing with an inert gas or by boiling. The sulphuric acid reagent may contain enough volatile acids to interfere with the measurement of low CO_2 concentrations, in which case it should be flushed with nitrogen before use.

Slow response

The most likely cause is incorrect assembly (see above), but the presence of an interfering substance in the sample may be the reason.

Drift

A drift towards an increasingly positive millivolt reading shows that a hole is developing in the membrane and acid is reaching the inside of the electrode. Normally about two hours are required before a steady potential of about 200 mV is reached. The electrode takes up to two hours to reach thermal equilibrium and variations in temperature may give rise to drift. Connecting the water jacket to a thermostatically-controlled bath removes this source of trouble.

Medium-term fluctuations

When a standard solution is pumped at a much lower flow rate than the acid reagent, as in the low-level calibration procedure, the readings may deviate for a few minutes at a time several millivolts from the expected reading. The effect can be eliminated by inserting a glass mixing coil in the line, or often simply by increasing the length of tubing between the flow cell and the confluence of the two streams.

Bibliography

Jensen, C. R., S. D. Van Gundy, and L. H. Stolzy, 1965, 'Recording CO_2 in soil-root system with a potentiometric membrane electrode', *Soil Sci. Soc. America Proc.*, **29**, 631.

Midgley, D., 1975, 'Investigations into the use of gas-sensing membrane electrodes for the determination of carbon dioxide in power station waters', *Analyst*, **100**, 386.

Moore, G. W., C. E. Roberson, and H. D. Nygren, 1962, 'Electrode determination of the carbon dioxide content of sea water and deep-sea sediment', *US Geological Survey Prof. Paper 450-B*,

Severinghaus, J. W., 1968, 'Measurements of blood gases: PO_2 and PCO_2', *Ann. N.Y. Acad. Sci.*, **148**, 115.

Smith, A. C., and C. E. W. Hahn, 1969, 'Electrodes for the measurement of oxygen and carbon dioxide tensions', *Br. J. Anaesth.*, **41**, 731.

Determination of Nitrite and Nitrogen Oxides with a Gas-sensing Membrane Electrode

The only commercially-available electrode is the Orion 95–46, although an experimental model made by Electronic Instruments Ltd. has also been described (Bailey and Riley, 1975). The electrode may be considered to respond to nitrous acid formed when the nitrite sample is acidified, but consideration of the reactions and equilibria below will show that this is merely a formal picture of the real case. The electrode works by using a glass electrode to sense the pH change produced when nitrous acid or nitrogen oxides cross a membrane and dissolve in an internal filling solution containing sodium nitrite.

$$H^+ + NO_2^- \rightleftharpoons HNO_2$$
$$3\,HNO_2 \longrightarrow HNO_3 + 2\,NO + H_2O$$
$$2\,HNO_2 \longrightarrow NO + NO_2 + H_2O$$
$$2\,NO + O_2 \longrightarrow 2\,NO_2$$
$$2\,NO_2 \rightleftharpoons N_2O_4$$
$$2\,NO_2 + H_2O \longrightarrow HNO_2 + HNO_3$$

APPARATUS

Gas-sensing electrode; pH meter; magnetic stirrer

Note that no reference electrode is needed, as the gas-sensing electrode is a complete electrochemical cell and contains its own internal reference electrode.

REAGENTS

Water

Water with a specific conductivity of 0.2 μS cm^{-1} (measured immediately at the outlet of a mixed-bed deionization column) should be suitable. Alternatively, add one crystal each of potassium permanganate and calcium or barium hydroxide to 1 litre of distilled water and re-distil in an all-Pyrex apparatus, discarding the first 50 ml of distillate. Collect the distillate as long as it is free of permanganate.

Stock nitrite solution (approximately 1000 ppm nitrogen)

Dissolve 4.974 g of sodium nitrite ($NaNO_2$) in water and make up to the mark in a 1-litre calibrated flask. Store in a refrigerator.

$1\ ml \cong 1\ mg$ nitrogen

Fresh bottles of sodium nitrite are preferable for preparing the solution, as sodium nitrite is oxidized in the presence of moisture. Even the fresh solid assays at less than 99%, however, and for the most accurate work the solution must be standardized by adding an excess of standard potassium permanganate solution, discharging the colour with standard sodium oxalate solution and finally back titrating with standard permanganate solution (APHA, 1971).

Intermediate nitrite solution (approximately 50 ppm nitrogen)

Prepare daily by dilution of the stock nitrite solution.

Standard nitrite solutions

Prepare daily by dilution of the intermediate nitrite solution.

Acid buffer

This solution is for use with samples containing less than $0.1\ mol\ l^{-1}$ of dissolved particles when analysed with the standard Orion electrode. For use with samples containing high concentrations of dissolved matter, see *Osmotic Effect*.

Dissolve 190 g of anhydrous sodium sulphate in about 800 ml of water and slowly add 53 ml of analytical-reagent grade concentrated sulphuric acid (sp. gr. 1.83). Cool the mixture and then dilute it to 1 litre with water.

SAMPLE COLLECTION

Collect the sample in a glass or plastic container and analyse immediately if possible, as bacteria can convert the nitrite to ammonia or nitrate. The sample (in a *plastic* bottle) may be stored for 1—2 days in a deep-freeze at $-20\ °C$.

ASSEMBLY, CONDITIONING AND STORAGE OF THE ELECTRODE

Assemble the electrode according to the manufacturer's instructions, using tweezers to handle the membrane. Stand a newly assembled electrode in internal filling solution for 30 min before use. Between measurements, store the electrode with the tip immersed in a solution composed of 100 ml deionized water and 10 ml acid buffer solution. Do not store in air and never let the membrane dry out. If the electrode is out of service for extended periods, it is best to disassemble it, wash the components with deionized water and store them dry. A new membrane will be needed on reassembly.

CONCENTRATION RANGE AND UNITS

The optimum concentration range is 0.1–70 ppm nitrogen. At lower concentrations the sensitivity decreases and the response time increases. At higher concentrations the tendency of the dissolved gas to leave the solution may result in drifting potentials.

1×10^{-3} mol l^{-1} ≡ 14 ppm nitrogen ≡ 46 ppm nitrite or nitrogen dioxide

7.14×10^{-5} mol l^{-1} ≡ 1 ppm nitrogen ≡ 3.29 ppm nitrite or nitrogen dioxide

2.17×10^{-5} mol l^{-1} ≡ 0.304 ppm nitrogen ≡ 1 ppm nitrite or nitrogen dioxide

ANALYTICAL PROCEDURE

Step 1 Pipette 50 ml of a standard solution of concentration s_1 into a 100-ml beaker or conical flask containing a magnetic stirrer bar. The concentration s_1 should be chosen so as to be at the upper end of the expected concentration range.

Step 2 Remove the electrode from the solution in which it is immersed, rinse it with deionized water and then remove any drops of water adhering to the electrode with a tissue. Immerse the electrode in the solution to be analysed.

Step 3 Add 5 ml of acid buffer solution and start the stirrer. The same gentle stirring rate should be used for all the analyses. Check that there are no bubbles of air adhering to the membrane.

Step 4 When a steady reading has been reached, note the e.m.f., E_1.

Step 5 Repeat Step 1 with a second standard solution of concentration s_2 from the lower end of the expected concentration range and not more than half s_1.

Step 6 Repeat Steps 2 and 3.

Step 7 When a steady reading has been reached, note the e.m.f., E_2.

Step 8 *Either* calculate the calibration slope

$$k = \frac{E_1 - E_2}{\log s_1 - \log s_2} \simeq 58 \text{ mV per tenfold increase in concentration}$$

or

Prepare a temporary calibration graph by drawing a straight line through the points $(\log s_1, 0)$ and $(\log s_2, E_1 - E_2)$.

Step 9 Repeat Step 1 with the sample solution.

Step 10 Repeat Steps 2 and 3.

Step 11 When a steady reading has been reached, note the e.m.f., E_x.
Step 12 Calculate the e.m.f. difference, $\Delta = E_1 - E_x$.
Step 13 Either read the concentration of the sample from the calibration graph, *or* calculate it from

$$c = s_1 / \text{antilog}\,(\Delta/k)$$

Step 14 Repeat Steps 9–13 for each sample. The calibration should be checked every 1–2 h by repeating Steps 1–4.

Worked example

An electrode gives e.m.f. values of 60.3 and 30.5 mV in 1.0 and 0.3 ppm nitrogen standards, respectively. The e.m.f. obtained in a sample solution is 55.0 mV. The calibration slope $k = (60.3 - 30.5)/(\log 1 - \log 0.3) = 29.8/0.5229 = 57.0$ mV per tenfold increase in concentration. The sample contains

$$c = 1.0/\text{antilog}\,(5.3/57.0) = 0.81 \text{ ppm nitrogen.}$$

Preparation and use of a permanent calibration graph

If the conditions discussed on p. 117 are satisfied, Steps 5–8 of the analytical procedure may be omitted and a permanent calibration graph used instead. Prepare the graph by carrying out Steps 1–4 and then repeating Steps 9–12 with at least four other standard nitrogen solutions. As the internal filling solution of the electrode is liable to oxidation, with a consequent change in the response characteristics of the electrode, a permanent calibration graph should only be used after thorough testing has proved it to be practicable and a standard solution should always be analysed with each batch of samples to check for changes in the performance of the electrode.

SOURCES OF ERROR

Temperature

The temperature of the sample and standard solutions should not differ by more than 1 °C.

Effect of other substances

The substances that are likely to interfere directly with the electrode are dissolved gases that on acidification of the sample will cross the membrane and change the pH of the internal filling solution, e.g. carbon dioxide, sulphur dioxide, acetic acid. Bailey and Riley (1975) found that their experimental electrode was not affected by 10^{-3} mol l^{-1} of dissolved carbon dioxide at a concentration of 14 ppm nitrite–

nitrogen, but read about 0.5 ppm high at 1.4 ppm nitrite—nitrogen. Orion (1976) report that the following molar ratios of interferent to nitrite result in a 10% error in the apparent nitrite concentration: carbon dioxide x 30, formic acid x 0.2, acetic acid x 3, hydrofluoric acid x 1, lactic acid x 0.2, but the last two interferences appear only slowly and should not be significant in the normal time of analysis (1—2 min). Prolonged use in hydrofluoric acid solutions may damage the glass electrode inside the gas-sensing electrode itself. Carbon dioxide interference could be removed by bubbling nitrogen through a fritted glass dispersion tube immersed in 100 ml of the sample, the pH of which should not be above 5.5 (adjusted by the addition of 2.5 mol l^{-1} perchloric acid if necessary). Bubbling for 5 min at 1 litre min^{-1} should remove the interference while causing a loss of less than 5% of the nitrite.

Gases such as sulphur dioxide, chlorine and bromine react with nitrous acid and should not coexist with it in solution. Immersion of the electrode in samples containing these gases rather than nitrous acid may result in the gases' crossing the membrane and reacting with the internal filling solution, which will then need to be replaced before further measurements are made.

Involatile substances that do not react with nitrite should not interfere, thus Tabatabai (1974) found that 0.1 mol l^{-1} concentrations of the following species did not interfere with analyses of nitrite solutions containing 0.5, 1.0 or 5.0 ppm nitrite—nitrogen: NH_4^+, Na^+, K^+, Ag^+, Ca^{2+}, Mg^{2+}, Cu^{2+}, Hg^{2+}, NO_3^-, Cl^-, SO_4^{2-}, PO_4^{3-}, urea.

Effect of pH

The pH of the sample after the addition of the acid buffer should be 1.1—1.7. If the sample is very acidic or very alkaline, or is buffered, the acid buffer may not be adequate to bring about the desired pH; in this case the sample should be pre-treated with perchloric acid or sodium hydroxide to bring it to approximately the correct pH and then the acid buffer should be added. The results ahould be corrected for the dilution involved in the pre-treatment of the sample.

Effect of detergents

Exposure of the membrane to a detergent will enable water to cross the membrane and stable e.m.f. readings will not be obtained.

Osmotic effect

Osmotic pressure differences between the internal filling solution and the acidified sample can cause transport of water across the membrane, with a resultant change in the concentration of the internal filling solution and hence a change in the e.m.f. The acid buffer solution is designed to maintain the correct osmotic pressure in the acidified sample solution for an Orion electrode, provided that the sample does not contain high concentrations of solute. The sum of the concentrations of all the dissoved particles (the osmolarity) should be 0.5 mol l^{-1}, but Tabatabai (1974) has shown that accurate results can be obtained with samples containing up to 0.2 mol l^{-1} of dissolved particles in addition to those added with the acid buffer. Samples containing 1 mol l^{-1} of potassium chloride (final osmolarity 2.3 mol l^{-1}) gave results 1% and 10% high at nitrite-nitrogen concentrations of 10 and 0.1 ppm, respectively. Samples containing 2 mol l^{-1} potassium chloride gave much less accurate results. The simplest way of dealing with samples containing high concentrations of solute is to dilute them until the osmolarity of the sample is no more than 0.2, provided that the nitrite concentration of the diluted sample is still within the working range of the electrode. Alternatively, the internal filling solution may be modified by adding sodium sulphate to bring the osmolarity of the internal filling solution up to that of the acidified sample or the acid buffer prepared with less sodium sulphate. Both methods are encompassed by equation (1), which can be used to calculate the required degree of dilution of the sample or the weight of sodium sulphate to be added to the filling solution or acid buffer for the addition of 1 volume of buffer to 10 volumes of sample.

$$\left(\frac{3w_B}{142} + 2.9\right)\frac{1}{11} + \frac{10}{11}\left(\Sigma \frac{n_i w_i}{m_i}\right)\frac{V}{V_D} = 0.5 + \frac{3w_F}{142} \tag{1}$$

where w_B and w_F are the weights (in grams) of anhydrous sodium sulphate added per litre of the acid buffer and internal filling solution, respectively; w_i is the weight of the ith solute per litre of sample; m_i is the molecular weight of that solute; and n_i the number of particles released per molecule. V_D is the volume to which V ml of sample are diluted.

Note If the internal filling solution is modified to cope with samples of high osmolarity, the standard solutions should be prepared in such a way that they also contain $\Sigma n_i w_i/m_i$ particles per litre in addition to the sodium nitrite.

Worked example

A brine solution contains 0.6 mol l^{-1} sodium chloride. If the usual acid buffer is used (w_B = 190 g) and the sample is undiluted

$$\left(\frac{3 \times 190}{142} + 2.9\right)\frac{1}{11} + \frac{10}{11} \times 1.2 \times 1 = 1.72 = 0.5 + \frac{3 \times w_F}{142}$$

hence w_F = 57.7 g of sodium sulphate are added per litre of filling solution. Alternatively, if V_D = 10V and w_F = 0, the correct balance would be obtained by having w_B = 18.5 g. In practice, the usual acid buffer could be used with the diluted sample, even though the osmotic balance is not perfect (Tabatabai, 1974).

PRECISION

Tabatabai (1974) carried out six replicate analyses on each of several types of natural water and soil extracts containing 0.5–4 ppm nitrogen as nitrite. The relative standard deviations were between 1% for the lower concentrations and 0.4% for the higher concentrations.

ACCURACY

99.2–100.8% recovery of spikes added to soil extracts and natural water samples have been reported by Tabatabai (1974), who also obtained good agreement between results with the electrode and a colorimetric method.

RESPONSE TIME

The time taken for a steady e.m.f. to be reached is shorter for an increase in concentration than for a corresponding decrease. The response time for a given multiple change in concentration, e.g. tenfold, is reduced as the final concentration increases. Bailey and Riley (1975) found times of 30–40 sec for a tenfold increase to 14 ppm of nitrite nitrogen compared with 90–100 sec for a similar increase to 1.4 ppm, the results being obtained with an experimental electrode. At 0.14 ppm the electrode took 7–8 min to reach a steady e.m.f. Orion (1976) quote response times of less than 2 min for any tenfold change in concentration.

COMPARISON WITH OTHER METHODS

The colorimetric determination of nitrite as an azo dye after diazotisation of a primary aromatic amine is not suitable for turbid or coloured solutions and is subject to interference by metal ions such as mercuric and cupric ions. The electrode is not affected by any of these interferences and the procedure is much simpler and should be adaptable to continuous monitoring and measurement in the field.

Determination of nitrite by oxidation to nitrate and measurement with a nitrate-selective liquid ion-exchange electrode is not only more complicated but more prone to interferences (see p. 378).

OTHER APPLICATIONS

Nitrate could be determined as nitrite after reduction on a cadmium column. No results have been reported, but the advantages would be the same as for the determination of nitrite itself compared with conventional nitrate-reduction methods and reduced susceptibility to interferences compared with the direct determination of nitrate with a nitrate-selective electrode, although the latter would be quicker and more convenient.

Sherken (1976) determined nitrite in aqueous extracts of smoked fish, using the electrode in an addition titration. Recoveries in the range 97–101% were obtained in the range 200–2000 ppm nitrite in the extract, with relative standard deviations increasing from 1.6% at the highest concentration to 6.3% at the lowest.

TRACING FAULTS

See p. 127 for a discussion of the faults found with ion-selective electrodes in general. The following relate to particular features of the NO_x electrode:

Partial or complete loss of response

(1) Check that the acid buffer has been added and, if so, that the pH of the sample is 1.1–1.7 (see *Effect of pH*). (2) Check that the membrane does not have a hole in it — normally this would produce a very high constant positive millivolt reading. (3) Check that the sensing electrode and the membrane are in contact, i.e. that the top and bottom caps are screwed on far enough. (4) If the electrode has been exposed to air, the filling solution immediately behind the membrane may be oxidized. Unscrew the top cap, withdraw the inner electrode slightly to allow fresh solution to replace the old solution and then re-tighten the cap. (5) Check that the glass electrode inside the NO_x electrode is not broken or damaged by performing a buffer test as described in the fault-finding section for the determination of ammonia.

Reading slowly drifting in one direction

(1) The sample may not be at a steady temperature. (2) The sample contains too high a concentration of dissolved matter, causing a drift to higher nitrite readings and more positive e.m.f. values (see *Osmotic Effect*). (3) Internal filling solution is leaking into the sample — check that the bottom cap is screwed on properly. (4) Loss of NO_x from the sample. Avoid high temperatures and vigorous stirring and use beakers of a shape and size that minimize the surface area to volume ratio of the sample solution. (5) The membrane has been exposed to a detergent. Replace the membrane. (6) As points (2), (3) and (5) for loss of response.

Slow response

(1) Check that there are no air bubbles adhering to the membrane. Holding the electrode at 20–30° from the vertical helps to prevent this problem. (2) As for drift, above.

Bibliography

APHA, 1971, *Standard Methods for the Examination of Water and Wastewater*, 13th ed., American Public Health Association, Washington, D.C.

Bailey, P. L., and M. Riley, 1975, 'Performance characteristics of gas-sensing membrane probes', *Analyst*, **100**, 145.

Orion, 1976, 'Nitrogen oxide electrode model 95–46', *Form IM95–46/672*, Orion Research Inc., Cambridge, Mass.

Sherken, S., 1976, 'Ion selective method for the determination of nitrite in smoked fish', *J. Assoc. Off. Anal. Chem.*, **59**, 971.

Tabatabai, M. A., 1974, 'Determination of nitrite in soil extracts and water samples by a nitrogen oxide electrode', *Commun. Soil Sci. Plant Anal.*, **5**, 569.

Determination of Sulphur Dioxide with a Gas-sensing Membrane Electrode

Electrodes of this type are the Orion 95–64 and the Electronic Instruments Ltd. 8010–800. On acidification of the sample, sulphur dioxide is released from solution, crosses a gas-permeable membrane and changes the pH of the internal filling solution of the electrode. Measuring the pH with a glass pH electrode contained within the body of the gas-sensing electrode provides an e.m.f. that is proportional to the logarithm of the dissolved sulphur dioxide concentration, including that present as sulphite and metabisulphite.

$$SO_3^{2-} + 2\,H^+ \longrightarrow SO_2 + H_2O$$

$$S_2O_5^{2-} + 2\,H^+ \longrightarrow 2\,SO_2 + H_2O$$

APPARATUS

Electrode; pH meter; magnetic stirrer; conical flasks

Note that no reference electrode is needed as the gas-sensing electrode is a complete electrochemical cell with its own internal reference electrode. The Orion and EIL electrodes differ in some aspects of their performance. In general, the Orion electrode may have some advantages for concentrations below 1 ppm SO_2, while the EIL electrode may be more suitable for concentrations above 120 ppm SO_2 and for samples with a high dissolved-solids content.

REAGENTS

Water

Deionized water with a specific conductivity less than 0.2 $\mu S\ cm^{-1}$ as it leaves the deionization unit should be suitable for preparing standard and reagent solutions.

Acid buffer solution A (Orion)

Dissolve 190 g of anhydrous sodium sulphate in about 800 ml water then add carefully 53 ml concentrated sulphuric acid (sp. gr. 1.83). Make up to 1 litre with water and mix.

Acid buffer solution B (EIL)

Slowly add 53 ml of concentrated sulphuric acid (sp. gr. 1.83) *or* 130 ml of concentrated perchloric acid (sp. gr. 1.54) to 750 ml of water and make up to 1 litre with water.

Standard sulphite solution (1000 ppm SO_3^{2-})

Dissolve 1.575 g of sodium sulphite heptahydrate ($Na_2SO_3 \cdot 7H_2O$, analytical reagent grade) in 250 ml of water to which 5 ml of glycerol have been added. Make up to the mark in a 500-ml calibrated flask. Prepare fresh every day.

$$1 \text{ ml} \cong 1 \text{ mg } Na_2SO_3$$

For the most accurate work this solution should be standardized iodometrically. Prepare less concentrated standards by dilution of the above solution, including 1 ml of glycerol per litre of dilute standard.

Standard sulphur dioxide solution (1000 ppm SO_2)

Dissolve 1.968 g of sodium sulphite heptahydrate ($Na_2SO_3 \cdot 7H_2O$, analytical reagent grade) in 250 ml of water to which 5 ml of glycerol have been added. Make up to the mark in a 500-ml calibrated flask. Prepare fresh each day.

$$1 \text{ ml} \cong 1 \text{ mg } SO_2$$

For the most accurate work, this solution should be standardized iodometrically. Prepare less concentrated standards by serial dilution of the above solution, including 1 ml of glycerol per litre of dilute standard.

Iodometric standardization of sulphite and sulphur dioxide solutions

To S ml (5 or 10 ml) of the appropriate sulphite or sulphur dioxide solution add 50 ml water, 1 ml of dilute (1 + 1) sulphuric acid and 1 ml of starch indicator solution. Titrate with 0.0125N potassium iodide–iodate titrant (445.8 mg anhydrous potassium iodate, 4.35 g potassium iodide and 310 mg sodium hydrogen carbonate per litre) until a faint permanent blue colour is observed. Note the volume, V ml, of titrant added (APHA, 1971).

$$\text{ppm } SO_2 = V \times 0.0125 \times 32000/S$$

$$\text{ppm } Na_2SO_3 = \text{ppm } SO_2 \times 1.968$$

SAMPLE COLLECTION

Collect the sample with as little contact with the air as possible (the apparatus described on p. 109 is suitable). Samples should be analysed at once (or as soon as the temperature of sample is the same as that of the electrode). If samples have to be stored, add 1 drop of glycerol per 100 ml of sample and store in a tightly stoppered glass bottle.

ASSEMBLY, CONDITIONING AND STORAGE OF ELECTRODES

Assemble the electrodes according to the manufacturer's instructions. Conditioning of the electrodes before use is not essential, but it may be advantageous to immerse the electrode for 10–15 min in a buffered standard solution similar in concentration to the samples to be analysed, especially if the concentrations of these samples are at the lower end of the calibration range of the electrode. The membrane should not be blotted dry between measurements and should not be exposed to the air for longer than is necessary. Store the electrode with the tip immersed in the sodium sulphite storage solution. The internal filling solution of the EIL electrode should last for at least 30 days before renewal is necessary; that of the Orion electrode for 2–3 weeks, these times depending on the care taken in storage and on the nature of the samples.

CONCENTRATION RANGE AND UNITS

The Nernstian response of the EIL electrode is from 2 to 2000 ppm SO_2; that of the Orion electrode from 0.15 to 150 ppm SO_2. Both electrodes may be used outside these ranges even though the calibrations will be curved; the upper limits are about 6000 ppm SO_2 for the EIL electrode and 1000 ppm SO_2 for the Orion electrode. The lower limits are not well defined.

$$10^{-3} \text{ mol l}^{-1} \equiv 64.07 \text{ ppm } SO_2 \equiv 80.07 \text{ ppm sulphite}$$

$$1.56 \times 10^{-5} \text{ mol l}^{-1} \equiv 1.0 \text{ ppm } SO_2 \equiv 1.25 \text{ ppm sulphite}$$

$$1.25 \times 10^{-5} \text{ mol l}^{-1} \equiv 0.80 \text{ ppm } SO_2 \equiv 1.00 \text{ ppm sulphite}$$

$$1 \text{ mol l}^{-1} \text{ metabisulphite} \equiv 2 \text{ mol l}^{-1} \text{ sulphite} \equiv 2 \text{ mol l}^{-1} SO_2$$

ANALYTICAL PROCEDURE

Step 1 Pipette 50 ml of a standard solution of concentration s_1 into a 100-ml conical flask containing a magnetic stirrer bar. The concentration s_1 should be chosen so as to be near the upper end of the expected concentration range of the samples.

Step 2 Add 5 ml of acid buffer solution.

Step 3 Remove the electrode from the solution in which it was previously immersed, rinse it with deionized water but do *not* dry the membrane with a paper tissue, and immerse it in the solution. An O-ring around the body of the electrode should hold it so that the tip is held above the stirrer bar and also seal the top of the conical flask, thereby reducing losses of sulphur dioxide by evaporation.

Step 4 Start the stirrer. Check that there are no bubbles of air at the surface of the membrane. Stir gently so as not to create a vortex and keep the same stirring rate for all measurements.

Step 5 When a steady reading has been attained, note the e.m.f., E_1.

Step 6 Repeat Step 1 with a second standard solution of concentration s_2 such that s_2 is near the bottom of the expected concentration range and $s_2 \leqslant 0.5 s_1$.

Step 7 Repeat Steps 2–4.

Step 8 When a steady reading reading has been attained, note the e.m.f., E_2.

Step 9 *Either* prepare a temporary calibration graph by drawing a straight line through the points $(\log s_1, 0), (\log s_2, E_1 - E_2)$
or
calculate the calibration slope

$$k = \frac{E_1 - E_2}{\log s_1 - \log s_2} \simeq 58 \text{ mV per tenfold increase in concentration}$$

Step 10 Repeat Step 1 with the sample solution.

Step 11 If glycerol was added to the sample when it was collected, or if the standard solution did not contain glycerol, proceed to Step 12, otherwise add 0.5 ml of glycerol to the solution.

Step 12 Repeat Steps 2–4.

Step 13 When a steady reading has been attained, note the e.m.f., E_x.

Step 14 Calculate the potential difference, $\Delta = E_1 - E_x$.

Step 15 Either read off the concentration from the calibration graph or calculate it from $c = s_1/\text{antilog}(\Delta/k)$.

Step 16 Carry out Steps 10–14 for each sample. The electrode should be recalibrated (Steps 1–5) at least every second hour and preferably every hour.

Worked example

The potentials observed with 100 ppm and 20 ppm standard solutions and a sample solution were 302.0, 261.3 and 283.5 mV, respectively. The calibration slope, $k = (302.0 - 261.3)/(\log 100 - \log 20) = 58.2$ mV per decade. The concentration of the sample,

$c = 100/\text{antilog} [(302.0 - 283.5)/58.2]$

$= 48.1$ ppm

Preparation and use of a permanent calibration graph

Steps 6–9 of the analytical procedure may be omitted and a permanent calibration graph used instead, provided the conditions discussed on p. 117 are met. Prepare the graph by carrying out Steps 1–5 of the method and then repeating Steps 11–14 with at least four more standard solutions. As the internal filling solution is liable to oxidation, with a consequent change in the response characteristics of the electrode, a permanent calibration graph should only be used after thorough testing has proved it to be practicable and a standard solution should always be analysed with each batch of samples to check for changes in the electrode.

SOURCES OF ERROR

Temperature

The standard and sample solutions should have the same temperature within 1 °C. As the electrodes are fairly slow to reach thermal equilibrium, changing their temperature by more than a few degrees may result in drifting potentials for up to an hour until a new equilibrium is established. With the Orion electrode temperature differences can cause an osmotic effect (see below).

Effect of other substances

Ionic species cannot cross the membrane and, therefore, do not interfere unless they release acidic gases when the acid buffer is added. A fiftyfold molar excess of acetic acid and a thirtyfold molar excess of hydrofluoric acid are reported to be the maximum tolerable levels that will produce a bias of less than 10% in the sulphur dioxide concentration (Orion, 1975). Prolonged exposure to hydrofluoric acid at concentrations above 10^{-3} mol l^{-1} can damage the inner glass sensing electrode permanently. Carbon dioxide does not interfere at concentrations likely to be absorbed from the atmosphere. Strong (about 1 mol l^{-1}) hydrochloric acid solutions may interfere. Hypochlorite and nitrite should not coexist with sulphur dioxide, but if they are present in excess, their gaseous forms chlorine and nitrogen dioxide can cross the membrane and react with the internal filling solution, which will need to be renewed. Oxygen will also react with the filling solution, although more slowly, and exposure of the membrane to air should be kept to a minimum.

Osmotic effect

If the total concentration of dissolved particles in the sample solution is too high or if the sample and the electrode are at different temperatures, water can be transported from the internal filling solution of the Orion electrode, thus changing the concentration inside the electrode and causing the potential to drift. Addition of the Orion acid buffer solution will bring dilute samples (up to 0.2 osmolar) to approximately the correct osmolarity so that drift will be negligible. More concentrated samples should be diluted before analysis. The EIL electrode is not subject to the osmotic effect, at least up to an osmolarity of 6.0 (Bailey and Riley, 1975). The osmolarity is calculated by summing the molar concentrations of all the species in the solution, allowing for electrolytic dissociation, e.g. in a solution containing glucose and sodium sulphate, the osmolarity (m_0) is given by

$$m_0 = c_{Na^+} + c_{SO_4^{2-}} + c_{glucose} = 3\, c_{Na_2SO_4} + c_{glucose}$$

Henry's law effects

The partial pressure of sulphur dioxide in a solution is related to its concentration by the Henry's law constant, H.

$$P_{SO_2} = Hc_{SO_2}$$

H depends on temperature, the total concentration of dissolved particles and the nature of these particles. In solutions containing low concentrations of dissolved particles, H is virtually constant at a given temperature, but at high concentrations, the results may be biased if the electrode was calibrated with standard solutions containing a low concentration of dissolved particles. For this reason, it may be better to use a standard addition calibration for highly concentrated samples, or else to dilute them if the sulphur dioxide concentration after dilution is still within the range of the electrode. Note that this is a separate phenomenon from the osmotic effect, which is characterized by drifting potentials.

Effect of acidity

The proportion of dissolved sulphur dioxide, converted into the gaseous form to which the electrode responds, depends on the pH of the solution. EIL and Orion recommend final pH values of 0.7 and 1.2, respectively, after addition of the acid buffer. If the sample contains a high concentration of alkali or is buffered, the acid buffer solution may not be adequate to adjust the pH to the required value and a low result will be obtained. In such cases the pH should first be adjusted to approximately the correct value with additions of concentrated sulphuric acid. Note that the heat of reaction may raise the temperature of samples treated in this way; if so, the samples should be cooled as quickly as possible to the same temperature as the standard solutions.

PRECISION

Both manufacturers claim a precision of 2% in concentration for their electrodes.

ACCURACY

No straightforward determination of accuracy have been reported, but Krueger (1974) obtained 5% higher values for the sulphur content of jet fuels determined as

SO_2 by the electrode compared with the standard gravimetric determination as sulphate.

RESPONSE TIME

When responding to a tenfold increase in concentration, the electrodes should reach equilibrium within 1–2 min. The response to decreases in concentration will be longer (3–4 min).

OTHER APPLICATIONS

The electrode may be used as an indicator in the titration of sulphur dioxide with potassium dichromate. Take 50 ml of sample, add 0.5 ml of glycerol and 5 ml of acid buffer and titrate with standard dichromate solution. Each mole of dichromate reacts with three moles of sulphur dioxide. Potassium dichromate can also be used as the reagent in a known subtraction type of calibration.

$$3 SO_2 + K_2Cr_2O_7 + H_2SO_4 \rightarrow Cr_2(SO_4)_3 + K_2SO_4 + H_2O$$

For the determination of sulphur dioxide in gases, the sample is passed through a pair of impingers or fritted bubblers filled with an absorbing solution and connected in series. A suitable absorbing solution consists of 10 g mercury (II) chloride, 6 g potassium chloride, 0.05 g disodium EDTA, 3.5 g sodium dihydrogen phosphate, 3.5 g disodium hydrogen phosphate and 20 ml glycerol dissolved in water and made up to 1 litre. Krueger (1974) found almost 100% collection efficiency with two bubblers and about 95% with one only. Before analysis, combine the two absorption solutions and make up to 100 ml with deionized water, adding 0.5 g of sulphamic acid if nitrites are likely to be present. Analyse the combined solution according to the analytical procedure.

The electrode can be used to determine free and total sulphur dioxide in foods (EIL, 1976). In the latter case, the sample is treated with alkali to raise the pH to 12 in order to release sulphite from aldehyde–bisulphite compounds before the addition of the acid buffer.

TRACING FAULTS

The commoner faults of potentiometric systems are dealt with on p. 127, but the following apply particularly to the SO_2 electrode.

Loss of sensitivity

(1) A hole in the polymer membrane will allow acid to reach the glass electrode directly, causing the electrode to give an almost constant and highly positive potential. Renew the membrane and the internal filling solution. (2) The sensing electrode and the polymer membrane are not in sufficiently close contact. Check that the assembly instructions have been followed correctly. (3) The internal filling

solution is exhausted. Try pulling up the sensing electrode to allow fresh solution into the film between the electrode and the polymer membrane. If no useful improvement occurs, renew the filling solution completely. (4) Check the sensing electrodes exactly as when tracing faults in the procedure for the determination of ammonia.

Slow response

The causes are likely to be the same as causes (2)–(4) for loss of sensitivity, but check that no air bubbles are trapped at the tip of the electrode.

Drift

A persistent change of the potential in one direction may be caused by the osmotic effect (*q.v.*), temperature differences between successive solutions or previous exposure to a very strong solution of sulphur dioxide. If the solution being analysed is in an open beaker, sulphur dioxide may be lost during the period of measurement; use a conical flask as recommended in the analytical procedure.

Bibliography

APHA, 1971, *Standard Methods for the Examination of Water and Wastewater*, 13th ed., American Public Health Association, Washington, D.C.

Bailey, P. L., and M. Riley, 1975, 'Performance characteristics of gas-sensing membrane probes', *Analyst*, **100**, 145.

EIL, 1976, *EIL SO_2 Probe Instruction Manual*, Electronic Instruments Ltd., Chertsey, Surrey.

Krueger, J. A., 1974, 'Use of a potentiometric sulphur dioxide electrode for lamp sulphur determinations', *Analyt. Chem.*, **46**, 1338.

Orion, 1975, 'Instruction manual sulfur dioxide electrode model 95–64', *Form IM95–64/5721*, Orion Research Inc., Cambridge, Mass.

Determination of Ammonia with a Gas-sensing Membrane Electrode

Electrodes of this type are the EIL 8002−8 and the Orion 95−10.

APPARATUS

Electrode; pH meter; magnetic stirrer

Note that the electrode is really a complete cell and needs no external reference electrode. Before assembly, however, it is advisable to check a new glass electrode against a reference electrode, e.g., a saturated calomel electrode, in two buffer solutions. A sluggish or non-responsive electrode should be rejected. Rinse the inside of the plastic body with deionized water and assemble the electrode according to the manufacturer's instructions.

Any magnetic stirrer can be used. Insulate the beaker from the heat produced by the motor by means of a suitable thickness of non-conducting material. A centre-ridge stirrer bar should not be used.

REAGENTS

Water

Pass distilled water through a column (300 mm × 25 mm i.d.) of cation-exchange resin in the hydrogen form. Ammonia contents of 0.01 ppm or less can be obtained. Commercial mixed-bed deionization units can also produce water of adequate quality. Store the water in a stoppered glass vessel in an ammonia-free room.

1 mol l^{-1} Sodium hydroxide

Dissolve 40 ± 1 g of sodium hydroxide pellets in about 500 ml of water, allow to cool and dissolve 20 ± 1 g of ethylenediaminetetraacetic acid disodium salt in the solution. Make up to 1000 ± 50 ml with water. Store in a polyethylene bottle. The solution is stable for at least 2 months.

Standard solutions for ammonia determinations

Solution A

Dissolve 3.141 g (± 0.001 g) of ammonium chloride (analytical reagent grade, dried at 105–110°C) in deionized water. Make to the mark with water in a 1-litre calibrated flask and mix.

1 ml ≡ 1000 µg ammonia

Solution B

Pipette 50 ml of solution A into a 500-ml calibrated flask, make to the mark with water and mix.

1 ml ≡ 100 µg ammonia

Standard solutions for ammoniacal nitrogen determinations

Solution A

Dissolve 3.819 g (± 0.001 g) of ammonium chloride (analytical reagent grade, dried at 105–110°C) in water, make to the mark in a 1-litre calibrated flask and mix.

1 ml ≡ 1000 µg ammoniacal nitrogen

Solution B

Pipette 50 ml of solution A into a 500-ml calibrated flask, make to the mark with water and mix.

1 ml ≡ 100 µg ammoniacal nitrogen

Store the solutions in glass-stoppered glass bottles. The A solutions are stable for at least 20 weeks and the B solutions for 5 weeks. Prepare less concentrated standards by dilution of solution B. These solutions should preferably be prepared freshly each day, but are stable for at least one week if stored in sealed glass flasks in the dark.

SAMPLE COLLECTION

Samples can be collected in glass or polyethylene bottles of any convenient size (minimum 100 ml). Clean new bottles by washing them with distilled or deionized water. Ammonia may be lost from samples at temperatures above 50°C; collect samples at less than 40 °C, using a cooling coil between the bottle and the sampling point if necessary. The sample bottle should be completely filled and immediately stoppered. Analyse the samples as soon as possible; if delay is unavoidable, store the samples in a refrigerator at 5 °C.

CONDITIONING AND STORAGE OF THE ELECTRODE

When first assembled, it is desirable to condition the electrode for 30 min in a solution made up of 2 ml of sodium hydroxide solution and 20 ml of a standard solution with a concentration in the range of interest. When not in use, the electrode can be stored with the tip immersed in a similar solution for 5 days, but for longer periods store in 0.1 mol l^{-1} ammonium chloride solution. Never let the membrane dry out; for this reason close the gap between the top of the beaker and the electrode with sealing film, e.g., Parafilm, to prevent evaporation when the electrode is not in use.

Lifetime of the Electrode Assembly

The membrane should last at least 3 months before replacement is necessary. Very consistent results have been obtained for periods up to 6 months. The lifetimes of the other components are not known, but should be considerable.

CONCENTRATION RANGE AND UNITS

The electrodes have been tested successfully over the range 0.02–17 000 ppm, but below 0.2 ppm the calibration may be biased unless care is taken to prepare pure water for the standard solutions.

$$10^{-3} \text{ mol l}^{-1} \equiv 14.01 \text{ ppm nitrogen} \equiv 17.03 \text{ ppm ammonia}$$
$$5.872 \times 10^{-5} \text{ mol l}^{-1} \equiv 0.823 \text{ ppm nitrogen} \equiv 1 \text{ ppm ammonia}$$
$$7.138 \times 10^{-5} \text{ mol l}^{-1} \equiv 1 \text{ ppm nitrogen} \equiv 1.216 \text{ ppm ammonia}$$

ANALYTICAL PROCEDURE

In a general chemistry laboratory, contamination of samples or the electrode's internal filling solution by ammonia may occur; analyses should be carried out in an atmosphere as free from ammonia as possible.

Step 1 Remove the probe from the solution in which it is immersed, rinse it with deionized water and immerse the tip in a beaker of water stirred by means of a magnetic stirrer bar.

Step 2 Add by pipette 20 ml of a standard solution of concentration s_1 to a 50-ml beaker containing a magnetic stirrer bar. Choose the standard to be at the upper end of the expected concentration range of the samples.

Step 3 Remove the electrode from the water and dry it with a soft tissue. Care is necessary in the region of the membrane.

Step 4 Immerse the probe in the test solution in the 50-ml beaker. Check that no air bubbles are trapped at the membrane's surface.

Step 5 Add 2 ml of 1 mol l^{-1} sodium hydroxide solution to the solution in the beaker. Either a 2-ml pipette with the tip cut off to ensure speedy delivery or an automatic pipette is suitable.

Step 6 When the electrode gives a steady potential, note the reading E_1.

Step 7 Repeat Steps 1–5 with a second standard solution of concentration s_2, not more than half the concentration s_1.

Step 8 When a steady potential has been reached, note the reading E_2.

Step 9 *Either* calculate the calibration slope

$$k = \frac{E_2 - E_1}{\log s_2 - \log s_1} \simeq -58 \text{ mV per tenfold increase in concentration}$$

or

prepare a temporary calibration graph by drawing a straight line through the points $(\log s_1, 0)$ and $(\log s_2, E_2 - E_1)$.

Step 10 Repeat Step 1; if all the solutions in a batch are of similar composition, varying by no more than a factor of 5, this step may be omitted.

Step 11 Pipette 20 ml of sample into a 50-ml beaker containing a magnetic-stirrer bar. The temperature of the sample and the standard solution should not differ by more than 1 °C; it may be convenient to use a water bath to bring the samples to the correct temperature.

Step 12 Repeat Steps 3, 4 and 5.

Step 13 When the potential is steady, note the reading E_x.

Step 14 Calculate the e.m.f. difference, $\Delta = E_x - E_1$.

Step 15 Either read the concentration from the calibration graph or calculate it from

$$c = s_1 \times \text{antilog}\,(\Delta/k)$$

Step 16 If more samples are to be analysed, repeat Steps 10–15 each time. The frequency with which restandardization is required (Steps 1–6) depends on the desired precision of the analysis, but it is recommended that it be done at least twice in a full working day.

Step 17 If no more samples are to be analysed, rinse the electrode with deionized water and store as described above.

Worked example

The e.m.f. decreases by 60 mV for a tenfold increase in ammonia concentration. With a 2-ppm standard the electrode reads 10.0 mV and with the sample solution −5.0 mV. The concentration of the sample is given by:

$$c = 2 \times \text{antilog}\,\frac{-5 - 10}{-60} = 2 \times \text{antilog}\,(0.25) = 3.55 \text{ ppm}$$

Preparation and use of a permanent calibration graph

Provided the conditions discussed on p. 117 are satisfied, a permanent calibration graph may be used and Steps 7–9 omitted from the routine analytical procedure. Prepare the graph by carrying out Steps 1–6 of the analytical procedure then repeating Steps 10–14 with at least four other standard solutions. The graph should be linear at concentrations above 0.5 ppm, but may start to curve in the region 0.5–0.1 ppm because of ammonia in the water used to prepare the standards. If purer water cannot be prepared, extrapolate the linear portion of the graph to the lower concentrations and do not use the curved part.

SOURCES OF ERROR

Effect of temperature

The standard potential of the electrode has a large temperature coefficient, 1.5 mV per degree, and greater than average care must be taken to keep the standard and sample solutions at the same temperature. If the sample is analysed at a higher temperature than the standard solution the result will have a low bias. Short-term fluctuations in temperature can produce significant effects and the electrode should not be situated where it is subject to sudden draughts, e.g. near an air-conditioning unit. The calibration slope will also change with temperature, but only by the usual effect on the factor $RT \ln(10)/F$.

Effect of other substances

Ionic species do not interfere unless they form strong complexes with ammonia, e.g. mercurous ion, or form insoluble hydroxides that can block the membrane, e.g. magnesium. These interferences should be removed by the presence of EDTA, but in impure waters such as those with a total hardness greater than 300ppm, it may be necessary to increase the amount of EDTA added. The following have been tested and shown not to interfere at ammonia concentrations between 0.1 and 1 ppm; the levels quoted should not be interpreted as the maxima permissible.

Cations 160 ppm Na^+; 50 ppm K^+; 120 ppm Ca^{2+}; 50 ppm Mg^{2+}; 2 ppm each of Cr^{3+}, Cu^{2+}, Ni^{2+}, Zn^{2+}; 5000 ppm Fe^{3+}; 2400 ppm Al^{3+}.

Anions 250 ppm CO_3^{2-}; 5500 ppm HCO_3^-; 214 ppm Cl^-; 185 ppm SO_4^{2-}; 4100 ppm NO_2^-; nitrate (18 ppm as N); silicate (20 ppm as SiO_2).

Free chlorine can interfere by forming chloramines; if it is present, modify Step 8 of the analytical procedure by adding 0.5 ml of sodium sulphite (1800 ppm) or sodium thiosulphate (7000 ppm) solution to the sample in the beaker and allowing it to stand for 10 min before immersing the electrode.

Detergents are reported to have deleterious effects on the membrane, but no details have been given. 0.4 ppm of the filming amine octadecylamine produces interferences levels of 0.026 and 0.14 ppm at ammonia concentrations of 0.1 and 1 ppm, respectively.

Table 1 Amine interferences

Amine	Manufacturer[a]	Concentration (ppm)	Interferences (as ammonia) at ammonia concentrations of:	
			0.1 ppm	1 ppm
Cyclohexylamine	O	120	–	0.5
	E	4	0.107	0.08
	E	1	0.023	0.03
Ethylamine	O	10[b]	–	6.63
	O	1[b]	–	0.49
Hydrazine	O	2	0.005	0
	E	4	0.012	0.06
	E	1	0	0
Methanolamine	E	1.05[b]	0.12[c]	–
Methylamine	O	10[b]	–	7.15
	O	1[b]	–	0.54
Morpholine	E	10	0.002	0.03

[a]EIL results, E; Orion results, O.
[b]Concentration expressed as ppm of nitrogen.
[c]In 0.4 ppm N ammonia solution.

Table 2 Precision of analysis of standard solutions with an EIL electrode

Concentration (ppm)		Standard deviations (ppm)[b]		
True	Found[a]	Within-batch σ_w	Between-batch σ_b	Total σ_t
Ammonia determinations (Midgley and Torrance, 1972)				
2.00	2.01	0.03	n.s.	0.05
0.80	0.81	0.025	n.s.	0.028
0.50	0.50	0.010	n.s.	0.012
0.20	0.21	0.007	n.s.	0.008
0.10	0.12	0.005	0	0.005
0.05	0.06	0.004	n.s.	0.006
Ammoniacal nitrogen determination (Beckett and Wilson, 1974)				
3.91	3.80	0.10	0	0.10
0.790	0.774	0.006	0.032[c]	0.032
0.400	0.381	0.007	n.s.	0.009
0.088	0.106	0.012	0	0.012
0.049	0.065	0.003	n.s.	0.004

[a]Mean of 10 results.
[b]Estimates of σ_w and σ_b have 5 and 4 degrees of freedom, respectively; n.s. = not significant at the $P = .05$ level.
[c]Statistically significant at the $P = .01$ level.

Table 3   Precision of analysis of natural waters

Type of sample	Electrode[a]	Ammonia concentration (ppm N)	Within-batch standard deviation (ppm N)	Degrees of freedom
River	E	1.7–2.4	0.042	10
River	E	0.8–0.9	0.026	9
Lake	O^1	1.57	0.008	5
River	O^1	2.19	0.008	5
Well	O^1	2.84	0.010	5
Rain	O^1	3.19	0.017	5
River	O^2	1.002	0.038	6
River	O^2	0.017	0.017	6
River	O^2	0.188	0.007	6
River	O^2	0.132	0.003	6

[a]E, EIL (Beckett and Wilson, 1974); O^1, Orion (Banwart, Tabatabai and Bremner, 1972); O^2, Orion (Thomas and Booth, 1973).

Nitrogenous organic compounds such as urea, amino acids, amides, purines, pyrimidines and hexosamines do not interfere. Volatile amines are the main source of interference and the reported effects are summarized in Table 1.

PRECISION

Precision tests on synthetic standard solutions have been carried out for ammonia and ammoniacal nitrogen determinations. The results are given in Table 2. The precision of analysis of natural waters for ammoniacal nitrogen is given in Table 3. Extracts of animal slurries containing 180–330 ppm ammoniacal nitrogen gave relative within-batch standard deviations of 1% (Byrne and Power, 1974).

ACCURACY

The recovery of spikes (0.1–1 ppm) detected by both EIL and Orion electrodes in boiler feed water, soil extracts, natural waters, sewage and trade effluents has been in the range 99–101%. 5 and 20 ppm spikes were recovered in the range 98–104% from water extracts of animal slurries (Byrne and Power, 1974). Analyses of samples by the electrode have been compared with those found by other methods; the results are shown in Table 4.

RESPONSE TIME

The electrode takes about 2–5 min after addition of the sodium hydroxide solution to reach equilibrium at a higher ammonia concentration than before. The change in the reverse direction may take twice as long, especially at levels below 1 ppm.

Table 4 Comparison of results by different methods

Type of sample	Concentration (ppm N)	Mean difference[a] (%)	No. of samples	Reference method
Boiler feed water	0.2–0.6[b]	+0.01	16	Cation glass electrode[c]
Boiler feed water	0.5–1.6[b]	−0.5	14	Nessler[d]
Untreated river water	1.7–2.4	−2.4	9	Indophenol blue[e]
Treated river water	0.8–1.2	−2.3	9	Indophenol blue[e]
Treated river water	1.4–1.9	−4.6	6	Indophenol blue[e]
River	0.03–0.38	+0.8	15	Indophenol blue[f]
Sewage	0.64–0.84	−4.6	5	Indophenol blue[f]
Potable water, sewage, trade wastes	0.5–25	−8.3	9	Nessler[g]
Potable water, sewage, trade wastes	0.5–25	−0.5	11	Indophenol blue[g]

[a]Difference = 100 × $(E - R)/E$, where E is the concentration indicated by the electrode and R that given by the reference method.
[b]Concentrations as ppm NH_3.
[c]Midgley and Torrance (1972); [d]Mertens, Van den Winkel and Massart (1974); [e]Beckett and Wilson (1974); [f]Thomas and Booth (1973); [g]Evans and Partridge (1974).

OTHER APPLICATIONS

Apart from the direct determination of ammonia, the electrode can be used for the determination of nitrate after reduction (McKenzie and Young, 1975; Mertens et al., 1975; Dewolfs et al., 1975), the determination of total nitrogen after Kjeldahl digestion (q.v.) and the determination of albuminoid nitrogen after distillation from alkaline permanganate (Evans and Partridge, 1974). Applications in clinical chemistry have been summarized by Bailey and Riley (1975) and Gilbert and Clay (1973) have made measurements in sea water.

COMPARISON WITH OTHER METHODS

Ammonium-sensitive glass electrode

The performance characteristics of the two types of electrode are generally very similar, but although the glass electrode is cheaper and does not have the complication of having to be periodically reassembled, it is subject to interference from alkali metal ions and the range of samples for which it is suited is much smaller (Goodfellow and Webber, 1972; Headridge and Long, 1976). Suitable glass electrodes are the EIL GKN 33 and Philips G15K.

Ammonium-selective liquid membrane electrode

Mertens et al. (1974) obtained results of essentially equal precision and accuracy for the determination of 0.25–1.6 ppm ammonia in boiler feed water with both gas-sensing and liquid-membrane electrodes. Dewolfs *et al.* (1975) found that the gas-sensing electrode had a lower limit of detection and was less subject to interferences, particularly from alkali metal ions (although the liquid membrane electrode would be better in this respect than ammonium-sensitive glass electrodes). Liquid membrane electrodes of the neutral carrier type with selectivity for ammonium ions are made by Philips (IS 550-NH$_4$) and Metrohm (EA301NH$_4$).

Absorptiometric methods

The indophenol blue and Nessler methods are capable of lower limits of detection and better precision at low concentrations (less than 0.2 ppm). The agreement between analyses using the different methods is good. The time required for analysis is about the same, but the electrode requires only one simple and stable reagent to be prepared beforehand. The electrode has the great advantage of being able to make direct measurements over a wide concentration range and in many kinds of samples, whereas the absorptiometric methods may need a diluted or distilled sample.

Distillation—titration

Byrne and Power (1974) compared the direct determination with the electrode in extracts of animal slurries with distillation of the extract and titration of the ammonia collected in boric acid solution. In the range 180—330 ppm ammoniacal nitrogen in the extract, the two methods agreed well (correlation coefficient 0.992) and had very similar standard deviations, but the electrode method took less time.

TRACING FAULTS

In addition to the common faults of ion-selective electrode systems, the following should also be noted:

Wildly fluctuating readings

Check that there is enough filling solution in the electrode to cover the internal reference electrode. If the membrane has dried out, the electrode will probably need reassembling with a new one, but try withdrawing the glass electrode slightly to allow fresh solution between it and the membrane.

Complete loss of response

The most likely cause is the appearance of a hole in the membrane; the glass electrode is thus directly affected by the sodium hydroxide added to the sample and a constant high pH (very negative e.m.f.) reading will be recorded. Reassemble the electrode with a new membrane, but first check that the glass electrode is not broken, as a constant e.m.f. can arise from this cause also.

Partial loss of sensitivity and slow response

(1) The gap between the membrane and the glass electrode may be too wide. Carefully tighten the electrode slightly against the membrane. (2) The membrane may be clogged with precipitates or bacterial growth. Carefully wipe the membrane with a tissue, then rinse with dilute (ca. 0.1 mol l^{-1}) hydrochloric acid, or reassemble with a new membrane. (3) The glass and reference electrodes inside the ammonia electrode may not be working properly. Check this by immersing them together in each of two modified buffer solutions, noting the e.m.f. in each buffer and comparing the difference between the two readings with the theoretical value. Both buffers must contain the same concentration of chloride ion — 0.1 mol l^{-1} is suggested — so that the theoretical difference between the e.m.f. values is given by $(58 \pm 2) \Delta$ pH, where Δ pH is the difference in the pH values of the buffers. The most suitable buffers to use are pH 4 (0.05 mol l^{-1} potassium hydrogen phthalate) and pH 6.9 (0.025 mol l^{-1} each of potassium dihydrogen phosphate and disodium hydrogen phosphate), to each of which has been added 0.58 g of sodium chloride per 100 ml of solution. Proprietary pH 4 and pH 7 buffer tablets, powders or solutions should be adequate provided sodium chloride is added as above. The difference between the readings obtained in these two modified buffers should be in the range 165–180 mV. If the e.m.f. difference is smaller, try rejuvenating the glass electrode by leaving it overnight in 0.1 mol l^{-1} hydrochloric acid solution. If the reference electrode is of the silver–silver chloride type, it may in time be stripped of its coating of silver chloride; re-chloridize it according to the instructions on p. 341. Orion electrodes have a solid-state reference electrode and should not be replated. With EIL electrodes, replating is possible, but care should be taken with the latest design, in which the reference and glass electrodes form a single unit. In this case take care not to include the glass electrode in the electrolysis circuit; the connection to the glass electrode is the central pin of the co-axial plug supplied with the electrode, while the reference electrode is connected via the screening.

Bibliography

Bailey, P. L., and M. Riley, 1975, 'Performance characteristics of gas-sensing membrane probes', *Analyst*, **100**, 145.

Banwart, W. L., M. A. Tabatabai, and J. M. Bremner, 1972. 'Determination of ammonium in soil extracts and water samples by an ammonia electrode', *Comm. Soil Sci. Plant Anal.*, **3**, 449.

Beckett, M. J., and A. L. Wilson, 1974, 'The manual determination of ammonia in fresh waters using an ammonia-sensitive membrane electrode', *Water Res.*, **8**, 333.

Byrne, E., and T. Power, 1974, 'Determination of ammonium nitrogen in animal slurries by an ammonia electrode', *Comm. Soil Sci. Plant Anal.*, **5**, 51.

Dewolfs, R., G. Broddin, H. Clysters, and H. Deelstra, 1975, 'Comparison of two electrodes for the determination of ammonia and nitrogen compounds', *Z. Anal. Chem.*, **275**, 337.

Evans, W. H., and B. F. Partridge, 1974, 'Determination of ammonia levels in water and wastewater with an ammonia probe', *Analyst*, **99**, 367.

Gilbert, T. R., and A. M. Clay, 1973, 'Determination of ammonia in aquaria and in sea water using the ammonia electrode', *Anal. Chem.*, **45**, 1757.

Goodfellow, G. I., and H. M. Webber, 1972, 'The determination of ammonia in boiler feed-water with an ammonium-selective glass electrode', *Analyst*, **97**, 95.

Headridge, J. B., and G. D. Long, 1976, 'The determination of mobile nitrogen in steel using an ammonium ion-selective electrode', *Analyst*, **101**, 103.

McKenzie, L. R., and P. N. W. Young, 1975, 'Determination of ammonia-, nitrate- and organic nitrogen in water and waste water with an ammonia gas-sensing electrode', *Analyst*, **100**, 620.

Mertens, J., P. Van den Winkel, and D. L. Massart, 1974, 'On-stream determination of ammonia in boiler feed-water with an ammonium ion selective electrode and an ammonia probe', *Bull. Soc. Chim. Belg.*, **83**, 19.

Mertens, J., P. Van den Winkel, and D. L. Massart, 1975, 'Determination of nitrate in water with an ammonia probe', *Anal. Chem.*, **47**, 522.

Midgley, D., and K. Torrance, 1972, 'The determination of ammonia in condensed steam and boiler feed-water with a potentiometric ammonia probe', *Analyst*, **97**, 626.

Thomas, R. F., and R. L. Booth, 1973, 'Selective electrode measurement of ammonia in water and wastes', *Environ. Sci. Technol.*, **7**, 523.

Determination of Total Nitrogen Using a Gas-sensing Ammonia Electrode after Kjeldahl Digestion

Suitable electrodes are the Electronic Instruments Ltd. 8002–8 and the Orion 95–10. The normal procedure for Kjeldahl digestion may be followed, but there is no need for the distillation step before determination of the ammonia produced.

APPARATUS

Electrode; pH meter; magnetic stirrer; Kjeldahl apparatus

See *Determination of Ammonia*

REAGENTS

Water

Pass distilled water through a mixed-bed deionization unit. Collect enough water in one batch to prepare all the solutions required. Store the water in a stoppered glass vessel in an ammonia-free room.

Concentrated sulphuric acid

Analytical reagent grade, sp. gr. 1.84.

Digestion mixture A (selenium catalyst)

Dissolve 110 g of anhydrous sodium sulphate or 135 g of potassium sulphate (analytical reagent grade) and 1.0 g of selenium dioxide in 600 ml water. Slowly add 200 ml of concentrated sulphuric acid and make up to 1.0 litre with water and mix.

Digestion mixture B (mercury catalyst)

Dissolve 110 g of anhydrous sodium sulphate or 135 g of potassium sulphate (analytical reagent grade) in 600 ml water. Take 200 ml of concentrated sulphuric acid, slowly add 190 ml to the sodium sulphate solution and add the remainder to 20 ml of water in a 100-ml beaker. Dissolve 2 g of red mercuric oxide in the dilute sulphuric acid in the beaker and then add the contents to the sodium sulphate–sulphuric acid solution. Make up to 1.0 litre with water and mix. Keep above 14° C to prevent crystallization.

pH adjustment solution

Dissolve 90.3 ± 0.3 g of sodium hydroxide and 5 ± 0.25 g of disodium EDTA (ethylenediaminetetraacetic acid disodium salt dihydrate) in water and dilute to 1 litre in a calibrated flask. If digestion mixture B is used, it may be necessary to include 30 g of sodium iodide to complex the mercury. Store in a plastic bottle.

Electrode filling solution

The procedure described below produces a concentration of 0.44 g ions l^{-1} of dissolved particles in the final solution. The reference solution inside the electrode should contain the same number of particles. The filling solution supplied by Orion can be used without modification, but that provided with EIL electrode should be replaced by one containing 0.22 mol l^{-1} (11.77 g l^{-1}) of ammonium chloride. If the procedure is varied it may be necessary to change the filling solution of both types of electrode for one containing 11.77 g l^{-1} of ammonium chloride and a suitable concentration of sodium or potassium sulphate (see the section on the osmotic effect).

Stock ammonia solution A

Dissolve 3.819 + 0.001 g of ammonium chloride (analytical reagent grade, dried at 105 °C) in water, make up to the mark in a 1-litre calibrated flask and mix. Store in a glass-stoppered glass bottle. This solution should be stable for at least 20 weeks.

   1 ml $\equiv$ 1 mg nitrogen

Stock ammonia solution B

Pipette 50 ml of stock ammonia solution A into a 500-ml calibrated flask and make up to the mark with water. Store in a glass-stoppered glass bottle. The solution should be stable for at least 5 weeks.

   1 ml $\equiv$ 100 μg nitrogen

Standard nitrogen solutions

These should be prepared fresh, as required, by dilution of stock ammonia solutions A and B. Add 10 ml of the appropriate digestion mixture per 100 ml of the final volume before making up to the mark.

SAMPLE COLLECTION

Samples may be collected in glass or polyethylene bottles. Clean new bottles by washing them with distilled or deionized water. Ammonia and volatile amines may be lost from samples at temperatures above about 50 °C; collect samples at less than 40 °C, using a cooling coil between the bottle and the sampling point if necessary. The sample bottle should be completely filled and immediately stoppered. Analyse the samples as soon as possible; if delay is unavoidable, store the samples in a refrigerator at 5 °C (polyethylene bottles may be kept deep frozen).

CONDITIONING AND STORAGE OF ELECTRODES

When first assembled, the electrode should be conditioned for 30 min in a solution typical of the samples to be analysed: use a standard solution as in Steps 2–6 of the procedure. Between batches of analyses, store the electrode with the tip immersed in 20 g l^{-1} sodium sulphate solution and never allow the membrane to dry out.

CONCENTRATION RANGE AND UNITS

Measurements in the range 0.1–800 ppm nitrogen have been made with the electrode.

10^{-3} mol l^{-1} ammonia $\equiv$ 14.01 ppm nitrogen

1 ppm nitrogen $\equiv 7.138 \times 10^{-5}$ mol l^{-1} ammonia

ANALYTICAL PROCEDURE

Kjeldahl Digestion

Add a portion, volume v_R ml, of sample solution containing 0.1–1.0 mg of nitrogen to a suitably sized Kjeldahl flask or boiling tube. Add 5 ml of digestion mixture A or B and two or three anti-bump granules and heat gently on the digestion rack until all the water has boiled off. Reflux the residue for 1 h, allow it to cool and add 5 ml of water. Transfer the digest to a 50-ml calibrated flask, using three 5-ml rinses with water and make up to the mark with water.

Determination of the ammonia in the digest

   Step 1 Remove the probe from the solution in which it is immersed, rinse it with deionized water and immerse the tip in a beaker of sodium sulphate solution (20 g l^{-1}) stirred by a magnetic stirrer bar.

Step 2 Add by pipette 10 ml of a standard solution of concentration s_1 ppm to a 150-ml beaker containing a magnetic stirrer bar. Choose the standard to be at the upper end of the expected concentration range of the samples.

Step 3 Add 80 ml of water to the solution in the beaker.

Step 4 Remove the electrode from the sodium sulphate solution and dry it with a paper tissue.

Step 5 Immerse the tip of the electrode in the solution in the beaker. Check that no bubbles are trapped at the surface of the electrode.

Step 6 Add 10 ml of pH adjustment solution.

Step 7 When the electrode gives a steady reading, note the potential E_1.

Step 8 Repeat Step 1.

Step 9 With a second standard solution of concentration s_2 ppm chosen to be near the lower end of the expected concentration range, repeat Step 2.

Step 10 Repeat Steps 3–6.

Step 11 When the electrode gives a steady reading, note the potential E_2.

Step 12 *Either* calculate the calibration slope

$$k = \frac{E_2 - E_1}{\log s_2 - \log s_1} \simeq -58 \text{ mV per tenfold increase in concentration}$$

or

prepare a temporary calibration graph by drawing a straight line through the points $[\log (s_1/20), 0]$ and $[\log (s_2/20), E_2 - E_1)]$. Note that this gives the nitrogen content in milligrams of the digest on the x-axis.

Step 13 Repeat Step 1. If all the solutions in the batch vary by no more than a factor of 5 in concentration this step may be omitted.

Step 14 Pipette 10 ml of the digest solution into a 150-ml beaker containing a magnetic stirrer bar.

Step 15 Repeat Steps 3–6.

Step 16 When the electrode gives a steady potential, note the reading E_x.

Step 17 Calculate the potential difference $\Delta = E_x - E_1$.

Step 18 Either read the nitrogen content of the digest, c_D mg, off the calibration graph, or calculate it from

$$c_D = \frac{s_1}{20} \times \text{antilog} (\Delta/k) \text{ mg, where } s_1 \text{ is in ppm.}$$

Step 19 Calculate the nitrogen concentration in the sample from

$$c = 1000 c_D / v_R \text{ ppm}$$

Step 20 If more samples are to be analysed, repeat Steps 13–19 each time. The frequency with which recalibration is necessary depends on the precision desired, but Steps 1–7 should be done at least twice in a full working day.

Step 21 If no more samples are to be analysed, store the electrode with its tip immersed in sodium sulphate solution (20 g l^{-1}).

Worked example

The e.m.f. of the electrode decreases by 58 mV for a tenfold increase in nitrogen concentration. With a 10 ppm standard solution the electrode reads 2.4 mV and with a digest of a 100-ml portion of sample it reads 9.5 mV. The digest therefore contains

$$c_D = \frac{10}{20} \times \text{antilog} \frac{9.5 - 2.4}{-58} = 0.377 \text{ mg}$$

and the concentration of the sample is 1000 × 0.377/100 = 3.77 ppm nitrogen.

Preparation and use of a permanent calibration graph

If the conditions discussed on p. 117 are satisfied, a permanent calibration graph may be used and Steps 8–12 omitted from the analytical procedure. Prepare the graph by carrying out Steps 1–7 of the analytical procedure and then repeating Steps 13–17 with at least four other, more dilute, standard solutions. The graph should be linear, but may show curvature below about the 0.25 mg level (5 ppm standard solution) because of ammonia in the water used for diluting the samples and preparing the standards. As almost the same amount of water is added to both standard and sample solutions, no significant bias will be present and the curved calibration can be used directly (cf. the determination of ammonia, where this is not so).

Note that in the above procedure a 20 ppm standard solution should give rise to the same e.m.f. as 1 mg of nitrogen in the v_R ml of sample taken for digestion.

SOURCES OF ERROR

Temperature

See the method for ammonia (p. 283). The standard and sample solutions should not differ in temperature by more than 1 °C.

Effect of other substances

See the method for ammonia. The amine interferences in that method are not relevant here, as the amines will be converted to ammonia during digestion. Note that the mercury added in digestion mixture B will interfere by forming ammonia complexes unless the pH adjustment solution contains sufficient complexing agent. If the EDTA in the normal pH adjustment solution is inadequate, include 30 g l^{-1} sodium iodide also.

Absorption of ammonia

In a general chemistry laboratory, contamination of the samples or the internal filling solution of the electrode may occur; analyses should be carried out in an atmosphere as free from ammonia as possible.

Osmotic effect

If the total concentration of dissolved particles in the final test solution differs from that in the electrode filling solution, an osmotic pressure difference arises and water will be transported across the membrane as well as ammonia. If the difference is big enough, the e.m.f. may drift for over an hour (Bailey and Riley, 1975) and accurate measurements become impossible. Any major variations in the composition of the digestion mixture, the pH adjustment solution or the electrode filling solution or in the volumes of sample or reagents taken must be compensated by a change in one of the other solutions or in one of the other volumes. The relationship between the parameters is given by

$$\frac{1}{v_t}\left\{\frac{v_d}{v_D}\left[Rv_R - 3.68v_A + 3 \cdot v_A \cdot \frac{w_s}{m}\right] + v_b[0.05w_b + 0.0027w_e + 0.0133w_I]\right\} = 0.03738w_N + 3\frac{w_G}{m} \quad (1)$$

The meaning of the symbols is as follows: v_A = volume of digestion mixture taken; v_R = volume of sample taken for digestion; v_D = volume to which digest is diluted; v_d = volume of digest taken for analysis (Step 14); v_b = volume of pH adjustment solution added (Step 6); v_w = volume of water added (Step 3); $v_t = v_d + v_b + v_w$; w_s = weight (in grams) of sodium or potassium sulphate per litre of digestion mixture; w_b = weight (in grams) of sodium hydroxide per litre of pH adjustment solution; w_e = weight (in grams) of disodium EDTA dihydrate per litre of pH adjustment solution, w_I = weight (in grams) of sodium iodide per litre of pH adjustment solution; w_N = weight (in grams) of ammonium chloride per litre of electrode filling solution; w_G = weight (in grams) of sodium or potassium sulphate per litre of electrode filling solution; the molecular weight, m = 142.05 for sodium sulphate or 174.26 for potassium sulphate. $R = \Sigma v_i C_i$ is the number of particles per litre of sample, where C_i is the molar concentration of the ith solute and v_i is the number of moles of particles formed on dissolution of one mole of solute.

A further condition that must be taken into account is the need for a sufficient excess of sodium hydroxide to give an effectively constant pH in the final solution.

$$\frac{1}{v_t}\left\{0.05v_b \cdot w_b - 7.36\frac{v_A \cdot v_d}{v_D} - 0.054v_b \cdot w_e\right\} \simeq 0.1 \quad (2)$$

Variations of 10–20% in the total number of dissolved particles are unlikely to produce a detectable effect and therefore the dissolved content of the sample can be neglected in many cases, particularly with freshwater samples, i.e. R is approximated to zero. On the other hand, a major change in the procedure, such as omitting the dilution of the digest, must be allowed for by changing the electrode filling solution (Stevens, 1976).

PRECISION

Relative standard deviations of 4% have been reported (Buckee, 1974; Stevens, 1976) for samples as different as river water (0.43 ppm N), wort (781 ppm N) and beer (473 ppm N). Bremner and Tabatabai (1972) found relative standard deviations of only 0.4–0.9% in soil digests. Deschreider and Meaux (1973), using a known addition calibration, reported a mean relative standard deviation of 1.2% in the determination of nitrogen in food products.

ACCURACY

McKenzie and Young (1975) obtained 90–105% recovery of 2 ppm spikes in samples varying from sewage effluent (3.4 ppm) to estuarine water (0.1 ppm). Bremner and Tabatabai (1972) recovered 99–100% of the nitrogen added to soil digests as ammonium chloride. Very close linear correlations between results obtained with electrodes and those from distillation–titration and distillation–Nesslerization have been reported by Bremner and Tabatabai (1972) and Stevens (1976), respectively. Buckee (1974) analysed pure acetanilide and glycine as reference materials and found relative errors of -0.3 and -0.1%, respectively.

COMPARISON WITH OTHER METHODS

The elimination of the distillation step means that results are obtained much quicker than in conventional Kjeldahl analysis. The agreement between results from the electrode and distillation–titration and distillation–Nesslerization has already been noted. The relative standard deviation for distillation–Nesslerization in the analysis of river water containing 0.4–0.5 ppm nitrogen was 11% compared with only 4% for the electrode method (Stevens, 1976).

AUTOMATION OF THE METHOD

The time-consuming nature of Kjeldahl analyses makes their automation very desirable. Buckee (1974) and Stevens (1976) have developed semi-automatic procedures in which the samples are digested, made up to a fixed volume and placed in an automatic sampler, from which the sample is pumped to an electrode fitted with a flow-through end-cap. Rates of 10–60 samples per hour were achieved. It is convenient if the dilution of the sample (Step 3 of the manual procedure) can be omitted. Buckee pumped a relatively dilute pH adjustment solution at a much higher rate than his samples, the resulting osmolality being the same as in the manual procedure. Stevens preferred to add a very concentrated pH adjustment solution, thereby diluting the sample as little as possible, as was desirable with freshwater samples containing very little nitrogen. In this case, the ionic content of the electrode's internal filling solution was increased by the inclusion of potassium sulphate in order to avoid the osmotic effect. In practice, the degree of dilution will be set by the performance of the pump and the required concentration of sodium hydroxide in the pH adjustment solution can be calculated from equation (1), where v_d, v_b, v_w and v_t now refer to the respective flow rates.

PROCEDURAL VARIATIONS

The details of the Kjeldahl digestion are largely irrelevant to the subsequent analysis of the digest and most of the variations developed in individual laboratories can be accommodated. Thus, sodium sulphate may be replaced by potassium sulphate, mercuric oxide by mercuric sulphate, selenium by selenium dioxide and the concentrations may also be changed. Equation (1) should be used to calculate the value of w_b or w_G needed to avoid the osmotic effect. If it is preferred to add concentrated sulphuric acid and solid catalyst mixture directly to the sample the following equation should be used instead

$$\frac{1}{v_t}\left\{\frac{v_d}{v_D}\left[R \cdot v_R - 18.4v_C + 3000\frac{w'_s}{m}\right] + v_b[0.05w_b + 0.0027w_e + 0.0133w_1]\right\} = 0.03738w_N + 3\frac{w_G}{m} \quad (3)$$

where v_C is the volume of concentrated sulphuric acid and w'_s the weight (in grams) of sodium or potassium sulphate added to the sample and the other symbols are as before.

TRACING FAULTS

See *Determination of Ammonia*, but note that drifting potentials and slow response times may also be caused by the osmotic effect.

Bibliography

Bailey, P. L., and M. Riley, 1975, 'Performance characteristics of gas-sensing membrane probes', *Analyst*, **100**, 145.

Bremner, J. M., and M. A. Tabatabai, 1972, 'Use of an ammonia electrode for determination of ammonium in Kjeldahl analysis of soils', *Comm. Soil. Sci. Plant Anal.*, **3**, 159.

Buckee, G. K., 1974, 'Estimation of nitrogen with an ammonia probe', *J. Inst. Brew.*, **80**, 291.

Deschreider, A. R., and R. Meaux, 1973, 'Utilisation d'une électrode ionique spécifique pour le dosage de l'azote par la méthode de Kjeldahl', *Analusis*, **2**, 442.

McKenzie, L. R., and P. N. W. Young, 1975, 'Determination of ammonia-, nitrate- and organic nitrogen in water and waste water with an ammonia gas-sensing electrode', *Analyst*, **100**, 620.

Stevens, R. J., 1976, 'Semi-automated ammonia probe determination of Kjeldahl nitrogen in freshwaters', *Water Res.*, **10**, 171.

Determination of Free and Total Cyanide Using a Cyanide-selective Electrode

Cyanide electrodes have membranes consisting of either silver iodide or, more commonly, a mixture of silver iodide and silver sulphide. In the presence of cyanide, the following reaction takes place

$$AgI + 2\ CN^- \rightleftharpoons Ag(CN)_2^- + I^-$$

in which iodide from the membrane is released into the solution. The potential of the electrode, which responds to the iodide concentration, can then be written in terms of this reaction

$$E = E^\circ - k \log a_{I^-}$$

$$= E^{\circ\prime} - k \log \frac{(a_{CN^-})^2}{a_{Ag(CN)_2^-}}$$

As the equilibrium (above) is far to the right, the concentration of the silver cyanide complex in the vicinity of the electrode is effectively half that of the initial cyanide concentration in the bulk of the solution. A theoretical treatment of this electrode has been given by Koryta (1975) and for most practical purposes the electrode can be considered to respond to cyanide ion according to

$$E = \text{constant} - k \log a_{CN^-}$$

It can be seen that the membrane will gradually be depleted of silver iodide and its lifetime will be dependent on the concentrations of the cyanide solutions to which it has been exposed. Orion (1974) recommend that their 94–06 electrode should only be used intermittently in solutions whose cyanide content is greater than 260 ppm, since at these levels there is rapid depletion of the sensing element. The estimated lifetime of this electrode is 200 h in 260 ppm cyanide solution and approximately 10^3 h in 0.26 ppm solution. The lifetime of the electrode will therefore be extended if concentrated samples are diluted to lie in the range 0.3–30 ppm before analysis.

Electrodes suitable for these analyses are the Orion 94–06A, Philips IS 550-CN,

Simac CN/1C, Leeds & Northrup 117404, Activion 003 15 005, EDT Supplies EE-CN, Tacussel PCN 2 and Radiometer F1042CN and F1542CN. Harzdorf (1976) prepared a silver—silver iodide electrode by iodizing a silver rod and demonstrated its satisfactory use for the determination of cyanide.

APPARATUS

Cyanide electrode; reference electrode; pH meter; magnetic stirrer; water-bath; hot plate; safety pipette or pipette pump.

REAGENTS

Water

Distilled water which has been passed through a mixed bed deionization unit such that its specific conductivity is $<0.2\ \mu S\ cm^{-1}$ is suitable for the preparation of standard and reagent solutions.

EDTA solution

Dissolve 7.44 g of analytical reagent grade disodium ethylenediaminetetraacetic acid dihydrate in approximately 900 ml of water and adjust to pH 4 by addition of acetic acid (analytical reagent grade), following the pH change on addition of acid with a pH electrode. Make up to 1 litre with water and store in a polyethylene bottle.

Phosphate buffer and ionic strength adjustment solution

Dissolve 35.5 g of disodium hydrogen orthophosphate and 101.1 g of potassium nitrate (both analytical reagent grade) in approximately 500 ml of water and add either 111 ml of 1N sodium hydroxide solution or 4.4 g of analytical reagent grade sodium hydroxide. Make up to 1 litre with water and store in a polyethylene bottle.

Standard cyanide solution A (1000 ppm)

Dissolve exactly 2.505 g of analytical reagent grade potassium cyanide in 500 ml of water and add 10 ml of 1N sodium hydroxide solution. Make up to 1 litre in a calibrated flask. Store this solution in a polyethylene bottle, where it should be stable for at least one week.

1 ml ≡ 1 mg cyanide

Additional standard solutions can be prepared by the appropriate dilution of standard solution A with water. When standard solutions containing less than 100 ppm cyanide are prepared, add sodium hydroxide solution in the proportion of

10 ml of 0.1N sodium hydroxide per litre of standard. For the most precise results, these additional standard solutions should be prepared daily. Care must be taken that the polyethylene bottles used to store these solutions are free from traces of any heavy metals. Bottles can be conditioned by washing carefully with deionized water, then allowing them to stand full of EDTA solution ($\sim 10^{-3}$ mol l^{-1}, pH 10) for a few days. Finally, they should be washed thoroughly with deionized water and used only for storing standard cyanide solutions.

METHOD A. TOTAL CYANIDE

The following procedure is for samples containing heavy metals, such as nickel and copper, which form cyanide complexes. In order to determine the total cyanide present, the complexed cyanide is liberated by displacement by EDTA. The procedure is intended for heavy-metal concentrations in the sample of the order of $10^{-4} - 10^{-3}$ mol l^{-1} and successful analyses have been reported in the presence of cadmium, zinc, nickel(II), copper(II) and chromium(III).

Step 1A Pipette 50 ml of a standard solution, whose concentration s_1 is representative of the upper end of the sample range, into a 100-ml beaker containing a magnetic stirrer bar. It is essential to use a hand-operated pipette pump to avoid ingestion of cyanide solution.

Step 2A Add 5 ml of EDTA solution to the beaker.

Step 3A Stand the beaker in a fume hood on a hot plate whose temperature control is set to heat the contents to 50 °C. Heat for exactly 5 min, remove from the hotplate and immediately add 5 ml of phospate buffer. Allow the contents of the beaker to cool to calibration temperature before continuing to Step 4A. The most convenient way to do this is to stand the beaker for approximately 20 min in a water bath set at the calibration temperature.

Step 4A Place the electrodes in a suitable holder, rinse them with water and then remove surplus water with a tissue.

Step 5A Remove the beaker from the water bath and place it on a magnetic stirrer. Immerse the electrodes in the solution and stir at a moderate rate (use the same speed for all solutions analysed).

Step 6A When a steady reading has been reached, note the e.m.f., E_1.

Step 7A Repeat Steps 1A—6A using a second standard whose concentration, s_2, is representative of the lower part of the expected sample range. Note the steady e.m.f., E_2.

Step 8A *Either* calculate the calibration slope

$$k = \frac{E_1 - E_2}{\log s_1 - \log s_2} \simeq -58 \text{ mV per tenfold increase in concentration}$$

or prepare a temporary calibration graph by drawing a straight line through the points ($\log s_2, E_2 - E_1$).

Step 9A Pipette 50 ml of sample solution into a 100-ml beaker containing a magnetic stirrer bar.

Step 10A Repeat Steps 2A and 3A, then remove the electrodes from the solution in which they are standing, rinse them with water and remove the surplus water with a tissue.

Step 11A Repeat Step 5A.

Step 12A When a steady reading has been reached, note the e.m.f., E_x.

Step 13A Calculate the difference, $\Delta = (E_x - E_1)$ mV

Step 14A Calculate the concentration of cyanide in the sample, c_x, from

$$c_x = s_1 \times \text{antilog}\left[\frac{\Delta}{k}\right]$$

or read it directly off the calibration graph.

Step 15A Repeat Steps 9A–14A for all samples and include a standard solution with every batch of 6–8 samples.

METHOD B. FREE CYANIDE

The following procedure does not include the EDTA complexation step and is intended for samples in which metal cyanide complexes are absent or are fully dissociated under the analytical conditions, e.g., zinc or cadium cyanide.

Step 1B As Step 1A.

Step 2B Add 5 ml of phospate buffer to the beaker.

Step 3B As Step 4A.

Step 4B Place the beaker on a magnetic stirrer, immerse the electrodes in the solution and stir at a constant rate.

Step 5B When a steady reading has been obtained, note the potential E_1.

Step 6B Repeat Steps 1B–4B with a second standard solution of concentration s_2.

Step 7B When a steady reading has been obtained, note the potential, E_2.

Step 8B As Step 8A.

Step 9B Pipette 50 ml of sample solution into a 100-ml beaker containing a magnetic stirrer bar.

Step 10B Add 5 ml of phosphate buffer to the beaker.

Step 11B Remove the electrodes from the solution in which they are standing, rinse them with water, then remove surplus water with a tissue.

Step 12B Repeat Step 4B.

Step 13B As Steps 12A–13A.

Step 14B As Steps 14A–15A.

Use and preparation of a permanent calibration graph

Under some conditions (p. 117) the second standard solution can be omitted (Steps 7A and 8A or 6B–8B) and a permanent calibration graph used. The latter can be prepared by carrying out Steps 1A–6A and repeating Steps 9A–13A with at least four other standard solutions for the total cyanide (Steps 1B–5B and 9B–13B for free cyanide) and plotting the graph as described on p. 118. The graphs should be linear at concentrations above 0.3 ppm.

CYANIDE ANTIDOTE

The following antidote is recommended (Muir, 1971) for those cases in which the victim is conscious following ingestion of cyanide solution:

> Two solutions should be made up and left ready for *immediate* use:
> **A.** 158 g ferrous sulphate ($FeSO_4 \cdot 7 H_2O$) and 3 g citric acid in a litre of distilled water (the solution *must* be inspected regularly and be replaced if any deterioration occurs).
> **B.** 60 g anhydrous sodium carbonate (Na_2CO_3) dissolved in a litre of distilled water.
> 50 ml of solution A is placed in a 170-ml (6-oz) wide-necked bottle with a plastic cap and labelled clearly 'CYANIDE ANTIDOTE A'. 50 ml of solution B is similarly bottled and labelled 'CYANIDE ANTIDOTE B'.
> Both bottles should bear the legend *'Mix the whole contents of bottles* 'A' *and* 'B' *and swallow the mixture'*.

The ferrous hydroxide suspension that is swallowed is likely to induce vomiting while at the same time forming insoluble non-toxic iron complexes with the cyanide.

It must be stressed that a doctor, who can administer an approved injection, should be summoned in cases of cyanide poisoning. Kits for the intravenous treatment of cyanide poisoning by doctors only are available, and one should be at hand in any laboratory where cyanides are handled regularly.

SOURCES OF ERROR

Effect of temperature

The temperature at which the samples are analysed should be within 1 °C of the temperature at which the calibration graph was determined. Additional caution is required in analysing samples following a heating stage (Step 3A) and it is useful to set the temperature of the water-bath to that at which the calibration was determined.

Effect of other substances

Ions that interfere with the electrode are those that form sparingly soluble silver salts. Their interference effects follow the inverse order of their solubility products, i.e., sulphide > iodide > bromide > thiocyanate > chloride. Orion (1974) state that there is no effect on the electrode or loss of accuracy of the measurement if the chloride and bromide concentrations are equal to or less than 10^6 and 5×10^3 times that of the cyanide, respectively. Similarly, the iodide concentration must be less than one-tenth of the cyanide concentration. Tóth and Pungor (1970) reported selectivity coefficients, K_{CNX} for the Radelkis electrode of $10^{-5} - 10^{-6}$ (chloride), $10^{-3} - 10^{-4}$ (bromide) and 1 (iodide).

Sulphide must be removed from all samples before analysis by precipitation as an insoluble metal sulphide. Riseman (1972) recommended the addition of an excess of lead nitrate and Clysters, Adams and Verbeek (1976) found that a tenfold excess was necessary at a sulphide level of 2×10^{-5} mol l^{-1} though even then the lead sulphide precipitate had to be removed by filtration. Sekerka and Lechner (1976) reported that the addition of a slight excess of lead did not effectively remove the sulphide interference and much more satisfactory results were obtained after addition of an excess of bismuth nitrate. The latter could be present in up to a thousandfold excess without affecting the measurement of cyanide.

Presence of metal cyanides

In many cases, the cyanide present in industrial effluent is accompanied by metal ions in the form of cyanide complexes, in which case it is necessary to liberate the complexed cyanide before direct potentiometric determination of total cyanide can be made. The most widely reported decomplexation technique involves exchange of EDTA for the complexed cyanide (Frant, Ross and Riseman, 1972) though this is known to be ineffective for cyanides of iron(II), iron(III) and cobalt (Clysters et al., 1976). Samples containing those metals are best treated in a specially constructed apparatus (Clysters et al., 1976) by an acidification procedure, which liberates hydrogen cyanide which is then carried over in a gas stream and trapped in 0.1 mol l^{-1} sodium hydroxide solution whose cyanide content is then measured. Alternatively, the method used by Sekerka and Lechner (1976) can be followed (see below).

In the analytical procedure, a loss of cyanide can occur during the heating and decomplexation stage (Step 3A) due to inadvertent evolution of hydrogen cyanide. Clysters et al. (1976) analysed 100 and 50 ppb cyanide solutions after acidification, EDTA addition and heating of the solutions at 50 °C for 5 min. They found the cyanide contents had dropped to 74 and 41 ppb, respectively. On the other hand, Frant et al. (1972) observed no loss in cyanide even though the heating stage was carried out in an open beaker. It is recommended in the analytical procedure that both the standard and sample solutions are treated identically with respect to time and temperature during Step 3A and this should minimize the errors brought about by loss of hydrogen cyanide.

In an alternative and more efficient decomplexation technique (Sekerka and Lechner, 1976) the metal cyanides are decomposed by irradiation with ultra-violet light (Goulden, Afghan and Brooksbank, 1972). The liberated metal ions are precipitated as their sulphides by sodium sulphide added before the irradiation stage. The excess of sulphide remaining at the end of this procedure is removed by the addition of an excess of bismuth nitrate. By contrast with the acid EDTA treatment, Sekerka and Lechner (1976) reported excellent results for solutions containing iron and colbalt cyanides. If the necessary equipment is available the following could be substituted for Steps 1A–3A of the analytical procedure:

Step 1A' Add by pipette 50 ml of either sample or standard solution to a fused silica tube. Add 0.6 ml of 50% orthophosphoric acid and 0.25 ml of 10^{-2} mol l^{-1} sodium sulphide solution.
Step 2A' Irradiate for 2 minutes with a 400 watt mercury lamp.
Step 3A' Add 1.25 ml of 10^{-2} mol l^{-1} bismuth nitrate solution, mix thoroughly, then add 5 ml of phosphate buffer solution.

PRECISION

Tóth and Pungor (1970) reported the reproducibility for the direct determination of cyanide by a silver iodide-based electrode as being within ±0.05 pa_{CN} (pa_{CN} = $-\log a_{CN}$) for the range $10^{-1}-10^{-5}$ mol l^{-1} potassium cyanide. At concentrations of 26, 2.6 and 0.26 ppm cyanide, the reproducibilities were ± 1.3, ± 0.13, and ± 0.03 ppm, respectively. Sekerka and Lechner (1976), using an Orion 94–06A cyanide electrode, quoted standard deviations of 2.7, 2.7 and 2.4 ppb for synthetic lake water samples containing 100, 200 and 500 ppb, respectively, but it must be noted that these workers used an analytical procedure which differed in detail from the one quoted above. Orion (1974) claim a precision of ± 2% with frequent calibration, if consideration is given to control of temperature, and stirring rate.

ACCURACY

György et al. (1969) determined the cyanide contents of nine fruit brandies by direct potentiometry using a silver iodide-based electrode and compared the results with those obtained by a standard argentimetric titration. The percentage differences between the cyanide contents as estimated by each technique varied from 0 to 44% for cyanide contents in the range 2–40 ppm. Sekerka and Lechner (1976) analysed a series of standard cyanide solutions by both the electrode technique and a conventional colorimetric procedure (ASTM, 1970). At 200 and 500 ppb the agreement was better than 2%.

RESPONSE TIME

Tóth and Pungor (1970) reported that the Radelkis cyanide electrode reached equilibrium in stirred solution within approximately 20 sec and even in dilute solutions (0.3 ppm) the response time did not exceed one minute. Response times for an Orion 94–06A electrode varied from several seconds at 26 ppm to several minutes at 26 ppb (Sekerka and Lechner, 1976). These workers recommended that an electrode used for the determination of low levels of cyanide should be kept exclusively for this purpose and should not be exposed to concentrations greater than 260 ppb.

COMPARISON WITH OTHER METHODS

Cyanide can be determined by titration with silver nitrate, using rhodanine as an indicator, down to concentrations of about 1 ppm, but at low concentrations the colour change at the end-point is difficult to detect. Standard colorimetric methods, e.g. chloramine-T, frequently involve a distillation of hydrogen cyanide from acidified sample solution and are therefore considerably more lengthy than the potentiometric technique.

Both the cyanide and the silver sulphide electrodes can be used as end-point detectors in cyanide titrations with silver nitrate as titrant. This procedure is suitable down to cyanide levels of approximately 1 ppm, but at the lower concentrations greater accuracy is obtained with the silver sulphide electrode.

Bibliography

ASTM, 1970 *ASTM Standards* 2B, D2306, American Society for Testing and Materials, Philadelphia.

Clysters, H., F. Adams and F. Verbeek, 1976, 'Potentiometric determinations with the silver sulphide membrane electrode. Part I. Determination of cyanide', *Anal. Chim. Acta*, 83, 27.

Frant, M. S., J. W. Ross, Jr., and J. H. Riseman, 1972, 'Electrode indicator technique for measuring low levels of cyanide', *Anal. Chem.*, 44, 2227.

Goulden, P. D., B. K. Afghan, and P. Brooksbank, 1972, 'Determination of nanogram quantities of simple and complex cyanides in water', *Anal. Chem.*, 44, 1845.

György, B., L. André, L. Stehli, and E. Pungor, 1969, 'Direct potentiometric determination of cyanide in biological systems', *Anal. Chim. Acta,* 46, 318.

Harzdorf, C., 1976, 'Silver/silver iodide electrodes of the second kind as sensors for cyanide', *Anal. Chim. Acta*, 86, 103.

Koryta, J., 1975, *Ion-selective Electrodes*, Cambridge University Press, Cambridge.

Muir, G. D., (ed.), 1971, *Hazards in the Chemical Laboratory*, The Royal Institute of Chemistry, London.

Orion, 1974, *Instruction Manual for Cyanide Activity Electrode, Model 94–06*, Orion Research Inc., Cambridge, Mass.

Riseman, J. H., 1972, 'Electrode techniques for measuring cyanide in waste waters', *American Laboratory*, 4, 63.

Sekerka, I., and J. F. Lechner, 1976, 'Potentiometric determination of low levels of simple and total cyanides', *Water Res.*, 10, 479.

Tóth, K., and E. Pungor, 1970, 'Determination of cyanides with ion-selective membrane electrodes', *Anal. Chim. Acta*, 51, 221.

Determination of Low Levels of Total Cyanide Using a Silver Sulphide Electrode

This method for the determination of low levels of cyanide is based on the silver indicator technique (Frant, Ross and Riseman, 1972) in which a small volume of $Ag(CN)_2^-$ solution is added to the sample prior to the addition of known concentrations of cyanide solution. The silver sulphide electrode responds to the activity of silver ions in solution and, assuming constant ionic strength throughout the analysis, this is related to the cyanide concentration through the overall formation constant, β_2

$$[Ag^+] = \frac{[Ag(CN)_2^-]}{\beta_2 [CN^-]^2} \tag{1}$$

Neglecting any dilution of the sample brought about by the addition of the $Ag(CN)_2^-$ solution, let the initial concentration of $Ag(CN)_2^-$ in a volume V_S of sample solution be A. If the original concentration of cyanide in the sample is c_S and a volume V_A of cyanide solution of concentration c_A is added, then the concentration of silver is

$$[Ag^+] = \frac{AV_S}{\beta_2 \left(\dfrac{V_S c_S + V_A c_A}{V_S + V_A}\right)^2 (V_S + V_A)} \tag{2}$$

The argentodicyanide ion is an extremely strong complex; therefore its concentration can be considered to be constant over a wide concentration range of added cyanide ion. Consequently, the only changes in A that need be considered are those brought about by dilution. The potential of the silver sulphide can be written in terms of expression (2) as

$$E = E^\circ + k \log \left[\frac{AV_S(V_S + V_A)}{\beta_2 (V_S c_S + V_A c_A)^2} \right] \tag{3}$$

where k is the calibration slope for the silver electrode (~59 mV per tenfold increase in concentration). Rearranging equation 3 and taking antilogarithms

$$(V_S + V_A)^{1/2} \text{ antilog} \left[\frac{E^\circ - E}{2k} \right] = \frac{\beta_2^{1/2}(V_S c_S + V_A c_A)}{A^{1/2} V_S^{1/2}} \quad (4)$$

The volume term $(V_S + V_A)^{1/2}$ on the left-hand side of equation 4 can be considered to be constant for volume additions $V_A \leqslant 5\% \, V_S$. Collecting constant terms in equation 4 and rearranging gives,

$$\text{antilog} \left[\frac{-E}{2k} \right] = M \left(c_S \frac{V_A c_A}{V_S} \right) \quad (5)$$

where M is a constant which includes the E° term of the previous equation. When the left-hand side of equation 5 is plotted on the y-axis versus the volume added, V_A, on the x-axis, the resulting plot is linear with an intercept $-V_e$ on the x-axis, where V_e can be regarded as the volume of added cyanide required to produce the initial concentration c_S. The concentration c_S is then

$$c_S = \frac{c_A V_e}{V_S}$$

In order that the correct conditions are maintained, the concentration c_A of the added cyanide solution used in the additions should be such that the final concentration of cyanide in solution, after all the additions have been made, should be about twice the initial concentration in the sample.

Suitable electrodes having silver sulphide membranes are the Orion 94-16, Beckman 39610, Leeds & Northrup 117408, Metrohm EA306S/Ag, Philips IS 550S, Activion 003 15 004, Coleman 3-805, EDT Supplies EE-S Tacussel PS 3, Simac S/1C and Radiometer F3001.

APPARATUS

Silver sulphide electrode; double-junction reference electrode; pH meter; magnetic stirrer; hotplate, water bath; safety pipette or pipette pump

The outer compartment of the double-junction reference electrode should be fitted with a solution containing 0.1 mol l^{-1} potassium nitrate and 10^{-3} mol l^{-1} potassium hydroxide.

REAGENTS

Water

Distilled water which has passed through a mixed-bed deionization unit such that its specific conductivity is 0.2 μS cm^{-1} is suitable for preparing standard and reagent solutions.

Argentocyanide indicator $(Ag(CN)_2^-)$ Solution

Prepare a 10^{-3} mol l^{-1} cyanide solution by dissolving 0.745 g of analytical reagent grade potassium cyanide in water and making up to 1 litre. Prepare a 10^{-2} mol l^{-1} silver nitrate solution by dissolving 0.425 g of analytical reagent silver nitrate in water and making up to 250 ml.

Titrate a 50-ml portion of the cyanide solution with the silver nitrate solution, following the course of the titration with the same electrodes as used in the analytical procedure. Plot the titration curve and note the millivolt reading corresponding to 98–99% of the volume required for the first end-point, which indicates the formation of $Ag(CN)_2^-$. Titrate the remaining bulk of the cyanide solution to the 98–99% value of e.m.f. This solution should be stable for several months when stored in a polyethylene bottle.

Standard cyanide solutions

Prepare the standard solutions as described in the previous method.

Phosphate buffer and ionic strength adjustment solution

Prepare as in the previous method.

CONCENTRATION RANGE

The following method can be used for the determination of either free or total cyanide in the range 30–3000 ppb. For samples which contain heavy metals that form cyanide complexes, a decomplexation stage is included (Step 3), involving the addition of EDTA. This procedure is intended for heavy-metal concentrations that are of the same order as the cyanide concentration. The free cyanide in the sample can be determined if Step 3 is omitted.

ANALYTICAL PROCEDURE

Step 1 Pipette 100 ml of sample into a beaker containing a plastic-coated magnetic stirrer bar. It is essential to use a hand-operated pipette pump to avoid the risk of ingestion of cyanide (*CYANIDE ANTIDOTE SEE p.* 302). Add 1 ml of cyanide indicator solution.

Step 2 If the sample is known to be free of metal cyanide complexes or if only the free cyanide is required, add 5 ml of phosphate buffer and proceed to Step 4. For other samples, add 5 ml of EDTA solution.

Step 3 Stand the beaker in a fume hood on a hotplate, the temperature control of which is set to heat the contents to 50 °C. Heat for exactly 5 min, then remove from the hotplate and immediately add 5 ml of phosphate buffer. Allow the contents of the beaker to cool to room temperature before continuing to Step 4. The most convenient way to do this is to stand the beaker for about 20 min in a water bath at room temperature. Swirl the contents of the beaker occasionally.

Step 4 Place the electrodes in a suitable holder, rinse them with water, then remove surplus water with a tissue.

Step 5 Place the beaker on the magnetic stirrer, immerse the electrodes in the solution and stir at a moderate rate (keep this stirring rate constant throughout the analysis).

Step 6 When a steady reading has been reached, note the potential, E_S.

Step 7 Pipette 1.0 ml of standard cyanide solution of concentration c_A into the beaker and note the new steady potential E_1.

Step 8 Repeat Step 7 for at least 3 further increments of 1.0 ml and note their steady readings E_2, E_3 and E_4.

Step 9 Calculate antilog (E_x/k), where k is the slope of the calibration graph and is equal to about -116 mV per tenfold increase in cyanide concentration. Plot this term on the y-axis against the total volume of standard cyanide solution added to give the reading E_x on the x-axis. Repeat for each of the known additions.

Alternatively, Gran plot paper without volume correction, such as that which is available from Orion Research Inc., can be used: plot the potential E_x on the vertical axis, noting that a major division corresponds to 10 mV, against the volume of cyanide solution added, V_A, on the x-axis.

Step 10 Draw a straight line through the points and extrapolate to the horizontal axis. The concentration c_S in the sample can be calculated from the intercept, $-V_e$.

$$c_S = \frac{V_e c_A}{100}$$

PREPARATION AND USE OF THE CALIBRATION GRAPH

In order to calculate each point on the Gran plot, a value must be chosen for k and this is best obtained from a calibration graph plotted at the same temperature as that at which the samples were analysed and covering the expected range of the samples. For a calibration graph covering the range 300–3000 ppb, carry out Steps 1–6 (omitting Step 3) of the analytical procedure with at least four standard cyanide solutions. Plot the values of e.m.f. on the y-axis versus the logarithms of the corresponding concentrations on the x-axis. The slope of this graph should be checked once a week with two standards.

Because of the uncertainty of sub-ppm cyanide solutions, the calibration graph for the range 30–300 ppb is best obtained by a titrimetric procedure. Prepare by dilution of the 1000 ppm cyanide solution a 50-ppm standard. Pipette 100 ml of deionized water into a beaker and add 1 ml of indicator solution and 5 ml of buffer solution. Set up the electrodes as described in Steps 4 and 5 and make at least four additions of the 50 ppm standard solution from a 1- or 2-ml microburette, noting the steady readings after each addition (the total volume added should not exceed 2 ml). Calculate the concentration c_x after each addition from the total volume added, V_x

$$c_x = \frac{V_x 50}{100 + V_x}$$

Plot the calibration graph as described for the range 300–3000 ppb. It is not necessary to plot this graph with each batch of analyses but the slope of the graph should be checked at two levels once a week if the electrode is in general use. The slope of the calibration graphs should be about -116 mV per tenfold increase in concentration at 25 °C.

SOURCES OF ERROR

Effect of temperature

Changes in temperature affect the slope of the calibration graph and the analyses should be carried out at temperatures that are within 1 °C of that at which the calubration graph was determined.

Effect of other substances

Sulphide ion must be absent from the analytical solution and methods for its removal, either by addition of an excess of lead nitrate (Frant et al., 1972) or bismuth nitrate (Sekerka and Lechner, 1976) have been discussed on p. 303. The most comprehensive report of anionic interferences was given by Hofton (1976), some of whose results for an Orion 94-16 electrode are given in Table 1.

Table 1 Effect of other substances on the determination of 50 ppb cyanide

Other substances	Concentration of other substance (ppm)	Concentration of cyanide determined (ppb)
Cl^-	5000	50
SO_4^{2-}	1000	50
SCN^-	500	60
$S_2O_3^{2-}$	100	55
NO_3^-	500	45

The effect of iodide and bromide ions was investigated by Clysters, Adams and Verbeek (1976) using an Orion 94-16A electrode. No interference effect was observed over a range, 10–100 ppb, of cyanide in the presence of 200 ppm iodide. In the absence of cyanide, solutions containing 500, 1000 and 10 000 ppm of iodide gave apparent cyanide concentrations of 21, 104 and 221 ppb, respectively. The interference from bromide was much less and in solutions containing no cyanide, 100, 1000 and 10 000 ppm bromide gave apparent cyanide concentrations of 4, 3 and 10 ppb, respectively.

The effect of metal ions that form cyanide complexes on the determination of cyanide has been discussed on p. 303. Both the analytical method using the cyanide electrode and the method under discussion include a decomplexation stage using EDTA at a concentration intended for heavy metals at concentrations less than 10^{-4} mol l^{-1} in the analytical solution. If metals are present in greater concentrations than the prescribed concentration of EDTA, the latter requires to be increased until it exceeds that of the total metal present approximately one hundredfold. This increase in EDTA can affect the final pH of the analytical solution, which should be adjusted to pH 11.3 with concentrated potassium hydroxide solution. Frant et al. (1972) tested the recovery of 2 ppm cyanide using a final concentration of 5×10^{-2} mol l^{-1} EDTA in solutions containing 10^{-3} mol l^{-1} of copper(II), nickel(II) or zinc ions. The final pH of each solution was adjusted to pH 11 and in each case the cyanide recovery was greater than 97%. The determination of cyanide in plating bath solutions has been discussed by Frant (1971).

ACCURACY

Frant et al. (1972) determined the cyanide content of potassium cyanide solutions and found the experimentally determined values agreed exactly with the nominal concentrations of 26, 2.6, 0.4 and 0.26 ppm. Values determined for 100 and 50 ppb solutions were 2 and 20% low, respectively. Clysters et al. (1976) carried out recovery tests on pond and river waters and 90% recovery of 10 ppb and 100% recovery of 100 ppb were reported. The same authors reported that concentrations of 1.0 and 0.103 ppm were determined, following a hydrogen cyanide evolution step, in standard solutions which contained 1.0 and 0.1 ppm, respectively.

PRECISION

Clysters et al. (1976) reported relative standard deviations of 1.1 and 9.5% at cyanide levels of 100 and 10 ppb, respectively.

COMPARISON WITH OTHER METHODS

The silver indicator electrode technique can be used for the potentiometric determination of cyanides at concentrations that were previously only accessible to sensitive colorimetric techniques (ASTM, 1970). It has the usual advantages of potentiometric analyses: speed, flexibility over a wide concentration range and suitability for *in situ* measurement. Until recently, this electrode technique was considered to be more sensitive, by a factor of ten, than that using the cyanide electrode but this may have to be reconsidered in the light of the results obtained by Sekerka and Lechner (1976) using an Orion 94-06A cyanide electrode where excellent agreement with colorimetric analyses was reported down to concentrations as low as a few parts per billion.

Bibliography

ASTM, 1970, *ASTM Standards 2B,* D2306, American Society for Testing and Materials, Philadelphia.

Clysters, H., F. Adams, and F. Verbeek, 1976, 'Potentiometric determinations with the silver sulphide membrane electrode. Part 1. Determination of cyanide', *Anal. Chim. Acta,* **83**, 27.

Frant, M. S., 1971, 'Application of specific ion electrodes to electroplating analyses', *Plating,* 686.

Frant, M. S., J. W. Ross, Jr., and J. H. Riseman, 1972, 'Electrode indicator technique for measuring low levels of cyanide', *Anal. Chem.,* **44**, 2227.

Hofton, M., 1976, 'Continuous determination of free cyanide in effluents using silver ion selective electrode', *Environ. Sci. Technol.,* **10**, 227.

Sekerka, I., and J. F. Lechner, 1976, 'Potentiometric determination of low levels of simple and total cyanides', *Water Res.,* **10**, 479.

Determination of Fluoride

Currently-available fluoride electrodes include the Orion 94-09A, Beckman 39600, Metrohm EA306F, Activion 003 15 008, Leeds & Northrup 117405, Tacussel PF 4, Coleman 3-803, Radiometer F1052 and EDT Supplies type EE-F. In addition, there are two combination fluoride electrodes, the Orion 96-09 and the Beckman 39650. In all these electrodes the membrane consists of a single crystal of lanthanum fluoride, LaF_3, which permits the transport of fluoride ions only. In the Orion electrodes, the crystal is doped with europous fluoride, EuF_2, in order to reduce its electrical resistance. Manufacturers claim a lifetime of 1–2 years for the electrode, but high temperatures and continuous operation will reduce the working life. All the work referred to in the rest of this section has been done with Orion electrodes, apart from the studies of Parthasarathy *et al.* (1974) who also used the two Beckman electrodes and found that the single examples tested had higher limits of detection than an Orion electrode.

APPARATUS

Fluoride electrode; reference electrode; pH meter; magnetic stirrer; plastic beakers; plastic bottles

Both standard and sample solutions should be stored in plastic bottles, especially if very dilute. If possible, prepare very dilute solutions by weight in plastic bottles. There may be an advantage in using a reference electrode with a plastic, rather than glass, body, e.g. the Orion 90-01 silver chloride electrode and the Beckman 40459 calomel electrode.

REAGENTS

Water

Distilled water that has been passed through a mixed-bed deionization unit and has a specific conductivity no greater than $0.2\ \mu S\ cm^{-1}$ at room temperature is suitable for the preparation of standard and TISAB solutions.

Standard fluoride solution A (100 ppm)

Dissolve 0.2210 g of anhydrous sodium fluoride (analytical reagent grade, dried at 120 °C) in water and dilute to 1 litre in a calibrated flask. Store in a polyethylene container. This solution should be stable for at least 3 months.

$$1 \text{ ml} \equiv 100 \text{ } \mu g \text{ fluoride}$$

Standard fluoride solution B (10 ppm)

Pipette 50 ml of standard solution A into a 500-ml calibrated flask and make up to the mark with water. Store in a polyethylene bottle. This solution should be stable for at least 3 months.

$$1 \text{ ml} \equiv 10 \text{ } \mu g \text{ fluoride}$$

TISAB (Total Ionic Strength Adjustment Buffer)

Add 58 g of sodium chloride, 57 ml of glacial acetic acid and 4.5 g of CDTA (cyclohexane-1,2-diamine-N,N,N',N'-tetraacetic acid) to 500 ml of water in a one-litre beaker. Slowly add, with stirring, about 120 ml of 5 mol l^{-1} sodium hydroxide solution and stir until all the solids have dissolved. Cool the beaker to room temperature by placing it in a water-bath and then immerse a calibrated pH electrode and its reference electrode in the solution Slowly add more 5 mol l^{-1} sodium hydroxide, with stirring, until the pH is 5.0–5.5. About 30 ml of sodium hydroxide are required for this step, making about 150 ml in all. Remove the electrodes and transfer the cool solution to a 1-litre calibrated flask. Make up to the mark with water.

CONCENTRATION RANGE AND UNITS

The electrode has a Nernstian response over the range 0.2–2000 ppm, but the non-linear response in the range 0.02–0.2 ppm is reproducible and can be used if a suitable calibration graph is prepared. At least four standard solutions in this range should be run with each batch of analyses if reliable measurements are to be made. Baumann (1971) has shown that in solutions containing 0.095–1.9 ppm total fluoride in the presence of cations that form strong fluoro-complexes, e.g. thorium, zirconium and lanthanum, the electrode responds in a Nernstian manner to the uncomplexed fluoride down to 10^{-9} mol l^{-1}, i.e. 19 ng l^{-1}. Warner and Bressan (1973) devised a technique for determining concentrations in the 0.3–20 ppb range with an accuracy of 10–20%, but the full implications of this method for general application are unknown.

$$10^{-3} \text{ mol l}^{-1} \equiv 19.0 \text{ ppm}; \quad 1 \text{ ppm} \equiv 5.263 \times 10^{-5} \text{ mol l}^{-1}$$

ANALYTICAL PROCEDURE

Step 1 Pipette 25 ml of a standard solution of concentration s_1, chosen to be at the upper end of the expected concentration range of the samples, into a 100-ml plastic beaker containing a plastic-coated magnetic stirrer bar.

Step 2 Add 25 ml of TISAB solution.

Step 3 Remove the electrodes from the solution in which they have been immersed, rinse them with deionized water, dry them with a paper tissue, and immerse them in the solution. Start the magnetic stirrer.

Step 4 When a steady reading has been obtained, note the potential E_1.

Step 5 With a standard solution of concentration s_2 chosen to be near the lower end of the sample concentration range, but not less than 0.2 ppm and not more than 0.5 s_1, repeat Step 1.

Step 6 Repeat Steps 2 and 3.

Step 7 When a steady reading has been obtained, note the potential E_2.

Step 8 *Either* calculate the calibration slope

$$k = \frac{E_1 - E_2}{\log s_1 - \log s_2} \simeq -58 \text{ mV per tenfold increase in concentration}$$

or

prepare a temporary calibration graph by plotting points $(\log s_1, 0)$ and $(\log s_2, E_2 - E_1)$.

Step 9 With a 25-ml portion of the sample solution, repeat Step 1. The temperature of the sample should not differ from that of the standard solutions by more than 1 °C.

Step 10 Repeat Step 2. If the sample contains aluminium ions it may be necessary to wait 10–15 min before proceeding to the next step (see the section on interferences).

Step 11 Repeat Step 3. When a steady reading has been reached, note the potential, E_x.

Step 12 Calculate the potential difference $\Delta = E_x - E_1$.

Step 13 Either read the concentration off the calibration graph or calculate it as follows:

$$c = s_1 \times \text{antilog}\,(\Delta/k)$$

Step 14 Repeat Steps 9–13 for each sample solution. The frequency with which restandardization is required depends on the precision and accuracy desired, but Steps 1–8 should be carried out each day and Steps 1–4 should be repeated at least once in a full working day.

Worked example
The e.m.f. decreases by 58.5 mV for a tenfold increase in fluoride concentration. With a 10 ppm standard the electrode reads −52.1 mV and with the sample

solution it reads -21.5 mV. The concentration of the sample is given by

$$c = 10 \times \text{antilog} \left\{ \frac{-21.5 - (-52.1)}{-58.5} \right\} = 10 \times \text{antilog}(-0.523) = 3.0 \text{ ppm}$$

Use and preparation of a permanent calibration graph

Provided the conditions discussed on p. 117 are satisfied, a permanent calibration graph may be used and Steps 5–8 omitted from the routine analytical procedure. Prepare the graph by carrying out Steps 1–4 of the analytical procedure and then repeating Steps 9–12 for each of at least four standard solutions. See p. 117 for the general precautions required. The graph should be linear at concentrations above 0.2 ppm.

SOURCES OF ERROR

Effect of temperature

Changes in temperature affect the slope of the calibration graph, producing an error of approximately 0.8% in the concentration of fluoride per degree.

Effect of other substances

The only species that is known to interfere with the electrode itself is hydroxide ion, for which the electrode has a selectivity coefficient of about 0.1. By adding TISAB to the sample, the pH is adjusted to about 5.2 and hydroxide interference can be neglected. If the sample is very alkaline, or very acidic, the composition of the TISAB added should be changed so that the final pH is in the range 5.0–5.5 and standard solutions should be prepared containing approximately the same concentration of acid or alkali.

As the electrode responds only to free fluoride, substances that form complexes with fluoride, e.g. H^+, Al^{3+}, Fe^{3+}, La^{3+} and other rare earth ions, ZrO^{2+} and Be^{2+}, would cause a negative bias in the determination of free fluoride in an untreated sample, but the inclusion of a much stronger complexing agent (CDTA) in the TISAB added to the sample releases the fluoride from its metal complexes and as TISAB also contains a pH buffer, the proportion of fluoride present as hydrofluoric acid is kept constant. In these circumstances a reliable calibration in terms of total fluoride can be made. The following substances have been tested and shown to produce an error of 0.01 ppm or less in the determination of 1 ppm fluoride; the levels quoted are the highest reported values and should not be interpreted as the highest permissible ones:

Phosphate, 75 ppm; sulphate, 3000 ppm; chloride, 3000 ppm; boric acid, 570 ppm; silica, 100 ppm; glucose, 180 g l^{-1}; urea, 6 g l^{-1}; hydrogen peroxide, 1.8 g l^{-1}; Ca^{2+}, 500 ppm; Mg^{2+}, 250 ppm; TiO^{2+}, 480 ppm Ti; Fe^{2+}, 560 ppm.

Although boric acid and silica do not interfere, fluoride present as fluoroborate or fluorosilicate will not be recovered completely by the standard procedure (Bock and Strecker, 1968). As fluoride is often added to drinking water as sodium fluorosilicate or fluorosilicic acid, its recovery from such solutions is important. Jordan (1970) added five drops of 1% phenolphthalein indicator per 20 ml of sample, then concentrated (sp. gr. 0.88) ammonia solution dropwise until the colour changed to pink. After the addition of two more drops of ammonia solution, the sample was allowed to stand for 1 min before the buffer was added and the analysis carried out in the normal way; 100% recovery of fluoride was obtained.

The most serious interferents are ferric and aluminium ions, 10 ppm of ferric iron produces a negative bias of 0.01–0.02 ppm in the determination of 1 ppm fluoride (Harwood, 1969; Erdmann, 1975). With the original formulation of TISAB which contained citric acid instead of CDTA (Frant and Ross, 1968), 10 ppm of ferric iron produced an error of −0.03 ppm at 1 ppm fluoride. 2 ppm of aluminium causes a negative of 0.2–0.3 ppm in the determination of 1 ppm fluoride and 5 ppm of aluminium a bias of 0.6–0.9 ppm (Harwood, 1969; Erdmann, 1975). The citrate-based TISAB cannot mask more than about 0.1 ppm aluminium (Harwood, 1969; Crosby, Dennis and Stevens, 1968). For samples containing much higher concentrations of aluminium, TISAB has been replaced by a fairly concentrated sodium citrate solution. Edmond (1969) diluted 20 ppm fluoride solutions containing up to 50 ppm aluminium with an equal volume of 1 mol l^{-1} sodium citrate and found no interference. Ingram (1970) found no interference with 200 ppm aluminium in 0.04–8.00 ppm fluoride solutions when 1 mol l^{-1} sodium citrate was added (1:1) and less than 5% error when the citrate concentration was 0.2 mol l^{-1}. Ingram included 0.2 mol l^{-1} potassium nitrate in the citrate solutions to reduce drift at low fluoride levels. Shiraishi et al. (1973) showed that for a given ratio of fluoride to aluminium the interference decreased as the solution became more dilute, e.g. a solution containing 380 ppm fluoride and 100 ppm aluminium gave only a 26% recovery when analysed in the normal way, but 100% after the sample solution had been diluted to one-hundredth of its original concentration. In each case, the fluoride solution was diluted 1:1 with 1 mol l^{-1} citrate buffer before measurement. At fluoride concentrations above 100 ppm, the hydroxide interference at pH 12 is negligible, while the hydroxide masks the aluminium (up to 500 ppm at least) more effectively than 1 mol l^{-1} citrate (Oliver and Clayton, 1970). This is particularly useful in the analysis of fluoride-rich solids, e.g. cryolite, fluorspar, as 0.5 g of solid digested with 6 g of sodium carbonate and made up to 250 ml with water needs no further treatment before measurement.

PRECISION

The precision of the analysis of synthetic standard solutions is better than 5% and figures of 1–2% can be achieved with care (Table 1). Better precision is obtained when TISAB is used to control interferences than when citrate buffer is used. Relative standard deviations of 1% or less have been reported for the analysis of real samples: 5–40 g l^{-1} of hydrofluoric acid in pickling baths (Entwistle, Weedon and

Table 1 Precision of analysis of synthetic standard solutions

Fluoride concentration (ppm)	Relative standard deviation (%)	Sample treatment	Reference
0.025–2.5	5	TISAB	MacLeod and Crist (1973)
2.5–25	2	TISAB	
1.7	3.5	Citrate[a]	Elfers and Decker (1968)
4–95	1.5	Citrate[a]	
0.05–0.1	1–2	TISAB	Louw and Richards (1972)
0.05–0.1	9–10	Citrate[a]	
1–10	1	TISAB	
1–10	5	Citrate[a]	
0.75–0.85	3.6–4.8[b]	TISAB	APHA (1971)
0.90	2.9[c]	TISAB	

[a] 1 mol l^{-1} Citrate buffer (pH 6).
[b] Results from 111 laboratories.
[c] Results from 13 laboratories.

Hayes., 1973) and about 2.5 ppm fluoride solutions of dissolved bone ash (Singer and Armstrong, 1968).

ACCURACY

100% recovery of 1 ppm spikes from three river waters has been reported by Crosby et al. (1968). Erdman (1975) obtained recoveries of 96–102% (mean 99.2%) for 0.5 ppm spikes in 13 natural waters (0.6–1.7 ppm). MacLeod and Crist (1973) found a mean recovery of 97% for a 20 ppm spike in 14–30 ppm solutions of soluble stack emissions from fertilizer and aluminium reduction plants.

Elfers and Decker (1968) reported errors of no more than 0.5% in the analysis of synthetic standard solutions (4–95 ppm). Inter-laboratory tests gave mean errors of 0.2 and 0.7% for the analysis of 0.75 and 0.85 ppm synthetic standard solutions, but in the presence of 0.5 ppm aluminium the error for a 0.9 ppm fluoride standard was 4.9%.

RESPONSE TIME

Response times of 3 min are commonly quoted for stirred solutions containing 0.5–5 ppm fluoride, although times as short as 20 sec and as long as 5 min have been reported, the response being slower in the more dilute solutions. Times of 60 min have been found for concentrations below 1 ppm, but these were obtained in unbuffered solutions and should be exceptional. At concentrations above 20 ppm only a few seconds are required. Vigorous stirring reduces the response time, especially at low concentrations. Sawyer and Foreman (1969) and Erdmann (1975) raised the solution temperature to 55 °C in order to analyse 20–40 samples per hour automatically. If aluminium is present, the response time may be increased, because of the slow release of fluoride from its aluminium complexes. Warner and Bressan (1973) set aside the sample solution for 15 min after adding the TISAB in order to ensure that all the fluoride was released. By preparing the samples (Step 9) before standardizing the electrode (Steps 1—8), the time taken to complete a batch of analyses should hardly be affected even if such a period of waiting is found necessary.

COMPARISON WITH OTHER METHODS

Crosby et al. (1968) made an extensive comparative study of the electrode and four spectrophotometric methods (alizarin, eriochrome cyanine R, SPADNS and alizarin complexone). The recovery of spikes of fluoride added to three river waters was much poorer for all the spectrophotometric methods, a mean of 92% compared with 100% for the electrode. The spectrophotometric methods were also more prone to interferences, including some common ions such as chloride, sulphate, phosphate and calcium that do not interfere with the electrode. Even aluminium interfered less with the electrode than with the other methods. The electrode may also be used directly in coloured or turbid samples whereas the spectrophotometric methods require a distillation step which may introduce uncertainty into the detection of very low levels of fluoride and considerably increases the time of analysis, 70 min for SPADNS—distillation compared with 7 min for the electrode, according to MacLeod and Crist (1973). Frant and Ross (1968) also reported that the electrode method gave better recoveries of spikes from natural-water samples than the SPADNS technique. In general, however, the electrode and SPADNS methods agree closely over the range 0.5—250 ppm fluoride. Inter-laboratory tests (APHA, 1971) showed that the standard deviation of the determination of synthetic sample solutions was consistently smaller than for the SPADNS and alizarin spectrophotometric methods, whether the latter incorporated a distillation step or not. In general, the electrode also gave more accurate mean results.

APPLICATIONS

The electrode may be used in the final stage of the analysis of almost any kind of sample. Among its applications are the analysis of air and stack gases (MacLeod and Crist, 1973; Elfers and Decker, 1968; Thompson et al., 1971); of vegetation and other biological samples after leaching (Louw and Richards, 1972; Jacobson and Heller, 1975), digestion and distillation (Ke, Regier and Power, 1969) and oxygen bomb combustion (Levaggi, Oyung and Feldstein, 1971); and of rocks and minerals (Oliver and Clayton, 1970; Ingram, 1970; Edmond, 1969). Entwistle et al. (1973) used a fluoride electrode in conjunction with a quinhydrone electrode to determine the free hydrofluoric acid in pickling baths.

The electrode may also be used in the determination of fluoride by potentiometric titration or known addition methods. The most useful titrants are lanthanum and thorium nitrates, and the sensitivity of the method is increased if the sample is adjusted to pH 5—6 with strong acid or base (not a buffer solution) and diluted 1:1 with ethanol. Studies of the factors affecting fluoride titrations have been made by Lingane (1967 and 1968) and Eriksson and Johansson (1970). Anfält and Jagner (1969 and 1970) have studied the effects of acetate and other buffer solutions on the lanthanum—fluoride titration and shown that they are injurious. TISAB, the function of which is to destroy metal—fluoride complexes, should not be added to samples for titration. Before use, the metal titrant solution should be standardized against a standard fluoride solution of similar concentration to the sample. Light and Mannion (1969) used the thorium titration to determine the fluorine content of organic compounds after combustion in an oxygen flask. For seven compounds containing 9.6—76% fluorine the difference between the theoretical and experimental compositions was always less than 0.3%.

Liberti and Mascini (1969) applied the addition titration technique (p. 91) to the determination of 0.1—200 ppm fluoride and obtained more accurate results than by direct potentiometry. Gyllenspetz, Kitchen and Rees (1973) devised a simple graphical calculation technique for known addition potentiometry with the fluoride electrode and Durst (1969) used it for an analate addition method (20—200 ppm samples). Durst (1968) used linear null-point potentiometry to determine fluoride at concentrations as low as 0.04 ppm with an error of less than 1% and a relative standard deviation of no more than 1%. The sample was made 0.1 mol l^{-1} in potassium nitrate and the standard fluoride solution was added to a 0.1 mol l^{-1} solution of potassium nitrate joined to the sample by a salt bridge.

Orenberg and Morris (1967) used the electrode to follow the titration of fluoride with tetraphenylantimony sulphate, achieving accuracies of better than 99% at concentrations of 20—1000 ppm. The sample is neutralized and then an equal volume of chloroform added in order to extract the complex formed during the titration.

The electrode can be used to determine thorium and the rare earth metals by titration with fluoride. Baumann (1968) has used it to determine concentrated (7.5—75 g l^{-1}) lithium solutions by diluting the sample with ethanol so that the final solution is about 5% aqueous and titrating with standard fluoride solution. The electrode has been applied in various titrimetric procedures for determining aluminium (q.v.).

Manahan (1970) has recommended the fluoride electrode as a reference electrode for use with the nitrate-selective electrode. The fluoride electrode is immersed in the sample solution, to which sodium fluoride has been added, and, therefore, there is no liquid junction potential and no interfering chloride ions are released, as from the usual reference electrodes. The nitrate concentration is determined by the known addition method.

TRACING FAULTS

The fluoride electrode is one of the most reliable ion-selective electrodes and faults are more likely to occur in the other parts of the system (p 127). Two faults could originate, however, in the composition of the fluoride sample solutions.

Loss of response

Check that the solution in which the electrode is immersed is not too alkaline. If TISAB is added this condition should be very rare.

Slow response

This may be observed in unbuffered solutions of low fluoride content and will normally be cured by adding TISAB. The slow release of fluoride from aluminium complexes may also retard the response; if necessary, add the TISAB 15 min before making the measurement to allow all the fluoride to be released.

Bibliography

Anfält, T., and D. Jagner, 1969, 'Effect of acetate buffer on the potentiometric titration of fluoride with lanthanum using a lanthanum fluoride membrane electrode', *Anal. Chim. Acta*, **47**, 483.

Anfält, T., and D. Jagner, 1970, 'Effect of carboxylic acid buffers on the potentiometric titration of fluoride with lanthanide nitrates using a lanthanum fluoride membrane electrode', *Anal. Chim. Acta*, **50**, 23.

APHA, 1971, *Standard Methods for the Examination of Water and Wastewater*, 13th edn., American Public Health Association, Washington, D.C., p. 168.

Baumann, E. W., 1968, 'Determination of lithium by potentiometric titration with fluoride, *Anal. Chem.*, **40**, 1731.

Baumann, E. W., 1971, 'Sensitivity of the fluoride-selective electrode below the micromolar range', *Anal. Chim. Acta*, **54**, 189.

Bock, R., and S. Strecker, 1968, 'Direkte elektrometrische Bestimmung des Fluorid-Ions', *Z. Anal. Chem.*, **235**, 322.

Crosby, N. T., A. L. Dennis, and J. G. Stevens, 1968, 'An evaluation of some methods for the determination of fluoride in potable waters and other aqueous solutions', *Analyst*, **93**, 643.

Durst, R. A., 1968, 'Fluoride microanalysis by linear Null-point potentiometry', *Anal. Chem.*, **40**, 931.

Durst, R. A., 1969, 'Determination of fluoride by analate addition potentiometry', *Mikrochim. Acta*, 611.

Edmond, C. R., 1969, 'Direct determination of fluoride in phosphate rock samples using the specific ion electrode', *Anal. Chem.*, **41**, 1327.

Elfers, L. A., and C. E. Decker, 1968, 'Determination of fluoride in air and stack gas samples by use of an ion specific electrode', *Anal. Chem.*, **40**, 1658.

Entwistle, J. R., C. J. Weedon, and T. J. Hayes, 1973, 'The determination of hydrofluoric acid and nitric acid contents of pickling bath liquors using ion-selective electrodes', *Chem. Ind. (London)*, 433.

Erdmann, D. E., 1975, 'Automated ion-selective electrode method for determining fluoride in natural waters', *Environ. Sci. Technol.*, **9**, 252.

Eriksson, T., and G. Johansson, 1970, 'A study of the optimal conditions for potentiometric titration of fluoride with lanthanum and thorium in unbuffered media', *Anal. Chim. Acta*, **52**, 465.

Frant, M. S., and J. W. Ross, 1968, 'Use of a total ionic strength adjustment buffer for electrode determinations of fluoride in water supplies', *Anal. Chem.*, **40**, 1169.

Gyllenspetz, A. B., D. Kitchen, and T. D. Rees, 1973, 'A simple graphical calculation technique for known addition potentiometry', *Chem. Ind.* (London), 640.

Harwood, J. E., 1969, 'The use of an ion-selective electrode for routine fluoride analyses on water samples', *Water Res.*, **3**, 273.

Ingram, B. I., 1970, 'Determination of fluoride in silicate rocks without separation of aluminium using a specific ion electrode', *Anal. Chem.*, **42**, 1825.

Jacobson, J. S., and L. I. Heller, 1975, 'Collaborative study of a potentiometric method for the determination of fluoride in vegetation', *J. Assoc. Off. Anal. Chem.*, **58**, 1129.

Jordan, D. E., 1970, 'Determination of total fluoride and/or fluosilicic acid concentration by specific fluoride ion electrode potentiometry', *J. Assoc. Off. Anal. Chem.*, **53**, 447.

Ke, P. J., L. W. Regier and H. E. Power, 1969, 'Determination of fluoride in biological samples by a nonfusion distillation and ion-selective membrane electrode method', *Anal. Chem.*, **41**, 1081.

Levaggi, D. A., W. Oyung, and M. Feldstein, 1971, 'Microdetermination of fluoride in vegetation by oxygen bomb combustion and fluoride ion electrode analysis', *J. Air. Pollut. Control Assoc.*, **21**, 227.

Light, T. S., and R. F. Mannion, 1969, 'Microdetermination of fluorine in organic compounds by potentiometric titration using a fluoride electrode', *Anal. Chem.*, **41**, 107.

Liberti, A., and M. Mascini, 1969, 'Anion determination with ion-selective electrodes using Gran's plots', *Anal. Chem.*, **41**, 676.

Lingane, J. J., 1967, 'A study of the lanthanum fluoride membrane electrode for end point detection in titrations of fluoride with thorium, lanthanum and calcium', *Anal. Chem.*, **39**, 881.

Lingane, J. J., 1968, 'Further study of the lanthanum fluoride membrane electrode for potentiometric determination and titration of fluoride', *Anal. Chem.*, **40**, 935.

Louw, C. W., and J. F. Richards, 1972, 'The determination of fluoride in sugar cane by using an ion-selective electrode', *Analyst*, **97**, 334.

MacLeod, K. E., and H. L. Crist, 1973, 'Comparison of the SPADNS-zirconium lake and specific ion electrode methods of fluoride determination in stack emission samples', *Anal. Chem.*, **45**, 1272.

Manahan, S. J., 1970, 'Fluroide electrode as a reference in the determination of nitrate ion', *Anal. Chem.*, **42**, 128.

Oliver, R. T., and A. G. Clayton, 1970, 'Direct determination of fluoride in miscellaneous fluoride materials with the Orion fluoride electrode', *Anal. Chim. Acta*, **51**, 409.

Orenberg, J. B., and M. D. Morris, 1967, 'Potentiometric titration of fluoride with tetraphenylantimony sulphate', *Anal. Chem.*, **39**, 1776.

Parthasarathy, N., J. Buffle, and D. Monnier, 1974, 'Study of the behaviour of solid-state membrane electrodes. Part 1. Role of various factors on the limits of sensitivity of chloride and fluoride electrodes', *Anal. Chim. Acta*, **68**, 185.

Sawyer, R., and J. K. Foreman, 1969, 'The development of electrometric methods for automatic analysis', *Laboratory Practice*, **18**, 35.

Shiraishi, N., Y. Murata, G. Nakagawa, and K. Kodama, 1973, 'Enhancement of demasking of fluoride ion from its aluminium complexes by dilution in the determination with a fluoride ion-selective electrode', *Anal. Lett.*, **6**, 893.

Singer, L., and W. D. Armstrong, 1968, 'Determination of fluoride in bone with the fluoride electrode', *Anal. Chem.*, **40**, 613.

Thompson, R. J., T. B. McMullen, and G. B. Morgan, 1971, 'Fluoride concentrations in the ambient air', *J. Air Pollut. Control Assoc.*, **21**, 484.

Warner, T. B., and D. J. Bressan, 1973, 'Direct measurement of less than 1 part-per-billion fluoride in rain, fog and aerosols with an ion-selective electrode', *Anal. Chim. Acta.* **63**, 165.

Determination of Chloride Using Electrodes Based on Silver Chloride

Chloride electrodes were traditionally made by electrolytically depositing a thin layer of silver chloride on a silver metal substrate. This form of chloride electrode is still used as the internal reference electrode in most glass electrodes and a number of ion-selective electrodes but has been largely displaced as a sensing electrode by chloride membrane electrodes which have longer operational lifetimes and are less prone to redox effects. However, silver–silver chloride electrodes are suitable for most chloride analyses and have the advantage of being cheaper than membrane types, whether made in the laboratory from silver wire (Whitfield, 1971; Torrance, 1974) and chloridized as described on p. 341 or manufactured commercially e.g. Electronic Instruments Ltd., 8004-2.

Chloride membrane electrodes are available from many manufacturers. The following have membranes consisting of single crystals of silver chloride: Philips IS 550-Cl, Metrohm EA306Cl, Radiometer F1012Cl and F1512Cl. Electrodes with polycrystalline membranes of silver chloride and silver sulphide are the Orion 94-17A, Beckman 39604, Leeds & Northup 117402 and EDT Supplies EE-Cl. The Radiometer F3005 and Simac E12K3/1C can be impregnated to make them responsive to chloride. Other membrane electrodes are the Activion 003 15001, Tacussel PCL 3, Coleman 3-802 and Simac Cl/1C. The Orion 96-17 and Beckman 39652 are combination chloride and reference electrodes; the Orion electrode is only suitable for fairly high concentrations as the reference half of the electrode contains a solution of chloride ions.

CONCENTRATION RANGE AND UNITS

Three methods are necessary to cover the complete range of chloride concentrations. Those described in this section deal with high (>10 ppm) and intermediate (0.1–10 ppm) chloride levels and require no apparatus that is not readily available in most laboratories, but for determining low levels the special apparatus and techniques described in the next section are necessary.

$$10^{-3} \text{ mol l}^{-1} \equiv 35.5 \text{ ppm chloride} \equiv 58.4 \text{ ppm NaCl}$$
$$2.82 \times 10^{-5} \text{ mol l}^{-1} \equiv 1.0 \text{ ppm chloride} \equiv 1.648 \text{ ppm NaCl}$$
$$1.71 \times 10^{-5} \text{ mol l}^{-1} \equiv 0.606 \text{ ppm chloride} \equiv 1.0 \text{ ppm NaCl}$$

APPARATUS

Chloride electrode; mercury—mercurous sulphate reference electrode; pH meter; magnetic stirrer

It is preferable to use a reference electrode which has a ground-glass sleeve junction.

REAGENTS

Water

Distilled water which has been passed through a mixed-bed deionization unit is suitable for the preparation of all reagent and standard chloride solutions.

Standard chloride solution A (1000 ppm)

Dry a sample of analytical reagent or purified grade sodium chloride in an oven at 150 °C for 4 h. Weigh 1.649 g of the dried salt and transfer it, with washing, to a 1-litre calibrated flask. Make up to the mark with water.

$1 \text{ ml} \equiv 1 \text{ mg chloride}$

Standard chloride solution B (100 ppm)

Pipette 50 ml of solution A into a 500-ml calibrated flask and make up to the mark with water.

$1 \text{ ml} \equiv 0.1 \text{ mg chloride}$

Additional standard solutions can be prepared by dilution of standard solutions A and B.

Buffer solution

Dissolve 77.8 g of analytical reagent or purified grade ammonium acetate in approximately 250 ml of water and add 57 ml of analytical reagent grade glacial acetic acid (sp. gr. 1.05). Dilute the mixture to 1 litre with water.

SAMPLE COLLECTION

Collect sufficient sample in a clean glass or polyethylene bottle for at least duplicate analyses. Care is required in handling samples, particularly those of low chloride content, since chloride is a common contaminant present in airborne dust and on the surface of the skin.

CONDITIONING AND STORAGE OF ELECTRODES

For the best performance, silver—silver chloride electrodes should be stored in a small volume of deionized water at all times. Solid-state chloride electrodes can be similarly stored if used on a day-to-day basis but for longer periods it is recommended that they are stored dry, covered with protective plastic caps.

If the electrodes have been stored dry, immerse their sensing surfaces in approximately 35 ppm chloride solution for about 30 min before use.

ANALYTICAL PROCEDURES

Method A (10—350 ppm chloride)

This method operates in the range of the Nernstian response of electrodes based on silver chloride.

Step 1 Add by pipette 50 ml of standard solution of concentration s_1 to a 100-ml beaker containing a magnetic stirrer bar. Prepare this standard such that its concentration is representative of the lower end of the expected concentration range of the samples.

Step 2 Add by pipette 5 ml of buffer solution.

Step 3 Remove the chloride electrode from the solution in which it is immersed and place it in a holder together with the reference electrode. Rinse both electrodes with deionized water, remove the surplus water with a tissue and then immerse them in the test solution. Note the depth of immersion and keep this level constant throughout.

Step 4 Stir the solution at a moderate rate (note the stirrer setting and keep this constant throughout the procedure) and when the electrodes give a steady reading note the e.m.f., E_1.

Step 5 Repeat Step 1 with a second standard solution of concentration s_2. Preferably $s_2 = 10 s_1$ and should not be less than $2 s_1$.

Step 6 Repeat Steps 2-4, noting the steady reading E_2.

Step 7 *Either* calculate the calibration slope k, which should be about —58 mV per tenfold increase in concentration at 20 °C

$$k = \frac{E_1 - E_2}{\log s_1 - \log s_2}$$

or

prepare a temporary calibration graph by drawing a straight line through the points $(\log s_2, 0)$ and $(\log s_1, E_1 - E_2)$.

Step 8 Add by pipette 50 ml of sample solution to a 100 ml beaker containing a stirrer bar. The temperature of the sample should be within 1 °C of that of the standard solutions and it may be convenient to use a water bath to bring the samples to the required temperature.

Step 9 Repeat Steps 2–4, noting the steady reading E_x millivolts.

Step 10 Calculate the difference $\Delta = E_x - E_2$ and use this to calculate the concentration of chloride in the sample from $c_x = s_2 \times$ antilog (Δ/k) or read the concentration corresponding to Δ from the calibration graph.

Step 11 Repeat Steps 8–10 for all the samples to be analysed. The e.m.f. values of the two standard solutions should be measured with every batch of about ten samples or at least twice a day.

Worked Example

The steady readings noted for 15 and 50 ppm standard chloride solutions were -218.0 and -248.7 mV, respectively. The steady reading of the sample was -231.3 mV. The calibration slope k is first calculated

$$k = \frac{(-218.0 + 248.7)}{\log \frac{15}{50}}$$

$$= -58.7 \text{ mV per decade}$$

The concentration of chloride in the sample is then

$$c_x = 50 \times \text{antilog} \frac{(-231.3 + 248.7)}{-58.7}$$

$$= 25.3 \text{ ppm}$$

Method B (0.1–10.0 ppm)

In this concentration range, the normal Nernstian plot of e.m.f. versus the logarithm of the chloride activity is curved and if accurate results are required the calibration graph has to be defined by more than two standard solutions. An alternative calibration function, to be plotted against e.m.f., is $\log(c + 1)$, where c must be expressed in ppm of chloride (Torrance, 1974). This function, whose derivation is shown below, provides a linear calibration graph requiring definition by only two standard solutions for analyses carried out at 25 ± 3 °C.

The response of the silver–silver chloride electrode at chloride concentrations in the region of the solubility of silver chloride, $\sqrt{K_s}$, (where K_s is the solubility product of silver chloride) can be written as

$$E = E^\circ_{Ag} + k \log K_s - k \log \left[\frac{fm}{2\sqrt{K_s}} + \left(\frac{f^2 m^2}{4 K_s} + 1 \right)^{1/2} \right]$$

where m and s are the molar concentrations of chloride in the sample and that dissolved from the electrode, respectively, and f is the mean univalent activity coefficient of the silver and chloride ions. On binomial expansion of the term raised to the power of one-half, and considering the first two members only, the expression within the square brackets becomes

$$\left[1 + \frac{fm}{2\sqrt{K_s}} + \frac{f^2 m^2}{8 K_s} \right]$$

At values of m of less than $\sqrt{K_s}$, this expression tends to

$$\left[1 + \frac{fm}{2\sqrt{K_s}} \right]$$

which at 25 °C becomes

$$\left[\frac{fc}{0.89} + 1 \right]$$

where c is the chloride concentration in parts per million. At the ionic strength produced by the addition of the acetate buffer, the mean activity coefficient was calculated to be 0.8 and it was shown experimentally that the simple expression log $(c + 1)$ provided an essentially linear calibration graph when plotted against the e.m.f. (Torrance, 1974). Although the theoretical derivation was made considering a silver–silver chloride electrode, the calibration procedure is applicable to all types of silver chloride membrane electrodes.

Step 1 If the electrode has been stored dry, immerse the sensing surface in a 10 ppm chloride solution for about 30 min before use. The analyses should be carried out in a position that is screened from strong natural light to avoid any photoelectric effects on the chloride electrode. In order to obtain satisfactory precision, the temperature of the test solutions must be strictly controlled. Ensure that the beaker is protected from any heat produced by the magnetic stirrer by standing it on a pad of insulating material.

Step 2 Place the bottles containing the samples, 10 and 1 ppm standard chloride solutions and the acetate buffer solution in a thermostatically controlled water bath. A convenient thermostat setting is 2–3 °C above ambient temperature. Allow sufficient time for the temperature of the samples and standards to reach equilibrium.

Step 3 Add by pipette 50 ml of 10 ppm standard chloride solution to a 100-ml beaker containing a magnetic stirrer bar.

Step 4 Add by pipette 2 ml of buffer solution

Step 5 Remove the chloride electrode from the solution in which it is immersed and place it in a holder together with the reference electrode. Rinse both electrodes with deionized water, remove the surplus water with a tissue and then immerse them in the test solution. Note the depth of immersion and keep this level constant throughout.

Step 6 Commence stirring at a moderate rate (keep the stirrer setting constant throughout the procedure) and note the e.m.f., E_1, 2 minutes after starting the stirrer.

Step 7 Add by pipette 50 ml of 1 ppm standard chloride solution to a 100-ml beaker containing a magnetic stirrer bar.

Step 8 Repeat Steps 4–6 noting the e.m.f. after 2 min, E_2.

Step 9 On two-cycle semi-log paper, plot E_1 and E_2, on the linear axis, against the corresponding values, on the log axis, of the calibration function $\log(c + 1)$ i.e. plot $(\log 11, E_1)$ and $(\log 2, E_2)$. Join the points and extrapolate the linear calibration graph to the function value 1.1, corresponding to a concentration of 0.1 ppm chloride.

Step 10 Add by pipette 50 ml of sample solution to a 100-ml beaker containing a magnetic stirrer bar.

Step 11 Repeat Steps 3–5, noting the e.m.f. after 2 min, E_x.

Step 12 Read the value of the function $(c + 1)$ corresponding to the value E_x on the calibration graph and obtain the concentration of chloride in the sample c_x from

$$c_x = (c + 1) - 1$$

Step 13 Repeat Steps 10–12 for each sample to be analysed. The responses to the two standard solutions should be remeasured after at most ten samples.

SOURCES OF ERROR

Effect of temperature

The analytical procedures are particularly sensitive to changes in temperature, especially in the low concentration range (Method B), where there is increased dissolution of the silver chloride from the electrode surface. The effects of changes in temperature over a range 0–100 °C were measured by Jones and Kehoe (1959) for the range 1–8000 ppm chloride. Their results indicate that electrodes calibrated at 25 °C have an error of approximately +2% at the 100 ppm level for a 1 °C increase in temperature. Torrance (1974) reported errors of +4% at 26 °C and −4% at 24 °C when analysing samples containing 10 and 1 ppm chloride with electrodes which had been calibrated at 25 °C.

Effect of light

Changes in intensity of natural light have been reported to produce changes in the e.m.f. of a silver–silver chloride electrode (Jones and Kehoe, 1959; Moody, Oke and Thomas, 1969). No effects were noted when both a silver–silver chloride electrode (Torrance, 1974) and a Radelkis chloride electrode (Torrance, 1976) were used in a laboratory screened from strong natural light. Electrodes such as the Orion 94-17, whose membrane consists of a mixture of silver chloride and silver sulphide, are considered to be less affected by changes in the intensity of natural light.

Effect of other substances

Ions such as sulphide, iodide, bromide and thiocyanate interfere with the response of the electrode due to the formation of their insoluble silver salts on the electrode surface. Phosphate and hydroxide ions, which also form insoluble silver compounds, do not constitute interferences at the pH of the buffer solution ($\sim$pH 4.7). In electrodes whose membranes consist of single- or mixed-crystal discs, this contamination can be removed by polishing the membrane to expose a fresh crystal surface. Silver—silver chloride electrodes cannot be treated in this way and the electrodes require to be stripped of their silver salts and replated.

The presence of species, such as EDTA, which form complexes with silver will give falsely high chloride readings and raise the limit of detection of the method. The extent of the interference depends on the concentration of the species and the strength of its silver complex. A possible means of removing this type of interference is to add an excess of a second, more readily complexed, metal ion with the buffer solution. This form of interference should not damage the electrode surface.

Strongly reducing solutions, such as photographic developer, must be avoided since they form a layer of silver metal on the electrode's surface. Orion (1971) state that their model 94-17 electrode can be used in solutions containing oxidizing agents such as copper(II), iron(III) and permanganate ion. Duff and Stuart (1975) using an Orion 94-17A electrode, reported interference from copper(II), aluminium and to a less extent, iron(III) ions in chloride analyses of samples buffered with citric acid.

Table 1 Selectivity coefficients for chloride electrodes

Interfering ion	Selectivity coefficient,[a] K_{ClX}		
	Radelkis	Philips IS 550-Cl	Radiometer F1012Cl
CN^-	1	400	8
$S_2O_3^{2-}$	1	60	0.8
AsO_4^{3-}	2×10^{-4}	–	–
CrO_4^{2-}	–	1.8×10^{-3}	–
CO_3^{2-}	4.6×10^{-5}	3×10^{-3}	–
SO_4^{2-}	10^{-6}	–	–
SO_3^{2-}	0.2	–	–
PO_4^{3-}	4.8×10^{-5}	–	–
$C_2O_4^{2-}$	4.5×10^{-5}	–	–
OH^-	–	2.4×10^{-2}	5×10^{-3}
Br^-	–	1.2	2
I^-	–	86.5	2

[a]Selectivity coefficients calculated from
$E = E° - k \log (c_{Cl} + K_{ClX} c_X^{1/z_X})$ where z_X is the charge on X and $k = RT/F \ln(10)$.

Table 2 Effect of substances expected in boiler waters

Substance	Concentration of substance (ppm)	Apparent chloride concentration[a] (ppm) at real concentrations of:	
		1 ppm	10 ppm
Ammonia	} Tested together	1	
Cyclohexylamine		1	1.04(0.98)[b]	10.0 (10.3)
Morpholine		1		
Hydrazine		0.1		
Na_3PO_4		10	1.04 (1.02)	10.0 (10.0)
SiO_2		10	1.02 (1.00)	9.7 (10.1)
NaOH		20	1.03 (1.02)	10.0 (10.0)
Fe^{3+}		10	0.95	9.8
Na_2SO_4	} Tested together	10		
Ca^{2+}		10	0.96 (0.96)	9.7 (10.0)
Mg^{2+}		10		

[a] If other substances had no effect the results would be expected to fall within the following ranges: 1.00 ± 0.04 and 10.0 ± 0.3 ppm (silver—silver chloride); 1.00 ± 0.04 and 10.0 ± 0.4 ppm (Radelkis electrode) for 95% confidence limits.
[b] The figures in parentheses were obtained with a Radelkis chloride electrode (Torrance, 1976).

Table 1 contains selectivity coefficients for the Radelkis and Philips electrodes (Koryta, 1972) and the Radiometer electrode.

The effects on a silver—silver chloride electrode of ions that could be expected to be present in boiler water are shown in Table 2 (Torrance, 1974).

Only iron(III) had a significant effect at the 1 ppm level and this was possibly due to the redox potential at the metal electrode. The presence of even less than 1 ppm of the filming amine octadecylamine produced a drift in the signal in the direction of increasing chloride concentrations and a similar effect was noted with solutions containing small amounts of commercial preparations used as corrosion inhibitors (Torrance, 1974).

PRECISION

No detailed results of precision tests are available for the higher concentration range but they would be expected to be slightly better than those reported for the lower analytical range. Relative standard deviations of about 5% were reported by Selmer-Olsen and Øien (1973) for the analysis of samples in the range 4.8—18.8 ppm using an Orion 94-17 electrode. Results of precision tests using Method B in the range 0.1—10 ppm are given in Table 3 (Torrance, 1974 and 1976).

Accuracy

Recovery tests were carried out on boiler-water samples spiked with known additions of chloride and the recoveries are shown in Table 4. Selmer-Olsen and

Table 3 Precision of determination of chloride in the range 0.1–10.0 ppm

Concentration (ppm)		Standard deviations[c] (ppm)		
Added	Found[a,b]	Within-batch (σ_w)	Between-batch (σ_b)	Total (σ_t)
0.10	0.08 (0.08)	0.01 (0.01)	0.04 (0.03)	0.04 (0.03)
0.74	0.72 (0.70)	0.03 (0.03)	n.s. (n.s.)	0.04 (0.06)
1.76	1.77 (1.79)	0.04 (0.04)	n.s. (n.s.)	0.06 (0.07)
3.38	3.49 (3.51)	0.09 (0.03)	n.s. (0.18)	0.09 (0.09)
6.14	6.12 (6.16)	0.04 (0.03)	0.14 (0.18)	0.15 (0.19)
10.00	9.80 (10.00)	0.22 (0.09)	n.s. (0.24)	0.23 (0.26)

[a] Mean of 10 results.
[b] The first set of figures was obtained using a silver–silver chloride electrode; the figures in parentheses were obtained using a Radelkis chloride electrode (Torrance, 1976).
[c] σ_w and σ_b have 5 and 4 degrees of freedom, respectively. n.s. denotes not significant at the $P = .05$ level.

Table 4 Recovery tests on spiked boiler-water samples

Sample	Chloride found (ppm)	Recovery of chloride (%)
3.08 + 2.37 ppm spike	5.46	100.4
3.05 + 2.50 ppm spike	5.60	101.2
3.00 + 2.50 ppm spike	5.51	100.4

Øien (1973) reported recoveries in the range 99.5–101% for the analysis of spiked soil extracts whose chloride contents were in the range 10–100 ppm.

The results obtained using an Orion 94-17 electrode were compared with those obtained using a colorimetric procedure involving mercury(II) thiocyanate for the analysis of soil extracts and water samples (Selmer-Olsen and Øien, 1973). Students t-test gave no indication of a difference between the results by the two methods at a probability level $P > 0.05$. Back (1960) compared the results obtained with a Beckman silver–silver chloride electrode with those obtained by a Mohr titration and found that the electrode results agreed within ±4% for sample solutions whose chloride contents were in the range 10–200 ppm.

COMPARISON WITH OTHER METHODS

Argentometric titrations of chloride may be more precise at high levels (>200 ppm), but they are not very suitable for determining low concentrations. Titration with coloured indicator end-points have the additional disadvantage that the colour change may be difficult to judge in coloured samples, and even in the most favourable circumstances considerable expertise may be needed before consistent end-point detection is achieved. An absorptiometric method based on the colour developed by the mercury(II) thiocyanate-iron(III) system has a higher precision at chloride concentrations below 1 ppm but this method requires three separate reagents, is time dependent and the experimental conditions must be carefully adjusted to suit the sample concentration range.

OTHER APPLICATIONS

The electrodes may be used as indicators in the argentometric titration of chloride, although a silver wire would do as well. They could in principle be used to determine silver, but only at fairly high concentrations and other electrodes are better for this purpose.

Bibliography

Back, W., 1960, 'Electrode for simplified field determination of chloride in ground water', *J. Amer. Water Works Assoc.*, **52**, 923.

Duff, E. J., and J. L. Stuart, 1975, 'Potentiometric determination of chloride in inorganic orthophosphates in citrate-buffered media', *Talanta*, **22**, 901.

Jones, R. H., and T. J. Kehoe, 1959, 'A new continuous chloride ion analyzer', *Ind. Eng. Chem.*, **51**, 731.

Koryta, J., 1972, 'Theory and application of ion-selective electrodes', *Anal. Chim. Acta.*, **61**, 329.

Moody, G. J., R. B. Oke, and J. D. R. Thomas, 1969, 'Influence of light on silver—silver chloride electrodes', *Analyst*, **94**, 803.

Orion, 1971, *Instruction Manual for Halide Electrode*, Model 94-17, Orion Research Inc., Harvard, Mass.

Selmer-Olsen, A. R., and A. Øien, 1973, 'Determination of chloride in aqueous soil extracts and water samples by means of a chloride-selective electrode', *Analyst*, **98**, 412.

Torrance, K., 1974, 'A potentiometric method for the determination of chloride in boiler waters in the range 0.1 to 10 μg ml^{-1} of chloride', *Analyst*, **99**, 203.

Torrance, K., 1976, Unpublished results.

Whitfield, M., 1971, *Ion Selective Electrodes for the Analysis of Natural Waters*, Australian Marine Sciences Association, Sydney, N.S.W.

Determination of Low Levels of Chloride

Silver–silver chloride electrodes have been shown to be suitable for this analysis (Bardin, 1962; Tomlinson and Torrance, 1977). Electrodes of this type can be obtained from a number of manufacturers, e.g. EIL model 8004-2, or can be made in the laboratory (Whitfield, 1971; Torrance, 1974). Solid-state membrane electrodes should also be suitable (Florence, 1971).

APPARATUS

Electrode; mercury–mercurous sulphate reference electrode; pH meter reading to 0.1 mV; flow cell; peristaltic pumps or multichannel pump; thermocirculator; 3-m length of stainless steel tubing (approximately 3 mm o.d. and 0.7 mm wall thickness); light-proof box for flow cell; chart recorder

Temperature control system

A thermocirculator capable of maintaining the temperature to within ±0.1 °C is desirable. Insert a 10-litre plastic bottle fitted with watertight inlet and outlet connections in the circulating system immediately after the thermocirculator outlet (see Figure 1). The water in this bottle acts as a thermal buffer, smoothing out small fluctuations in the temperature of the water leaving the circulator. Connect the outlet of this bottle to the inlet of a second similar bottle whose neck is fitted with additional inlet and outlet holes for the stainless steel tubing, which is coiled so that it fits inside the bottle. Complete the circulation system by connecting the outlet of the second bottle to the inlet of the circulator. If a thermocirculator is not available, a water bath can be used as the heat sink for the stainless steel coil, provided the necessary temperature stability can be maintained. The stability may be improved by enclosing the water bath or by covering its surface with a layer of floating plastic balls. Set the temperature of the circulating system or water bath at about 2 °C above room temperature.

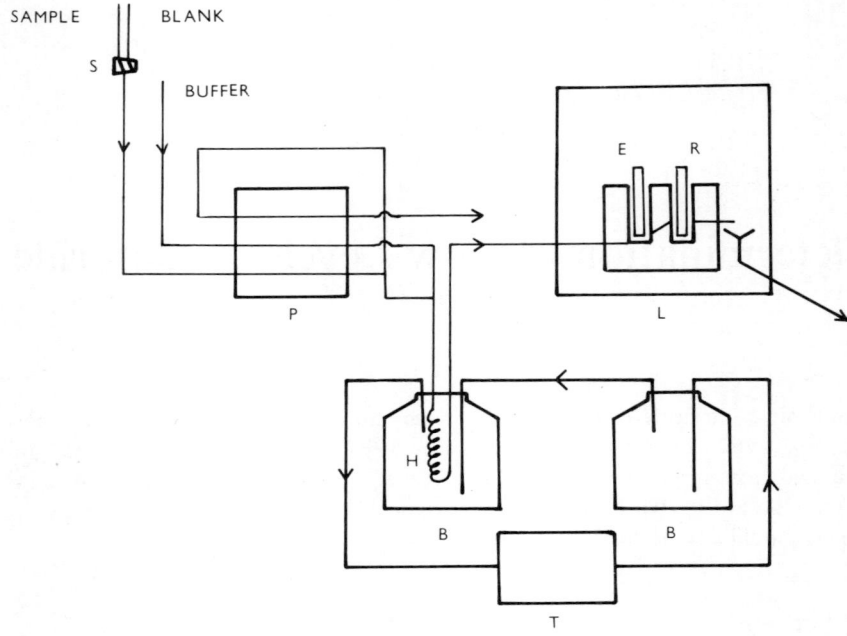

Figure 1 Analysis system for low-level chloride determinations. B = 10-litre bottle; E = silver–silver chloride electrode; H = heat-exchanger coil; L = light-proof box; P = pump; R = reference electrode; S = double-oblique stopcock; T = thermocirculator

Analysis system

The apparatus is shown schematically in Figure 1. Connect two short sample lines to the same side of an oblique two-way stopcock and connect the common outlet to a peristaltic pump capable of delivering approximately 8 ml min^{-1}. Immediately after the pump, insert a glass de-bubbling T-piece (such as is used in Technicon Autoanalyzer systems) in the sample line; abstract a subsidiary flow of about 3 ml min^{-1} and run this to waste. This is a precaution against air bubbles entering the flow cell since they can upset the stability of the electrodes for periods of as long as 15–20 min. Directly after the de-bubbler, combine the sample and buffer streams at a glass T-piece. The ratio of the sample flow to the buffer flow should be 10:1. The buffered analysis stream is connected to the inlet of the stainless-steel coil and the outlet is connected to the flow cell by the minimum length of narrow-bore plastic tubing.

Flow cells suitable for potentiometric analysis are available commercially, e.g. EIL model 24 8990 240, or they can be machined from blocks of perspex or plexiglass to a design such as that described by Webber and Wilson (1969). Place the mercury–mercurous sulphate reference electrode downstream of the chloride electrode in the flow cell. Reference electrodes fitted with ground-glass sleeve junctions are preferable to those with ceramic frit junctions, as the latter can develop a bias after prolonged exposure to chloride solutions. Arrange the outlet of the flow cell such that the effluent falls dropwise into a catchpot rather than flowing to waste in a continuous stream.

REAGENTS

Water

It is necessary to use deionized water whose specific conductivity is less than 0.1 μS cm^{-1} at 25 °C for the preparation of standard and buffer solutions. In many cases is not possible to obtain water of this quality by passing distilled water once through a mixed-bed deionization unit and consequently the water must be circulated through the deionization unit until its conductivity falls to the desired level. A suitable circulation system can be assembled along the following lines: Using nylon tubing, connect a 20-litre dense plastic aspirator to the top of the mixed-bed unit through a pump with a delivery between 100 and 1000 ml min^{-1}. At the outlet of the column, fit a conductivity cell followed by a three-way Teflon stop-cock. One outlet of the stop-cock returns the water to the aspirator while the other is used for removing water from the circuit as required. Connect the aspirator to the atmosphere through a soda-lime guard tube to avoid ingress of carbon dioxide.

Buffer solution

Add 114 ml of analytical-reagent grade acetic acid (sp. gr. 1.05) to about 250 ml deionized water. Slowly add, with stirring, 55 ml of concentrated ammonia solution (sp. gr. 0.88, analytical reagent grade) and allow the solution to cool before making the volume up to 1 litre. This solution should be prepared in a fume cupboard or under a fume hood.

Standard chloride solutions

Prepare standard solution A (1000 ppm) as described in the previous procedure for the higher analytical ranges (p.324).

Solution C (10 ppm)

Pipette 10 ml of solution A into a 1-litre calibrated flask and make up to the mark with water.

1 ml ≡ 10 μg chloride

Standard solutions for the concentration range 10–150 ppb are prepared by dilution of solution C.

SAMPLE COLLECTION

Samples are best collected in polyethylene aspirators. These aspirators should first be washed with deionized water and then allowed to stand full of water for 2–3 days before being finally rinsed prior to collection of samples. Great care is required in the collection and handling of samples, as at no time should the samples come into contact with the skin. Samples should be collected as quickly as possible and the bottles sealed and transported to the laboratory for analysis with the minimum delay.

CONCENTRATION RANGE AND UNITS

The method is suitable for concentrations up to about 150 ppb.

$$10^{-6} \text{ mol l}^{-1} \equiv 35.45 \text{ ppb chloride} \equiv 58.45 \text{ ppb sodium chloride}$$
$$2.82 \times 10^{-8} \text{ mol l}^{-1} \equiv 1 \text{ ppb chloride} \equiv 1.649 \text{ ppb sodium chloride}$$
$$1.71 \times 10^{-8} \text{ mol l}^{-1} \equiv 0.606 \text{ ppb chloride} \equiv 1 \text{ ppb sodium chloride}$$

ANALYTICAL PROCEDURE

The potential of a silver–silver chloride electrode in flowing solutions containing less than 200 ppb chloride is directly proportional to the chloride concentration (Bardin, 1962). In this concentration range, the response of the chloride electrode depends on both the chloride present in the sample and the chloride that dissolves from the electrode surface as silver chloride. The relationship between the two sources of chloride is governed by the solubility product of the sparingly soluble salt and it is such that at concentrations below 100 ppb, over 90% of the chloride in the vicinity of the electrode is derived from dissolution of the silver chloride. It is this latter condition that makes precise control of temperature necessary, as the solubility of silver chloride is highly temperature dependent.

Step 1 The control of temperature is so important that it is advisable to leave the thermocirculator running continuously or, if this is not practicable, to start it on the day before the analyses are to be carried out.

Step 2 Collect a supply of deionized water in a plastic aspirator and connect the tap to a sample input line. Enough water for at least 8 h continuous running is needed.

Step 3 Connect a bottle of buffer solution to the buffer input line to the peristaltic pump. The volume of buffer solution should be sufficient for 8 h continuous use.

Step 4 Select the deionized-water input line and start the pump(s). After all the air in the system has been displaced initially, check that no air bubbles are entering from leaky joints. It is also useful at this stage to check the stability of the temperature of the analysis stream in the flow cell. If the stability is worse than ±0.1 °C, check that the proportional band setting on the thermocirculator is set as finely as possible or that the circulation in the water bath is homogeneous and does not produce currents of warm water.

Step 5 Place the electrodes in the flow cell with the reference electrode downstream. Connect the electrodes to the pH meter and select the relative millivolt scale.

Step 6 Connect the pH meter to the recorder and select a scale on the latter such that the change in relative millivolts from 0 to −10 produces exactly a full-scale deflection. Adjust the calibration control on the pH meter until the reading is 0.0 mV and start the recorder chart. The electrodes take 2–3 h to stabilize after start-up and the reading on the meter should be readjusted, if necessary, after this period.

Step 7 When a steady trace has been recorded with the deionized water, select a sample bottle and wash the outside of the tap with water. Flush the tap with a small volume of sample before connecting it to the second input line to the two-way stop-cock.

Step 8 Turn the two-way stop-cock to select the sample and record the steady response for 15 min.

Step 9 Turn the two-way stop-cock to select the deionized water and record the steady response for 15 min.

Step 10 Measure the difference, Δ mV, between the steady response of the sample solution and the baseline response of the deionized water immediately preceding it.

Step 11 Calculate the concentration, c_x, of chloride in the sample from

$$c_x = \Delta/k \text{ ppb}$$

where k is the slope (mV per ppb) of the calibration graph.

Step 12 Repeat Steps 7–11 for each sample. It is advisable to include a standard chloride solution whose concentration is representative of those of the samples with every batch of about five samples to check the calibration. Any change in the calibration slope will produce a bias in the measured concentration of the standard solution.

Preparation of the calibration graph

By dilution of standard solution C, prepare 1 litre each of at least five solutions spanning the concentration range up to 150 ppb. Carry out Steps 7–10 of the analytical procedure with each solution and note the millivolt differences between their responses and that of deionized water. Plot the millivolt differences on the y-axis against the corresponding nominal chloride concentrations on the x-axis. The graph should be linear with a slope, k, of the order of −0.022 mV per ppb of chloride at 25 °C. A 50 ppb standard solution should be run once per day; if its value of Δ (Step 10) differs by more than 5% from the calibration value, the calibration graph should be redetermined.

SOURCES OF ERROR

Effect of temperature

As has been emphasised in the analytical procedure, the control of the temperature of the analysis stream is of primary importance for the success of the method. The temperature dependence may be judged from the slope of the calibration graph, which changes from -0.022 mV per ppb at 25 °C to -0.033 mV per ppb at 10 °C. The most common fault in the temperature control is the occurrence of short-term variations in the temperature of the water leaving the thermocirculator, causing a related fluctuation in the recorder trace. Suggestions for improving the control are given in Step 4 of the procedure. Variations in the ambient temperature can affect the analytical system by altering the potential of the electrodes inside the light-proof box. This particularly applies to the reference electrode, since the reference element may not be sufficiently insulated from changes in air temperature by the internal reference solution. A variation of ±1 °C inside the box can be tolerated, but should greater variations occur it may be impossible to obtain a steady response from the system. If this is the case, the light-proof box can be fitted with an air circulation fan which operates in conjunction with a heating element and a temperature controller.

Effect of other substances

Table 1 shows the effect of substances that can be found in power station waters (Tomlinson and Torrance, 1977). These results were obtained at 10 °C, but the interference effects are expected to be similar at the ambient temperatures assumed in the analytical procedure.

Florence (1971) investigated the effect of other substances, at a chloride level of 100 ppb, on a differential analytical method for the determination of chloride using an Orion 94-17A chloride electrode and found that the hydrazine content should be less than 40 ppm. The interference he reported from ammonia would not be significant in the acetic acid buffer solution used in the analytical procedure. Aleskovskii, Bardin and Bystritskii (1966) reported that there was no interference from low concentrations of ions other than those which formed sparingly soluble silver salts.

PRECISION

Precision tests have been carried out on standard chloride solutions at 25 °C with the results given in Table 2.

ACCURACY

Satisfactory recovery tests have been carried out on power station waters (Tomlinson and Torrance, 1977). 100.5% recovery of a 47 ppb spike was obtained

Table 1 Interference effect of substances in power station waters

Substance	Concentration (ppm)	Interference effect[a] (ppb) at chloride concentrations of:	
		0 ppb[b]	50 ppb
Na_2SO_4	75	1.7	3.1
Na_2HPO_4	50	1.4	2.7
Na_3PO_4	50	2.7	<1
NaOH	4	<1	<1
NaOH	40	4.1	3.8
Na_2CO_3	50	2.7	<1
NH_3	10	<1	<1
Morpholine	10	<1	<1
Cyclohexylamine	10	<1	<1
Hydrazine	1	<1	<1
H_2SO_4	10	<1	<1
Na_2SO_3	5	4.4	4.5
Fe^{2+}, Cu^{2+}, Zn^{2+}	1 + 1 + 1	−2.9	<1
Ca^{2+}, Mg^{2+}, K^+	2 + 2 + 20	1.5	<1
Oxygen	~4	<1	—

[a]If other substances had no effect the results would have been expected to fall within the limits of 0 ± 1.2 ppb and 50 ± 3.4 ppb at the 95 % confidence level.
[b]The chloride concentration of deionized water was taken nominally as 0 ppb.

Table 2 Precision of analysis of standard chloride solutions at 25 °C

Concentration of chloride (ppb)		Standard deviation[a] (ppb)
Added	Found[b]	
20	19.9	1.8
40	41.1	1.8
60	60.6	1.8
80	81.4	2.2
100	101.3	2.6

[a]The standard deviation for a single determination with 5 degrees of freedom.
[b]Each result is the mean of six determinations.

from boiler water containing 108 ppb chloride and 100% for a 5 ppb spike added to condensed steam containing less than 1 ppb chloride.

The chloride levels determined by this procedure are subject to a small negative bias, as there is no simple way to determine exactly how much chloride is present in the deionized water used for calibration. If the conductivity of the deionized water is in the range 0.06–0.07 $\mu S\ cm^{-1}$ at 25 °C and if the anionic contribution (other than hydroxide ions from the dissociation of water) is considered to be entirely from chloride ions, then the chloride levels in the deionized water would be in the range 0.7–3.0 ppb. Concentrations of this order have been found by Florence (1971) in distilled deionized water.

RESPONSE TIME

The response time of 5–8 min is largely due to the delay in the heat-exchange coil and the time taken to sweep out the previous solution from the flow cell.

COMPARISON WITH OTHER METHODS

Florence (1971) developed a manual differential potentiometric technique for the determination of chloride concentrations up to 350 ppb. The procedure involves titrations of both the sample and blank solutions and is, therefore, rather lengthy. A relative standard deviation of 3% for samples containing 100 ppb chloride is much the same as that shown in Table 2. An experimental electrode, based on a mixture of mercuric sulphide and mercurous chloride (Sekerka, Lechner and Wales, 1975) was more sensitive than silver chloride electrodes at low levels of chloride (the graph of e.m.f. against the logarithm of the concentration was linear down to 50 ppb), but the relative standard deviation of 14% at 100 ppb was poorer than the value of 2.6% for the corresponding level in Table 2.

The most popular non-electrochemical technique for trace chloride analysis is an absorptiometric method based on a mercury(II) thiocyanate–iron(III) system. Florence and Farrar (1971) obtained a limit of detection of 15 ppb and a relative standard deviation of 12% at the 50 ppb level for one version of this method. The limit of detection was extended by concentrating the chloride by co-precipitation with lead phosphate (Rodabaugh and Upperman, 1972); excellent recoveries were obtained at chloride levels down to 1 ppb and the standard deviations in the concentration range up to 40 ppb were less than 1 ppb. While this procedure appears to be the most accurate and precise that has been reported to date, like all techniques that involve a concentration stage, it is time-consuming, and results such as those reported can only be obtained by a skilled operator.

TRACING FAULTS

The common faults of solid-state and reference electrodes discussed on p. 127 mostly apply here also, but there may be additional problems caused by the flow system and the very low concentrations being determined.

Noise

The major problems of temperature control in this respect have been dealt with already and will not be considered further. A fluctuation of signal like that due to temperature variations has occasionally arisen from poor mixing of the sample and buffer streams at the glass T-piece. A Technicon T-piece with the buffer entering through a capillary side-arm gives good results and as an additional aid to mixing, the capillary should be placed vertically so that the more dense buffer solution is drawn down into the main sample flow.

Noisy signals can arise from earthing problems, particularly earth loops associated with the thermocirculator. There are no exact directions that can be given to eliminate this fault and trial-and-error procedures often produce the best results. A general improvement has been noted when the electrical earth connection on the circulator was disconnected and the sample solution earthed through the stainless steel coil, but this should not be done without considering the safety of the particular apparatus involved.

Lifetime of the silver—silver chloride electrode

The surface of the electrode is constantly being dissolved and, in time, insufficient silver chloride will remain to cover the electrode surface, resulting in an electrode that gives a noisy and drifting signal. Electrodes that are not covered evenly with a dark brown layer of silver chloride and show small lighter patches of metal substrate should be replaced immediately. An electrode may be re-chloridized, but first strip it of any residual silver chloride by washing it with ammonia solution (1 volume of sp. gr. 0.88 ammonia to 4 volumes of water). Rinse the electrode thoroughly with water and then lightly etch the silver metal surface with dilute nitric acid. Chloridize the electrode in 0.01 mol l^{-1} hydrochloric acid, versus a platinum cathode, for 24 h at a current density of 0.1—0.2 mA cm^{-2}. After electrolysis, remove the electrode from the acid solution, wash it with water and store in the dark, in water, for a week before use.

Bibliography

Aleskovskii, V. B., V. V. Bardin, and A. L. Bystritskii, 1966, 'The potentiometric determination of microquantities of chlorides in water', *Zavod. Lab.*, **32**, 148.

Bardin, V. V., 1962, 'A potentiometric method for determining small concentrations of chloride ions', *Zavod. Lab.*, **28**, 910.

Florence, T. M., 1971, 'Differential potentiometric determination of parts per billion chloride with ion-selective electrodes', *J. Electroanal. Chem.*, **31**, 77.

Florence, T. M., and Y. J. Farrar, 1971, 'Spectrophotometric determination of chloride at the parts-per-billion level by the mercury(II) thiocyanate method', *Anal. Chim. Acta*, **54**, 373.

Rodabaugh, R. D., and G. T. Upperman, 1972, 'Determination of parts per billion levels of chloride in high-purity waters by coprecipitation and spectrophotometry', *Anal. Chim. Acta*, **60**, 434.

Sekerka, I., J. F. Lechner, and R. Wales, 1974, 'Determination of chloride in water with a HgS/Hg$_2$Cl$_2$ electrode', *Water Res.*, **9**, 663.

Tomlinson, K., and K. Torrance, 1977, 'A potentiometric system for the continuous determination of low levels of chloride in high purity power station waters', *Analyst*, **102**, 1.

Torrance, K., 1974, 'A potentiometric method for the determination of chloride in boiler water in the range 0.1—10 μg ml^{-1} of chloride', *Analyst*, **99**, 203.

Webber, H. M., and A. L. Wilson, 1969, 'The determination of sodium in high purity water with sodium-responsive glass electrodes', *Analyst*, **94**, 209.

Whitfield, M., 1971, *Ion Selective Electrodes for the Analysis of Natural Waters*, Australian Marine Sciences Association, Sydney, N.S.W.

Determination of Bromide and Iodide

Many suitable electrodes are available for both ions.

Bromide

Activion 003 15 002, Beckman 39602 and 39651, EDT Supplies EE-Br, Leeds & Northrup 117400, Metrohm EA306Br, Orion 94-35, Philips IS 550-Br, Radiometer F1022Br, F1522Br and F3006, Simac Br/1C and E12K3/1C, Tacussel PBR 2.

Iodide

Activion 003 15 003, Beckman 39606 and 39653, EDT Supplies EE-I, Leeds & Northrup 117406, Metrohm EA306I, Orion 94-53, Philips IS 550-I, Radiometer F1032I, F1532I and F3007, Simac I/1C and E12K3/1C, Tacussel PI 2.

All of the above are solid-state membrane electrodes, mainly of mixed AgX/Ag_2S composition. The Philips IS 550-Br and Radiometer F1022Br and F1532Br have single-crystal membranes. The Beckman 39651 and 39653 are combination reference and ion-selective electrodes. The Radiometer F3006 and F3007 and Simac E12K3/1C are impregnated with a suitable sensitizing mixture to form the membrane. In addition, silver—silver bromide and silver—silver iodide electrodes may be prepared analogously to silver—silver chloride electrodes (p. 341) and as described by Ives and Janz (1961).

APPARATUS

Electrode; double-junction reference electrode; magnetic stirrer; pH meter

A micrometer syringe and a chart recorder are required for lower ranges. The outer compartment of the double-junction reference electrode should be filled with 0.1 mol l^{-1} potassium nitrate solution.

REAGENTS

Water

Deionized water whose specific conductivity is no greater than $0.2\ \mu S\ cm^{-1}$ is suitable for the preparation of reagents and standard solutions of concentration not less than 10^{-5} mol $^{-1}$ (0.8 and 1.27 ppm bromide and iodide, respectively). If calibration standards of lower concentration are required then water whose conductivity is less than $0.1\ \mu S\ cm^{-1}$ should be used.

Standard bromide solution A (1000 ppm)

Dissolve 1.489 g of analytical reagent grade potassium bromide in water and make up to the mark in a 1-litre calibrated flask.

   1 ml ≡ 1 mg bromide

Standard solutions of lower concentrations can be prepared by sequential dilution of solution A.

Standard iodide solution A (1000 ppm)

Dissolve 1.308 g of analytical reagent grade potassium iodide in water and make up to the mark in a 1-litre calibrated flask.

   1 ml ≡ 1 mg iodide

Standard solutions of lower concentrations can be prepared by sequential dilution of solution A. Standard solutions containing less than 1 ppm iodide should be stabilized by the addition of ascorbic acid antioxidant solution in the proportion of 200 ml per litre of standard solution (200 ml + 1 litre).

Ascorbic acid antioxidant solution

Dissolve 80 g of sodium hydroxide pellets in 500 ml of deionized water that has been deoxygenated either by boiling or by bubbling with nitrogen. Add slowly 320 g of sodium salicylate, stirring to avoid formation of flocculent clumps. When all the salt is dissolved, add 72 g of ascorbic acid, cool and make up to one litre. Store in a tightly-stoppered plastic bottle. This solution should be effective for at least two weeks.

Ionic strength adjustment solutions

Dissolve 101 g of analytical reagent grade potassium nitrate in 1 litre of water.

SAMPLE COLLECTION

Samples can be collected in clean plastic bottles. Samples containing iodide at concentrations below approximately 0.5 ppm should be stabilized by the addition of antioxidant solution added in the proportion of 200 ml per litre of sample. These samples should be kept in tightly stoppered bottles and analysed as soon as possible.

CONCENTRATION RANGE AND UNITS

Method A is intended for use with samples whose concentrations lie within the linear Nernstian range of the electrodes. The procedure is considered suitable for samples containing 200–0.5 ppm bromide and 300–0.5 ppm iodide. More concentrated samples should be diluted before analysis. The termination of the linear Nernstian range should be determined for each individual electrode by plotting a calibration graph so that the appropriate choice of method can be made.

Method B is intended for samples lying within the non-linear range of the electrodes and the procedure is considered to be suitable for the analysis of samples containing either 500–50 ppb of bromide or 500–12 ppb of iodide.

$$10^{-3} \text{ mol l}^{-1} \text{ bromide} \equiv 80 \text{ ppm}$$
$$1.25 \times 10^{-5} \text{ mol l}^{-1} \text{ bromide} \equiv 1 \text{ ppm}$$
$$10^{-3} \text{ mol l}^{-1} \text{ iodide} \equiv 127 \text{ ppm}$$
$$7.87 \times 10^{-6} \text{ mol l}^{-1} \text{ iodide} \equiv 1 \text{ ppm}$$

ANALYTICAL PROCEDURES

Method A

Step 1A Pipette 50 ml of standard solution into a 100-ml beaker containing a magnetic stirrer bar. The concentration of the standard solution s_1 should be representative of the upper end of the concentration range of the samples to be analysed. Add 5 ml of the ionic strength adjustment solution.

Step 2A Place the halide electrode together with the reference electrode in a suitable holder, rinse them with water and then remove the surplus water with a tissue. Connect the electrodes to the pH meter and select the millivolt scale.

Step 3A Immerse the electrodes in the solution in the beaker and stir at a moderate rate. Keep this stirring rate and the depth of immersion of the electrodes constant throughout.

Step 4A Note the steady e.m.f., E_1.

Step 5A Repeat Steps 1A–4A with a second standard solution whose concentration s_2 is representative of the lower end of the sample concentration range.

Step 6A Note the steady e.m.f., E_2.

Step 7A *Either* calculate the calibration slope

$$k = \frac{E_1 - E_2}{\log s_1 - \log s_2} \simeq -58 \text{ mV per tenfold increase in concentration}$$

or

prepare a temporary calibration graph by drawing a straight line through the points, $(\log s_1, 0)$ and $(\log s_2, E_2 - E_1)$.

Step 8A Pipette 50 ml of sample and 5 ml of ionic strength adjustment solution into a 100-ml beaker containing a magnetic stirrer bar.

Step 9A Repeat Steps 3A and 4A, noting the steady reading E_x.

Step 10A Calculate the potential difference $\Delta = E_x - E_1$.

Step 11A Calculate the halide concentration, c_x, in the sample from

$$c_x = s_1 \times \text{antilog}\left[\frac{\Delta}{k}\right]$$

or read the concentration corresponding to Δ directly off the calibration graph.

Step 12A Repeat Steps 8A–11A for all the samples to be analysed.

Preparation and use of a permanent calibration graph

Provided the conditions discussed on p. 117 are satisfied, a permanent calibration graph may be used and Steps 5A and 6A omitted from the analytical procedure. Prepare the graph as described on p. 117 by carrying out Steps 1A–4A of the analytical procedure and then repeat Steps 8A–10A with at least four standard solutions.

Method B

It is advisable to carry out this analysis in a laboratory sheltered from strong natural light to avoid photoelectric e.m.f.s. Alternatively, dark or dense plastic beakers will eliminate this effect.

Step 1 Rinse the electrodes with deionized water and remove the surplus water with a tissue.

Step 2 Pipette 50 ml of sample into a beaker containing a magnetic stirrer bar. To bromide samples only, add 5 ml of ionic strength adjustment solution.

Step 3 Immerse the electrodes in the solution and stir at a moderate rate. Keep the depth of immersion and stirring rate constant throughout.

Step 4 Note the steady e.m.f., E_x. (It may be convenient to follow the e.m.f. changes on a chart recorder since the time taken for the electrode to reach equilibrium can be as long as 10–15 min at the lowest concentrations.)

Step 5 Read the concentration of halide in the sample, c_x, directly from the calibration graph.

Step 6 Repeat Steps 1–5 for all the samples.

Preparation of Calibration Graphs for the Bromide Range 500–50 ppb and the Iodide Range 500–12 ppb

In these ranges, the calibration graphs for these electrodes are non-linear at the lowest levels. The most convenient method of preparing calibration curves is to add, sequentially, known volumes of a concentrated standard solution by means of a micrometer syringe of approximately 0.2-ml capacity. Pipette 50 ml of freshly collected deionized water and either 5 ml of ionic strength adjustment solution (for bromide calibration) or 10 ml of antioxidant solution (for iodide calibration) into a 100-ml beaker containing a magnetic stirrer bar. Immerse the electrodes in the solution as described in Step 3 of the analytical procedure. Fill a micrometer syringe with 250 ppm standard halide solution and place its tip below the surface of the solution after rinsing it thoroughly with water. In many cases it is convenient to introduce the standard through fine-bore PTFE tubing connected to the syringe. Make at least four additions covering the expected concentration range of the samples, noting after each addition the total volume added and the steady e.m.f.; the latter can conveniently be recognised from a recorder trace. Calculate the concentration, c_i ppb, after the addition of V_i ml from

$$c_i = 5 \times 10^3 \, V_i \text{ ppb}$$

Plot the e.m.f. values on the y-axis versus the logarithms of the corresponding concentrations on the x-axis. This graph should be determined each day the analysis is carried out if the highest accuracy is required, but it is possible that a fixed calibration curve can be used if the shape of the graph is found to be constant from day to day. In the latter case, a standard solution representative of the upper end of the sample concentration range should be analysed with each batch to correct for any shift of the standard potential and so to relate the e.m.f. values to those of the fixed calibration graph.

SOURCES OF ERROR

Effect of temperature

The temperature of standard and sample solutions should be kept within 1 °C of the temperature at which the calibration curve was determined.

Effect of other substances

Ions which form sparingly soluble silver salts interfere with the performance of bromide and iodide electrodes if they are present in sufficient concentrations to precipitate their salts on the membrane surface. Solutions containing strong reducing agents, e.g. photographic developer, cause malfunction due to the reduction of the silver ions on the membrane surface to silver. The principal ionic interferences are sulphide, cyanide and, to a less extent, hydroxide, chloride and bromide or iodide ions. For most analytical purposes, sulphide must be absent and if necessary it can be removed by precipitation with cadium nitrate. Table 1 (Orion, 1971) shows the maximum allowable concentrations of common interfering ions expressed as multiples of the halide concentration in the sample. Note that the concentrations in the ratio must be expressed as mol l^{-1}.)

Table 1 Maximum allowable ratio of interfering ion to halide ion for the Orion solid-state bromide and iodide electrodes

Interfering ion	Maximum ratio for electrode	
	Bromide	Iodide
OH^-	3×10^4	—
Cl^-	400	10^6
Br^-	—	5×10^3
I^-	2×10^{-4}	—
CN^-	8×10^{-5}	0.4

Onken et al. (1975) investigated the effects of a range of chloride concentrations (60–240 ppm) on the determination of bromide in soils using an Orion 94–35 bromide electrode. They found that the apparent bromide concentration could be successfully corrected for the presence of chloride by applying simple regression equations following a colorimetric determination of the apparent chloride content. For bromide concentrations in the range 6–100 ppm, the correction was less than 10%. Duff and Stuart (1975) measured the iodide interference in bromide analysis, using a Radiometer F1022 bromide electrode in a solution buffered at pH 2.5 with citric acid: interference occurs when the iodide to bromide ratio is greater than 1:500. This ratio can be increased to 1:10 if the sample is treated with hydrogen peroxide, which oxidizes the iodide to iodine, leaving the bromide ion concentration unchanged provided the temperature of the sample does not rise above room temperature.

Kontoyannakos, Moody and Thomas (1976) noted that the linear range of the Orion 94-53A iodide electrode was extended to about 12.7 ppb by either removing the dissolved oxygen by physical means or by the addition of ascorbic acid based antioxidant solution. They concluded that the removal of oxygen limited the oxidation of iodide to iodine.

Effect of pH

The interference effect of hydroxide (Table 1) indicates that there is negligible pH dependence in the case of the iodide electrode while the bromide electrode can be used over its entire working range up to pH 12. Paletta (1969) noted that the response time of the Orion iodide electrode was improved at a concentration of 127 ppb by the addition of a buffer solution.

Effect of light

Paletta (1969) reported that it was necessary to shield the Orion 94-53 electrode from strong natural light during the analysis of samples containing less than 127 ppb of iodide. A detailed account of the effects of light on a number of iodide electrodes has been given by Veselý (1974).

PRECISION

Duff and Stuart (1975) reported relative standard deviations of 33–2.5% for the determination of bromide in extracts of phosphate rock samples using a Radiometer F1022Br electrode.

Zeinalova, Morshina and Senyavin (1975) determined the total iodine content of natural waters using an experimental silver iodide–silver sulphide electrode to measure the iodide concentration, following reduction with sodium arsenite of any iodine and iodate present. A relative standard deviation of 10% was reported in model solutions in the concentration range 0.127–1270 ppm iodide. Hoover, Melton and Howard (1971) reported standard deviations, for five replicate analyses, of 0.9, 11.3 and 15.2 ppb at iodide levels of 14.3, 153 and 356 ppb, respectively, in extracts of feed and plant samples using an Orion 94-53 electrode.

Arino and Kramer (1968) reported a relative standard deviation of 1% for the determination of iodide in solutions containing ^{131}I for the concentration range 1.27–1,270 ppm using an electrode manufactured by the National Instrument Laboratories Inc.

ACCURACY

There are no reports of accuracy tests on analytical methods for either bromide or iodide in water samples.

Onken *et al.* (1975) reported a recovery of 96% for the addition of 100 ppm of bromide to a soil sample. Hoover *et al.* (1971) obtained recoveries in the range 87–110% for the determination of iodide in plant and feed samples but these recovery figures, and those of the examples given for the bromide in soil samples, will also include any errors from the extraction procedure.

RESPONSE TIME

The response times of the solid-state electrodes at concentrations of 10^{-3} mol l^{-1} are expected to be less than one minute (Philips, 1975; Orion, 1971). As the concentration of halide in the sample decreases, response times increase and Paletta (1969), using an Orion 94-53 iodide electrode, reported times of 15 and 40 min, respectively, for 10^{-5} and 10^{-7} mol l^{-1} solutions of iodide even though the solutions were stirred throughout.

COMPARISON WITH OTHER METHODS

A standard volumetric method for the determination of bromide and iodide above 50 ppm is based on an iodine titration with thiosulphate. This method is much slower than a direct potentiometric method since it requires a number of preparative steps before the final titration but it has certain advantages in its comparative freedom from interference. For the determination of bromide concentrations greater than 100 ppb, another method measures the colour developed after a specific time by phenol red in the presence of hypobromite either by a photometer or by comparison with standard solutions in Nessler tubes. The time dependence makes the comparison technique difficult. Methods for the determination of low levels of bromide (<100 ppb) are often lengthy and one standard method involves an inconvenient solvent extraction stage. Standard methods for the analysis of low levels of iodide are also complicated and the method using the catalytic properties of iodide in the oxidation of arsenite ion by ceric ion is highly time and temperature dependent.

OTHER APPLICATIONS

These halide electrodes can be used to detect the end-point in titrations using silver nitrate as titrant. This form of analysis may be preferred in cases where either difficulty is experienced with high and variable ionic strength of the sample or where the additional accuracy obtainable from a titration procedure is required.

Christova, Ivanova and Novkirishka (1976) used a Crytur iodide electrode for the indirect determination of each of the following species following their oxidation with an ethanolic solution of iodine: arsenite, sulphite, ascorbic acid, hydrazine and hydroxylamine. The reproducibility was found to be similar to that of direct potentiometric methods. At reductant levels above 10^{-5} mol l^{-1} the errors did not exceed 2–3% and at lower levels these increased to about 5%.

The results obtained by Zeinalova *et al.* (1975) indicated that it was possible to use an iodide electrode for the determination of iodate following its reduction with sodium arsenite, which was shown to have no deleterious effects on the electrode membrane. Similar systems may be applicable for the determination of bromate.

Popescu and Hălălău (1976) measured the concentration of iodide in the presence of a large excess of chloride by a Gran plot technique. The sample was a solution containing 200 g l^{-1} of kitchen salt containing a small amount of iodate which was first reduced to iodide by stannous chloride. A relative standard deviation of 0.6% was obtained in solutions containing 21 ppm of potassium nitrate.

Most iodide electrodes are suitable for the determination of cyanide as described on p. 298. The manufacturer's instructions should be consulted to verify that cyanide analysis is compatible with his product.

Riseman and Frant (1976) reported the use of the Orion 94-53 iodide electrode as a reference electrode in a monitor using a fluoride electrode for the continuous determination of fluoride in a sample stream to which a fixed concentration of iodide was added.

Bibliography

Arino, H., and H. H. Kramer, 1968, 'Determination of specific activity of ^{131}I solutions via an iodide electrode', *Nucl. Appl.*, **4**, 356.

Christova, R., M. Ivanova, and M. Novkirishka, 1976, 'Indirect potentiometric determination of arsenite, sulphite, ascorbic acid, hydrazine and hydroxylamine with an iodide-selective electrode', *Anal. Chim. Acta.*, **85**, 301.
Duff, E. I., and Stuart, J,L., 1975, 'Mutual interference effects during the successive determination of iodide, bromide, chloride and fluoride in a single sample using halide-selective electrodes', *Analyst*, **100**, 739.
Hoover, W. L., J. R. Melton, and P. A. Howard, 1971, 'Determination of iodide in feed and plants by ion-selective electrode analysis', *J. Assoc. Off. Anal. Chem.*, **54**, 760.
Ives, D. J. G., and G. J. Janz, 1961, *Reference Electrodes, Theory and Practice*. Academic Press, New York and London.
Kontoyannokos, J., G. J. Moody, and J. D. R. Thomas, 1976, 'The detection limit of the Orion iodide/silver ion-selective electrode', *Anal. Chim. Acta*, **85**, 47.
Onken, A. B., R. S. Hargrave, C. W. Wendt, and O. C. Wilke, 1975, 'The use of the specific ion electrode for determination of bromide in soils', *Soil Sci. Soc. Amer. Proc.*, **39**, 1223.
Orion, 1971, *Instruction Manual for Halide Electrodes*, Orion Research Inc., Cambridge, Mass.
Paletta, B., 1969, 'Direct electrometric determination of iodide and iodate ions', *Mikrochim. Acta*, **6**, 1210.
Philips, 1975, *Guide to the Use of Ion-selective Electrodes, Types IS 550/IS 561*, N. V. Philips' Gloeilampenfabrieken, Eindhoven.
Popescu, I. C., and M. Hălălău, 1976, 'Determinarea continutului de iodat din sarea de bucătărie iodata folosind un electrode-membrană I-selectiv si metoda adaosului cunoscut', *Rev. Chim. (Bucharest)*, **27**, 161.
Riseman, J., and M. Frant (Orion Research Inc.), 1976, US 3 964 988 (*Chem. Abs.*, 1976, **85**, 116287h).
Veselý, J., 1974, 'Iodide membrane electrodes', *Coll. Czech. Chem. Comm.*, **39**, 710.
Zeinalova, E. A., T. N. Morshina, and M. M. Senyavin, 1975, 'Potentiometric determination of the various forms of iodine in natural waters by means of ion-selective electrodes', *Zhur. Anal. Khim.*, **30**, 966.

Determination of Thiocyanate

Thiocyanate-selective electrodes include the Orion 94-58, Metrohm EA306SCN, Activion 003 15 006, Tacussel PSCN 1 and Radiometer F3008. All are solid-state electrodes with membranes typically comprising silver sulphide and silver thiocyanate. The Radiometer F3008 is of the Růžička Selectrode type in which the membrane is formed by impregnation of a graphite–PTFE electrode.

APPARATUS

Thiocyanate electrode; (double-junction or mercury–mercurous sulphate) reference electrode; pH meter; magnetic stirrer

Conventional calomel or silver–silver chloride reference electrodes should be adequate for most purposes, as a considerable amount of chloride would have to leak from the reference electrode before interference would occur. With very dilute thiocyanate solutions, however, a mercury–mercurous sulphate or double junction reference electrode should be used. The outer sleeve of the double-junction electrode should be filled with ionic strength adjustment solution (1 mol l^{-1} potassium nitrate) diluted 1 + 10 with water.

REAGENTS

Stock thiocyanate solution (ca. 10 000 ppm)

Dissolve 16.9 g of analytical reagent grade potassium thiocyanate in water and make up to 1 litre. As the analytical reagent grade material is only 98% KSCN, this solution should be standardized against mercuric nitrate. Prepare less concentrated standard solutions by serial dilution once the stock solution has been standardized.

Standardization of stock solution

Heat 100.3 g of dry mercury (analytical reagent grade) in a 600-ml beaker with 250 ml water (*in a fume cupboard*) and slowly add concentrated nitric acid until all the mercury has dissolved. Boil the solution to drive off all nitrous fumes, allow it to cool and then transfer it to a 1-litre calibrated flask. Make up to the mark with water and mix. This solution contains 0.5 mol l^{-1} mercury and needs no standardization. Prepare a 0.05 mol l^{-1} solution by dilution of the above.

Prepare a dilute nitric acid solution by adding 150 ml of concentrated acid to 400—500 ml of water, boiling to drive off nitrous gases and diluting to 1000 ml.

Prepare an iron alum indicator solution by dissolving 25 g of ammonium ferric sulphate $(NH_4Fe(SO_4)_2 \cdot 12 H_2O$, analytical reagent grade) in water and dilute to 500 ml.

Pipette 20 ml of the stock thiocyanate solution into a conical flask and dilute to 150—200 ml with water. Add 10 ml dilute nitric acid and 2 ml iron alum indicator. Titrate against standard 0.05 mol l^{-1} mercuric nitrate solution with constant stirring until the red colour just disappears.

1 ml 0.05 mol l^{-1} Hg(NO$_3$)$_2 \equiv$ 5.808 mg CNS$^-$

Ionic strength adjustment solution

Dissolve 101 g of analytical reagent grade potassium nitrate in water and make up to 1 litre in a calibrated flask.

SAMPLE COLLECTION

Collect samples in the same way as for bromide solutions (p. 344).

CONCENTRATION RANGE AND UNITS

The electrodes have linear responses at concentrations above 0.6 ppm. and may be used at concentrations down to 0.06 ppm.

10^{-3} mol l$^{-1} \equiv$ 58.08 ppm

1 ppm $\equiv 1.722 \times 10^{-5}$ mol l^{-1}

ANALYTICAL PROCEDURES

In the absence of reported analytical results for thiocyanate electrodes, it is recommended that the same procedures should be followed as for bromide (p. 345): use Method A for thiocyanate concentrations above about 0.6 ppm and Method B for lower concentrations.

SOURCES OF ERROR

Temperature

The temperature affects both the standard potential and the calibration slope of the electrode; the standard and sample solutions should not differ in temperature by more than 1 °C.

Effect of other substances

Approximately 100- and 2000-fold molar excesses of chloride and hydroxide ions, respectively, are tolerable, but the thiocyanate concentration should exceed those of bromide and iodide by factors of 10 and 5000, respectively. Sulphide ions cannot be tolerated. Complexing agents such as thiosulphate and cyanide ions will interfere unless there is a hundredfold excess of thiocyanate.

Transition metal ions form weak thiocyanate complexes and will therefore reduce the concentration of free thiocyanate in solution. It may be possible to mask these metal ions by working in the pH range 8–9. At pH < 2, protonation of the thiocyanate will reduce the concentration of free ion in solution. The electrode may be used in the pH range 2–11 for thiocyanate concentrations of 0.6–600 ppm.

PRECISION, ACCURACY, RESPONSE TIME

No results have been published.

Bibliography

Most of the information of the electrode has come from manufacturers' technical literature. The following may also be consulted:

Mascini, M., 1972, 'Preparation and analytical evaluation of a new thiocyanate solid-state heterogeneous membrane electrode', *Anal. Chim. Acta,* **62**, 29.

Moody, G. J., and J. D. R. Thomas, 1971, *Selective Ion Sensitive Electrodes*, Merrow, Watford.

Determination of Sulphide

Sulphide-selective electrodes are available from many manufacturers — Orion 94-16, Beckman 39610, Leeds & Northrup 117408, Radiometer F3001, Radiometer F1212S and F1712S, Metrohm EA306S/Ag, Philips IS 550-S, Simac S/1C, Activion 003 15 004, EDT Supplies EE-S, Coleman 3-805 and Tacussel PS 3. The Beckman 39654 is a combination sulphide and reference electrode. All are solid-state electrodes based on silver sulphide membranes, although the details of construction may vary considerably. A metal–insoluble salt electrode can be prepared by immersing a piece of silver wire in an aqueous solution of hydrogen sulphide or ammonium sulphide, by exposing the silver wire to hydrogen sulphide gas or by anodizing the silver in a solution of sodium sulphide. Ives (1961) has reviewed the preparation of silver–silver sulphide electrodes.

The electrode may be used either for direct potentiometry or for potentiometric titrations, the latter being more precise. The titrimetric procedure is also used to standardize the stock sulphide solution needed for preparing dilute standard solutions for the direct method.

APPARATUS

Sulphide electrode; double-junction reference electrode; pH meter; magnetic stirrer

Any type of double-junction reference electrode may be used. The outer chamber should be filled with 3 mol l^{-1} potassium chloride solution for the determination of sulphide by direct potentiometry or with 1 mol l^{-1} potassium nitrate solution for titrimetric applications.

REAGENTS

Water

Use deionized water with a specific conductivity of less than 0.2 μS cm^{-1} when freshly prepared.

SAOB (Sulphide Anti-oxidant Buffer) Stock Solution

Dissolve 80 g sodium hydroxide in about 500 ml of deionized water (preferably freshly boiled or degassed with nitrogen), then slowly add 320 g sodium salicylate, stirring to prevent the formation of flocculent clumps. When all the solid has dissolved, add 72 g ascorbic acid and, when it has dissolved, cool the solution rapidly to room temperature and make it up to 1 litre in a calibrated flask. Store in a tightly stoppered plastic bottle. This solution has a shelf life of about two weeks, depending on its exposure to air. Its life may be extended by storing it in a refrigerator, but it should be returned to room temperature before use. If the solution turns dark brown, it has been oxidized and should be discarded.

50% SAOB solution

Prepare daily by 1:1 dilution of the stock solution with deionized water.

25% SAOB solution

Prepare daily by 1:3 dilution of the stock solution with deionized water.

Stock sulphide solution (approximately 1000 ppm)

Dissolve about 0.75 g of sodium sulphide nonahydrate ($Na_2S \cdot 9H_2O$) in 100 ml of 25% SAOB solution. It is best to select large crystals and rinse them with deionized water before weighing. Determine the exact sulphide concentration by titrating a 25-ml portion with standard ($0.1\ mol\ l^{-1}$) lead or cadmium solution as described below. Store the solution in a tightly stoppered plastic bottle; it should be stable for at least one week.

$$1\ ml \cong 1\ mg\ sulphide$$

Dilute sulphide standards

Prepare these solutions daily by serial dilution. Pipette an appropriate volume of stock sulphide solution into a 100-ml calibrated flask and make up to the mark with 25% SAOB solution. Use this solution to prepare more dilute standards, and so on until the requisite range of concentration has been covered.

Lead titrant stock solution ($0.1\ mol\ l^{-1}$)

Dissolve 33.120 g of lead nitrate (analytical reagent grade) in water and make up to the mark in a 1 litre calibrated flask. A $0.1\ mol\ l^{-1}$ solution of lead perchlorate is available from Orion Research Inc.

$$1\ ml \equiv 3.2066\ mg\ sulphide$$

Cadmium titrant stock solution (0.1 mol l^{-1})

Dissolve 30.847 g of cadmium nitrate tetrahydrate (Cd(NO$_3$)$_2 \cdot$ 4 H$_2$O) or 25.650 g of cadmium sulphate (3 CdSO$_4$. 8 H$_2$O) in water and make up to the mark in a 1-litre calibrated flask.

1 ml ≡ 3.2066 mg sulphide

Dilute titrant solutions

Titrant solutions for determining low concentration of sulphide should be prepared by dilution of one of the above solutions. The concentration of the titrant should be about ten times the expected sulphide concentration.

SAMPLING

As far as possible, the sample must be kept out of contact with air. If the sample is taken from a closed vessel or pipe, use the technique described for collecting samples without atmospheric contamination (p. 109). If taking samples from open water, fill the vessel as quickly as possible and with a minimum of splashing. In either case, immediately dilute the sample 1:1 with SAOB stock solution and store the mixture in a tightly stoppered plastic bottle. 1 ppm samples stored in this way are stable for about a week.

CONCENTRATION RANGE AND UNITS

The electrode has a Nernstian response in the range 300 ppm to 100 ppb total sulphide. The extension of this response to 30 ppb has been reported, but not consistently achieved. Even if the response is non-Nernstian, however, useful readings can be obtained from a calibration graph down to 10 ppb total sulphide. The electrode will respond to as little as 10^{-14} mol l^{-1} free sulphide ion in the presence of a relatively large total sulphide concentration, e.g., during a titration or in solutions of hydrogen sulphide. This does not represent an analytical limit, however, and the electrode can only be calibrated if the pH and the equilibrium constants for the reactions of sulphide in the solution are known (Berner, 1963; Mor et al., 1975).

By using a pre-concentration technique, Baumann (1974) has extended the analytical range of the method to 2 ppb and an even greater degree of concentration may be possible.

10^{-3} mol l^{-1} ≡ 32.064 ppm

1 ppm ≡ 3.119 x 10^{-5} mol l^{-1}

ANALYTICAL PROCEDURE

Step 1 Rinse the electrodes with deionized water, dry them with a soft paper tissue and immerse them in about 50 ml of a standard solution in a beaker containing a magnetic stirrer bar. The concentration, s_1, of the standard solution should be chosen to be at the upper end of the expected concentration range.

Step 2 Start the stirrer and adjust it so that the solution is stirred briskly, but without forming a vortex. Use the same rate of stirring for all subsequent measurements.

Step 3 When a steady potential is reached, note the reading, E_1.

Step 4 Repeat Step 1 with a second standard solution with a concentration, s_2, at the lower end of the expected concentration range and not more than $0.5 s_1$.

Step 5 When a steady reading has been obtained, note the e.m.f., E_2.

Step 6 *Either* calculate the calibration slope

$$k = \frac{E_1 - E_2}{\log s_1 - \log s_2} \simeq -29 \text{ mV per tenfold increase in concentration}$$

or

plot a temporary calibration graph by joining the points ($\log s_1$, 0) and ($\log s_2$, $E_2 - E_1$).

Step 7 Repeat Step 1 with the sample solution to which the buffer has already been added.

Step 8 When a steady reading has been obtained, note the e.m.f., E_x.

Step 9 Calculate the e.m.f. difference, $\Delta = E_1 - E_x$.

Step 10 Either read the concentration (c) corresponding to Δ off the calibration graph or calculate it as follows

$$c = 2 s_1 \times \text{antilog} \, (-\Delta/k)$$

Step 11 Repeat Steps 7–10 for each sample solution. The value of E_1 (Steps 1 and 3) should be checked at least once during a full working day, or more often, depending on the required precision.

Worked example

The e.m.f. values observed with a 30 ppm standard solution, a 3 ppm standard solution and the sample were -800.0 mV, -771.5 mV, and -790.8 mV, respectively.

$$\Delta = -9.2 \text{ mV}$$

and

$$k = (-800.0 - (-771.5))/(\log 30 - \log 3) = -28.5 \text{ mV}$$

per tenfold increase in concentration. Therefore the concentration, c, is given by

$$c = 2 \times 30 \times \text{antilog} \, (+9.2/-28.5)$$

$$= 60 \times 0.475$$

$$= 28.5 \text{ ppm}$$

Preparation and use of a permanent calibration graph

By dilution of the sulphide stock solution with 25% SAOB solution, prepare at least five standard solutions covering the range of sulphide concentrations under investigation. With the most concentrated standard, carry out Steps 1–3 of the analytical procedure and with the others repeat Steps 7–9. Plot the e.m.f. differences, Δ, on the y-axis against the logarithm of twice the concentration on the x-axis. For the most concentrated standard, $\Delta = 0$. Thus, if the most concentrated standard is 10 ppm sulphide, it will be plotted as (log 20, 0). The factor of two allows for the dilution of the sample with SAOB solution.

If this calibration graph is used, Steps 4–6 may be omitted from the routine analytical procedure. Before this course is adopted, the factors discussed on p. 117 should be considered.

SOURCES OF ERROR

Effect of temperature

The temperatures of the standard and sample solutions should not differ by more than 1 °C, as changes in temperature affect both the calibration slope and standard potential of the electrode.

Effect of other substances

There are no direct interferences with the solid-state membrane electrodes, but the silver–silver sulphide electrode is affected by strong oxidizing agents. Species that form complexes with sulphide ion will interfere with the determination of the total sulphide concentration, e.g., elemental sulphur and tin, arsenic and antimony ions. The extent of their interferences has not been determined in the conditions of the analysis. Hydrogen ions will also form complexes with sulphide ions, and at constant total sulphide concentration the electrode has a linear dependence on pH over the range pH 9–12. At pH 12.7, only half the sulphide is present as free S^{2-} ion. For all but the most acidic samples, the SAOB solution should contain enough sodium hydroxide to keep the pH at an effectively constant high value. Hseu and Rechnitz (1968) found no interference from 480-fold molar excesses of chloride, bromide, iodide, thiocyanate or nitrate ions and 160-fold excesses of sulphate, oxalate or chromate ions. Light and Swartz (1968) found no significant interferences from thiosulphate, sulphite, carbonate, bicarbonate, fluoride and cyanide ions in addition to those above. Bock and Puff (1968) found no interference effects from the presence of 1 mol l^{-1} of urea, glucose, sodium sulphate, sodium chloride, disodium hydrogen phosphate or sodium aluminate in solutions containing 32 μg l^{-1}–32 g l^{-1} sulphide at pH 13.5 nor from tungstic (25 g l^{-1} H_2WO_4) or vanadic (18 g l^{-1} V_2O_5) acids in the same conditions.

Sulphide may be lost by volatilization from acidic samples or by oxidation. SAOB solution is added to the samples in order to prevent such losses. Without

SAOB solution, the limit of detection is as high as 3 ppm, even if the solutions are purged with nitrogen.

Dilution factor

The 1:1 dilution of sample with SAOB solution introduces a factor of 2 into the calculation in Step 10 of both the direct and titrimetric methods. Attention should be paid to this at all times.

PRECISION

Relative standard deviations of 4—5% have been reported for sulphide concentrations in the range 45—800 ppb (Morie, 1971; Baumann, 1974). The latter found relative standard deviations of 8.5 and 23% at sulphide concentrations of 18 and 9 ppb, respectively. An Orion electrode was used in all cases. Berner (1963) used a silver—silver sulphide electrode to determine the free sulphide in solutions of fairly low pH (3.8—6.9) and found that values of $pS = -\log [S^{2-}]$ were reproducible to ±0.2 units over the range pS 8—14.

ACCURACY

Allam, Pitts and Hollis (1972) obtained a correlation coefficient of 0.9965 between results by the electrode method and the standard methylene blue method over a range 0.1—6.5 ppm sulphide.

RESPONSE TIME

The equilibrium response times of Orion electrodes are less than 2 min at concentrations above 1 ppm, but may be as long as 15—20 min at levels of about 50 ppb (Baumann, 1974; Crombie, Moody and Thomas, 1975). The times depend also on the rate of stirring and the condition of the membrane.

PRE-CONCENTRATION TECHNIQUE

The following procedure is suitable for analysing samples containing 2—30 ppb sulphide. It may be possible to apply the technique to even more dilute samples by taking a larger volume for analysis, but no results have been reported.

Water for sulphide standards

Pass oxygen-free nitrogen through deionized water for 30 min, or boil deionized water for 10 min and allow it to cool while passing oxygen-free nitrogen through it. Use this deoxygenated water immediately.

Stock sulphide solution (approximately 500 ppm)

Dissolve 3.75 g of sodium sulphide nonahydrate ($Na_2S \cdot 9 H_2O$) in water and make up to 1 litre in a calibrated flask. Determine the exact concentration of the solution by titration (see below). The solution should be restandardized weekly.

Working sulphide solution (approximately 5 ppm)

Prepare daily by dilution of the stock sulphide solution with deionized water.

1 ml $\cong$ 5 µg sulphide

Calculate c_0, the concentration of the working solution in ppb, from the experimentally determined concentration of the stock solution and the dilution factor.

Zinc acetate solution ($1 \ mol \ l^{-1}$)

Dissolve 219.5 g of zinc acetate dihydrate ($Zn(OAc)_2 \cdot 2 H_2O$) in water and make up to 1 litre in a calibrated flask.

Sodium hydroxide solution ($1 \ mol \ l^{-1}$)

Dissolve 40 g of sodium hydroxide in water and make up to 1 litre in a calibrated flask. Store in a plastic bottle.

Alkaline antioxidant reagent (AAR)

Dissolve 120 g of sodium hydroxide and 186 g of disodium EDTA ($Na_2H_2EDTA \cdot 2H_2O$) in water and make up to 1 litre in a calibrated flask. Store in a plastic bottle. On the day of use, take a 100-ml portion of this solution and dissolve it in 7.2 g of ascorbic acid. This provides enough reagent for 20 samples.

Procedure

To a known volume, v_s ml, of freshly collected sample, e.g., 250 or 50 ml, add first 1 ml of zinc acetate solution then 1 ml of sodium hydroxide solution. Stir for two minutes, then allow the floc of zinc sulphide co-precipitated with zinc hydroxide to settle (or use a centrifuge) before siphoning off the supernatant solution. The precipitate is stable for about 24 h (Pomeroy, 1954). Note that no SAOB solution is added to the sample.

Dissolve the precipitate by adding to it 5 ml of AAR solution and leaving it for 5 min. Transfer the solution to a 25-ml calibrated flask and make up to the mark with deionized water. Analyse the sample by carrying out Steps 7—9 of the analytical procedure. Prepare two standard solutions by adding 0.5 ml of zinc acetate solution to suitable volumes, v_1 and v_2 ml, of working sulphide solution, dissolving the resultant precipitate in 5 ml of AAR solution and making up to 25 ml in a calibrated flask. With two such standard solutions, carry out Steps 1–5 of the analytical procedure.

Calculate the concentration c_1 and c_2 of the standard solution from the concentration, c_0, of the working sulphide solution.

$$c_1 = c_0 \cdot v_1/25 \quad \text{and} \quad c_2 = c_0 \cdot v_2/25$$

Carry out Step 6 of the analytical procedure and then calculate the original concentration, c, of the sample.

$$c = \frac{25}{v_s} \times c_1 \times \text{antilog}\,(-\Delta/k)\ \text{ppb}$$

Note that there is no factor of 2 in the calculation, as in the main method.

Baumann (1974) obtained 100% recoveries of 1.8 and 4.6 ppb solutions concentrated ten times with relative standard deviations of 6% in the measured concentration.

COMPARISON WITH OTHER METHODS

The standard methods for sulphide (APHA, 1971) have higher limits of detection than potentiometry and are more prone to interference. The iodometric method requires the sulphide to be stripped from the sample as hydrogen sulphide, but in these acidic conditions sulphite and thiosulphate can decompose to sulphide and therefore bias the result. The methylene blue method has a limit of detection of 50 ppb (comparator) or 20 ppb (spectrophotometer), but is not suitable for turbid or coloured samples, unless a separation step is included.

OTHER APPLICATIONS

The sulphide electrode is also used as a silver electrode and, with the addition of argentocyanide ion to the solution, as a cyanide-selective electrode. It may also be used as an indicator electrode for titrations of silver, sulphide, thiols, chloride, bromide and iodide.

TRACING FAULTS

Apart from the general points discussed on p. 127, tarnishing of the surface of silver sulphide membranes can cause a reduction in the calibration slope and long response times. If necessary, renew the surface of the membrane according to the manufacturer's instructions.

POTENTIOMETRIC TITRATIONS WITH THE SULPHIDE ELECTRODE

Although silver ion has been used to titrate sulphide, the titration curve is asymmetrical and the equivalence point does not coincide with the point of inflection. In consideration of this, Tamele, Irvine and Ryland (1960) titrated to a fixed potential of +50 mV for a silver—silver sulphide electrode versus a 0.1 mol l^{-1} KCl calomel electrode in a buffer solution of 0.1 mol l^{-1} sodium acetate. If the titration is to be accurate, oxygen must be excluded from the apparatus; SAOB solution cannot be used, as it reduces silver ions to silver metal and would bias the result. The end-point is obscured by the presence of other substances that form complexes or sparingly soluble salts with silver ion, but particularly by cyanide ion and, to a lesser extent, by iodide. Schmidt and Pungor (1971) have studied titration curves in the presence of several interferents.

Titration with cadmium or lead ions gives a symmetrical titration curve that is much less affected by the presence of other anions and, as lead and cadmium ions are not reduced by the SAOB solution, more complicated precautions against oxidation of the sulphide are unnecessary. Ehman (1976) and Green and Schnitker (1974) preferred cadmium for the titration of very dilute sulphide solutions.

Samples are collected and standards prepared as in the direct method.

Titration procedure

    Step 1 Rinse the electrodes with deionized water and dry them with a paper tissue.

    Step 2 Pipette a convenient volume, v_s ml, of the sample or standard solution into a beaker containing a magnetic stirrer bar.

    Step 3 Immerse the electrodes in the solution and start the stirrer

    Step 4 Fill a burette with titrant solution and adjust the level of solution until the meniscus exactly coincides with a scale division. Note the reading, R_o. The concentration of the titrant, m mol l^{-1}, should be about ten times that of the sample.

    Step 5 Slowly add about 50% of the expected equivalent volume of titrant solution.

    Step 6 When a steady potential has been reached, note the e.m.f., E_1, and the burette reading R_1.

    Step 7 Add further portions of titrant solution, each time noting the steady e.m.f., E_i, and the burette reading, R_i. For an equivalent volume of about 10 ml, the titrant should be added in increments of 0.5 ml initially, then of 0.1 ml in the vicinity of the end-point (indicated by the rapidly-changing e.m.f.), and then again of 0.5 ml after the end-point. Continue the titration until an excess of 30—40% of titrant has been added.

    Step 8 Plot the e.m.f. readings, E_i, on the y-axis against the volume addition, $R_i - R_o$, on the x-axis.

    Step 9 Obtain the equivalent volume, R_e, from the point of inflection of the curve.

    Step 10 Calculate the concentration of sulphide in the solution. For a standard solution:

$$c = m \cdot R_e/v_s \text{ mol l}^{-1} \equiv 3.2066 \times 10^4 \ m \cdot R_e/v_s \text{ ppm sulphide.}$$

For a sample, allowing for dilution with SAOB solution:

$$c = 2\, m \cdot R_e/v_s \text{ mol } l^{-1} \equiv 6.413 \times 10^4\, m \cdot R_e/v_s \text{ ppm sulphide}.$$

Concentration range

Sulphide concentrations between 120 ppb and 30 ppb have been determined with 10^{-3} mol l^{-1} titrant (Green and Schnitker, 1974; Ehman, 1976).

Precision

Relative standard deviations of about 2% have been reported for the titration of sulphide solutions with concentration of 30 ppb–3 ppm, including real samples (Green and Schnitker, 1974), as well as synthetic standards (Naumann and Weber, 1971).

Accuracy

Very good recoveries have been obtained over wide ranges of concentration by Ehman (1976) and Slanina *et al.* (1971).

Interferences

Slanina *et al.* (1971) found that thousandfold molar excesses of the following were tolerable for a lead titrant: chloride, bromide, iodide, phosphate, sulphate, acetate, cyanide. Naumann and Weber (1971) reported that $10^5 - 10^6$-fold molar excesses of halide ions, sulphite, thiosulphate and thiocyanate did not interfere. No comparable figures have been reported for a cadmium titrant, but no marked differences would be expected.

Bibliography

Allam, A. I., G. Pitts, and J. P. Hollis, 1972, 'Sulphide determination in submerged soils with an ion-selective electrode', *Soil Sci.*, **114**, 456.

APHA, 1971, *Standard Methods for the Examination of Water and Wastewater*, 13th edn., American Public Health Association, Washington, D.C.

Baumann, E. W., 1974, 'Determination of parts per billion sulphide in water with the sulphide-selective electrode', *Anal. Chem.*, **46**, 1345.

Berner, R. A., 1963, 'Electrode studies of hydrogen sulphide in marine sediments', *Geochim. Cosmochim. Acta*, **27**, 563.

Bock, R., and H.-J. Puff, 1968, 'Bestimmung von Sulfid mit einer Sulfidionen-empfindlichen Elektrode', *Z. Anal. Chem.*, **240**, 381.

Crombie, D. J., G. J. Moody, and J. D. R. Thomas, 1975, 'Observations on the calibration of solid-state silver sulphide membrane ion-selective electrodes', *Anal. Chim. Acta*, **80**, 1.

Ehman, D. L., 1976, 'Determination of parts-per-billion levels of hydrogen sulphide in air by potentiometric titration with a sulphide ion-selective electrode as an indicator', *Anal. Chem.*, **48**, 918.

Green, E. J., and D. Schnitker, 1974, 'The direct titration of water-soluble sulphide in estuarine muds of Montsweag Bay, Maine', *Marine Chem.*, **2**, 111.

Hseu, Tong-Ming, and G. A. Rechnitz, 1968, 'Analytical study of a sulphide ion-selective membrane electrode in alkaline solution', *Anal. Chem.*, **40**, 1054.

Ives, D. J. G., 1961, 'Oxide, oxygen, and sulphide electrodes', in *Reference Electrodes, Theory and Practice*, D. J. G. Ives and G. J. Janz, (eds.), Academic Press, New York and London, Ch. 7.

Light, T. S., and J. L. Swartz, 1968, 'Analytical evaluation of the silver sulphide membrane electrode', *Anal. Lett.*, **1**, 825.

Mor, E., V. Scotto, G. Marcenaro, and G. Alabiso, 1975, 'The use of membrane electrodes in the determination of sulphides in sea water', *Anal. Chim. Acta,* **75**, 159.

Morie, G. P., 1971, 'Determination of hydrogen sulphide in cigarette smoke with a sulphide ion electrode', *Tobacco Sci.*, **107**, 34.

Naumann, R., and C. Weber, 1971, 'Titration von Sulfid mit einer sulfidionensensitiven Elektrode', *Z. Anal. Chem.*, **253**, 111.

Pomeroy, R., 1954, 'Auxiliary pretreatment by zinc acetate in sulphide analyses', *Anal. Chem.*, **26**, 571.

Schmidt, E., and E. Pungor, 1971, 'Studies on sulphide selective membrane electrodes', *Anal. Lett.*, **4**, 641.

Slanina, J., E. Buysman, J. Agterdenbos, and B. Griepink, 1971, 'Die Bestimmung des Schwefels. III.', *Mikrochim. Acta,* 657.

Tamele, M. W., V. C. Irvine, and L. B. Ryland, 1960, 'Potentiometric determination of sulphide ions and the behaviour of silver electrodes at extreme dilution', *Anal. Chem.*, **32**, 1002.

Titrimetric Determination of Sulphate

Although sulphate-sensitive electrodes have been made (Rechnitz, Lin and Zamochnick, 1964; Mohan and Rechnitz, 1973), they have relatively poor selectivities and working ranges and they are not available commercially. Sulphate is determined titrimetrically using a lead-selective electrode (p. 243) as an indicator. As lead sulphate is moderately soluble in water, the sample is diluted with an organic solvent to improve the sensitivity of the technique.

APPARATUS

Lead-selective electrode; double-junction reference electrode; pH meter; magnetic stirrer; burette

Any of the lead electrodes used for the determination of lead (p. 243) can be used. The outer chamber of the double-junction reference electrode should be filled with 1 mol l^{-1} sodium nitrate solution.

REAGENTS

Water

Pass distilled water through a mixed-bed deionization unit. The specific conductivity of this water should be less than 0.2 μS cm^{-1} at the outlet of the deionization unit.

Methanol

Use analytical reagent grade.

Standard sulphate solution A (100 ppm)

Dissolve 1.4788 g of anhydrous sodium sulphate (analytical reagent grade, dried for 2 h at 110 °C) in water, transfer the solution to a 1-litre calibrated flask and make up to the mark with water. Store in a polyethylene or lead-free glass bottle.

> 1 ml $\equiv$ 100 μg sulphate $\equiv$ 33.38 μg sulphur

Standard sulphate solution B (10 ppm)

Pipette 50 ml of standard solution A into a 500-ml calibrated flask and make up to the mark with water. Store in a polyethylene or lead-free glass bottle.

1 ml ≡ 10 µg sulphate ≡ 3.338 µg sulphur

Other standards are prepared by dilution of standard solutions A and B.

Lead titrant solution A (0.1 mol l^{-1})

Dissolve 33.120 g of lead nitrate (analytical reagent grade) in water and make up to the mark in a 1-litre calibrated flask.

1 ml ≡ 9.6066 mg sulphate ≡ 3.2066 mg sulphur

Lead titrant solution B (0.01 mol l^{-1})

Pipette 50 ml of lead titrant solution A into a 500-ml calibrated flask and make up to the mark with water.

1 ml ≡ 960.66 µg sulphate ≡ 320.66 µg sulphur

Lead titrant solution C (0.002 mol l^{-1})

Pipette 10 ml of lead titrant solution A into a 500-ml calibrated flask and make up to the mark with water. Store in a polyethylene bottle.

1 ml ≡ 192.13 µg sulphate ≡ 64.13 µg sulphur

Other lead titrant solutions may be prepared as desired by dilution of titrant solution A. Lead perchlorate solutions would be more suitable titrants, but lead perchlorate of sufficient purity is not readily available. Ready-prepared 0.1 mol l^{-1} lead perchlorate solution is available from Orion Research Inc. and their reagents and may be used as the titrant solution A.

Note The theoretical sulphate and sulphur equivalence relationships for the titrant solutions are included only as a guide. The true values should always be determined by titration against a standard sulphate solution.

Ionic strength adjustor

Dissolve 7.0 g of sodium perchlorate ($NaClO_4 \cdot H_2O$, analytical reagent grade) in water. If phosphate is present in the samples dissolve also 2.1 g of lanthanum nitrate hexahydrate, $La(NO_3)_3 \cdot 6H_2O$. Transfer the solution to a 100-ml calibrated flask and make up to the mark with water.

Indicator solution

Dissolve 0.6 g of bromocresol green and 0.4 g of methyl red in water and make up to 1 litre with water.

SAMPLE COLLECTION

Collect 250 ml of sample in a washed polyethylene or lead-free glass bottle.

CONCENTRATION RANGE AND UNITS

The lower limit of the technique has been stated to be in the range 100–500 ppm sulphate (Orion, 1975; Eysenbach, Suttkus and Heller, 1975), but with careful attention to technique levels as low as 10 ppm may be determined (Selig, 1970; Hicks, Fleenor and Smith, 1974). Heistand and Blake (1972) found that even lower levels could be determined if the 25-ml portion of sample were spiked with 150 μg or sulphate.

$$10^{-3} \text{ mol l}^{-1} \equiv 32.064 \text{ ppm sulphur} \equiv 96.06 \text{ ppm sulphate}$$

$$1.041 \times 10^{-5} \text{ mol l}^{-1} \equiv 0.3338 \text{ ppm sulphur} \equiv 1 \text{ ppm sulphate}$$

$$3.119 \times 10^{-5} \text{ mol l}^{-1} \equiv 1 \text{ ppm sulphur} \equiv 2.996 \text{ ppm sulphate}$$

ANALYTICAL PROCEDURE

General

The electrodes should be conditioned and stored according to the manufacturers' instructions (see also *Determination of Lead*, p. 243). The lead titrant solution should be standardised against a standard sulphate solution of similar (±50%) concentration to the sample, as there is evidence that the reaction is not completely stoichiometric. The sample may be titrated first and the approximate concentration calculated from the theoretical concentration of the titrant. A standard solution of similar concentration is then prepared and the equivalent concentration of the titrant obtained by titrating it against the standard solution. Finally, the empirically obtained equivalent concentration is used to calculate the true concentration of the sample. The concentration of the titrant should be about ten times the concentration of the sample if a burette is used to make the additions, but a more concentrated titrant may be used if a high-precision micrometer syringe is available.

Titration procedure

Step 1 Pipette 25 ml of sample into a 100-ml beaker containing a magnetic stirrer bar, add 1 ml of ionic strength adjustor solution and 25 ml of methanol.

Step 2 Add four drops of indicator and add dilute perchloric acid or sodium hydroxide solution until the solution is slightly acid (yellow) to the indicator. If the sample has naturally a pH in the range 4–6, this step may be omitted.

Step 3 Remove the electrodes from the solution in which they have been stored, rinse them with deionized water and dry them with a tissue. Immerse the electrodes in the solution in the beaker and start the magnetic stirrer. The stirring speed should be high and uniform, but not so high as to create a vortex.

Step 4 Adjust the level of titrant in the burette so that the meniscus rests on the mark for an exact number of millilitres. Note the reading R_o.

Step 5 Clamp the burette with its tip about 30 mm above the surface of the solution in the beaker and as far from the electrodes as possible.

Step 6 Slowly add the first portion of titrant to the solution in the beaker. This portion should be 25–50% of the expected titre and the meniscus should finish opposite a marked division on the burette.

Step 7 Release the burette from its clamp (or the clamp from the clamp-stand) and touch the surface of the solution with the tip of the burette.

Step 8 Allow the e.m.f. to reach the steady value.

Step 9 Note the e.m.f. reading E_1 and the burette reading R_1.

Step 10 Add further portions of titrant, repeating Steps 7 and 8 in each case and noting the readings R_i and E_i. R_i should always coincide with a marked division on the burette. The size of the portions to be added varies with the proximity to the end-point. Near the end-point, add the smallest scale division on the burette. On either side of the end-point, where the increase in e.m.f. per millilitre of titrant is smaller, add in multiples (up to 5) of the smallest division.

Step 11 When the end-point has been exceeded by about 25%, stop adding titrant.

Step 12 Plot the e.m.f. values, E_i, on the y-axis against the volume additions $(R_i - R_o)$ on the x-axis.

Step 13 Find the equivalent volume of titrant, R_e, from the point of inflexion of the titration curve.

Step 14 Repeat Steps 1–12 with a standard solution of concentration $s \sim t \times R_e/25$, where t is the theoretical equivalent concentration (ppm sulphate) of the titrant.

Step 15 From the point of inflexion of the standard titration curve, find R_s, the volume of titrant equivalent to 25 ml of standard solution.

Step 16 Calculate, c, the concentration of sulphate in the sample, from

$$c = s \times R_e/R_s$$

Step 17 Repeat Steps 14 and 15 every time a new batch of titrant is prepared.

Variations in procedure

Samples that are low in sulphate may be diluted at Step 1 with 25 ml of dioxane or 50 ml of methanol in order to increase the sharpness of the end-point, but the chances of precipitating calcium sulphate are greater. Inconsistent results have been reported when dioxane is used (Goertzen and Oster, 1972), because of its tendency to decompose and form peroxide, which attacks the membrane. If available, lead perchlorate may be used instead of lead nitrate for the titrant solutions; stock standard solutions may be obtained from Orion Research Inc. Kister et al. (1976) found that water—isopropanol mixtures gave sharper end-points than water—acetone mixtures when the titrant was lead nitrate, but that the water—acetone mixture was better when lead perchlorate was the titrant. In each case, 25 ml of solvent was added per 10 ml of aqueous sample. Both the above mixtures were better than water—dioxane mixtures. The best results were obtained using isopropanol and lead nitrate.

SOURCES OF ERROR

Standardization

Selig (1970) found that the titration reaction appeared to be non-stoichiometric and the results of Eysenbach et al. (1975) may also be interpreted in this light. It is recommended, therefore, that the titrant solution be standardized with a standard sulphate solution of similar concentration (±50%) to the sample, using the same titration procedure as with the samples.

Effect of other substances

Anions that form less soluble lead salts than the sulphate will react preferentially with the titrant. Molybdate and tungstate are fairly rare, but phosphate will be a common interferent. When phosphate is present, add lanthanum nitrate to the ionic strength adjustor in order to precipitate the phosphate. Carbonate ions could interfere, but they can be removed by acidification and neutralization. Chloride ions at high concentrations could interfere by forming lead chloride. Mascini (1973) has described a procedure for removing chloride: Stir 50—100 ml of sample with 0.5—1 g of a strongly acidic cation-exchange resin in the silver form for 20 min, filter off the resin, without washing, and stir the filtrate for 20 min with a further portion of resin in the hydrogen form. Filter off the resin, without washing, and take a 25-ml portion of the filtrate for analysis. This procedure will also remove some, but not all, of the phosphate present.

Cations that interfere with the lead electrode (see p. 245) will prevent it responding to changes in lead concentration and no titration curve will be obtained. Silver, mercury and copper ions must be absent and cadmium present only in trace amounts. Sulphate present as high concentrations of calcium sulphate (760 ppm) may precipitate on addition of methanol. The ionic strength adjustor increases the solubility of calcium sulphate and should always be added before the methanol.

Kister et al. (1976) studied the effect of a tenfold molar excess of the chloride salts of common cations and sodium salts of common anions on the titration of 2×10^{-3} mol l^{-1} sulphate in various experimental conditions. 25 ml of either isopropanol or acetone were added per 10 ml of sample and the mixture titrated with either lead nitrate or lead perchlorate. In the presence of sodium, lithium, chloride, nitrate, perchlorate and acetate the sharpest end-points were obtained with isopropanol and lead nitrate, but acetone and lead nitrate were best if potassium or magnesium were present.

Effect of pH

In alkaline solutions, lead hydroxide may be precipitated, but if the solution is too acidic (<pH 4) the sharpness of the end-point decreases (Kister *et al.*, 1976). In either case, Step 2 of the analytical procedure should be followed.

Effect of temperature

If the temperature changes during the titration, the resultant changes in the slope factor and standard potential of the electrode and the solubility of lead sulphate will increase the time taken to reach a steady e.m.f. after each addition of titrant and will make the end-point less sharp. The effect on the accuracy will, however, be much smaller than in direct potentiometry. Magnetic stirrer motors are a common source of heat: insulate the beaker from the heat with a 1-cm thick layer of a material such as expanded polystyrene or plastic foam. Mixing equal portions of water and methanol at room temperature produces a perceptibility warm solution. It is advantageous, therefore, to carry out Steps 1 and 2 of the procedure well in advance of the others so that the solution has time to cool, which will be helped if the beaker is covered with a watch glass and placed in a water-bath at room temperature. A suggested procedure is to work one solution in advance, i.e. do Steps 1 and 2 for sample n and leave it to cool while carrying out Steps 3–12 for sample $n-1$, and so on.

Effect of stirring

Changes in stirring may produce changes in e.m.f. that reduce the sharpness of the end-point. Keep the rate of stirring constant.

TIME OF ANALYSIS

The time taken to complete a titration depends on the number of additions made and the speed of response of the electrode. The electrode should reach a steady e.m.f. within a minute of the addition, but longer times have been reported, especially with dilute reactants; ensure that the surface of the electrode is in good condition (see the section on conditioning and storage). The number of additions required depends largely on foreknowledge of the approximate end-point; thus it may be possible to extend Step 6 of the procedure to 70–80% of the end-point. Experience of a few titrations will often yield a 'starting e.m.f.' before which the electrode need not reach a steady e.m.f. and the volume added is not recorded. By using an automatic titrator with a constant-flow titrant delivery system and recording the potential continuously on a chart recorder, the whole titration from Steps 2 to 12 can be completed in less than 5 min (Goertzen and Oster, 1972; Hicks *et al.*, 1974).

PRECISION

Eysenbach *et al.* (1975) obtained a relative standard deviation of 0.80% from titrations of 0.05 mol l^{-1} sulphuric acid (4800 ppm sulphate). Ross and Frant

(1969) reported 0.2% for solutions containing more than 48 ppm sulphate and 1% for solutions of 10–48 ppm. Mascini (1973) reported relative standard deviations of 1–2% for synthetic sulphate standard solutions (100 and 1000 ppm), 4–5% for mineral water samples (50–75 ppm) and less than 2% for sea water samples (2300–2500 ppm).

Selig (1970) found a relative standard deviation of 0.06–0.6% for the analysis of sulphur in organic compounds after oxidation to sulphate (300–600 ppm sulphate in the solution titrated). Heistand and Blake (1972) obtained a relative standard deviation of 8.8% for the sulphur content of a sample of petroleum when different amounts of the sample were combusted, giving sulphate concentrations of 1–50 ppm in the solution titrated. Hicks et al. (1974) obtained a relative standard deviation of 2.1% for the titration of a standard sulphate solution.

ACCURACY

Mascini (1973) obtained recoveries of 98–100% (mean 98.8%) and 97–104% (mean 100.9%) with 100 and 1000-ppm sulphate standards, respectively. Goertzen and Oster (1972) found a bias of +1.1% compared with gravimetric determination as barium sulphate in waters and soil extracts containing 480–11 700 ppm sulphate. Hicks et al. (1974) analysed standard coal samples for sulphur by titrating the sulphate formed after combustion in a bomb calorimeter and obtained agreement within 1% of the theoretical values (330–1800 ppm sulphate in titrated solutions). In synthetic mixtures of carbon and flowers of sulphur or cystine, the same workers found an average relative error of −0.08% for mixtures giving 220–3300 ppm sulphate in the solution titrated. Heistand and Blake (1972) obtained good agreement between the titration and turbidimetric methods over a 25-fold concentration range. Selig (1970) found errors of 0.1–0.3% for the sulphur content of a range of organic compounds after oxidation to sulphate.

COMPARISON WITH OTHER METHODS

Hicks et al. (1974) compared the titration with gravimetric precipitation as barium sulphate. The precipitation method took 4 h compared with less than 5 min (using an automatic titrator) and had a standard deviation three times as large. Heistand and Blake (1972) used both the titration and turbidimetric methods and found that the mean algebraic difference between the results of the two methods was only 0.12 ppm in samples containing 2–50 ppm and that the titration method was faster and simpler. Sulphate can also be determined by an amperometric titration with lead ions, but the solutions must first be purged with nitrogen to remove oxygen.

OTHER APPLICATIONS

The sulphate titration can be used as the last stage in the determination of the sulphur content of many materials after combustion in a bomb calorimeter. Substances analysed in this way include petroleum (Heistand and Blake, 1972), coal

(Hicks *et al.*, 1974) and various organic chemicals (Selig, 1970). Reynolds (1971) used the lead electrode to determine low concentrations (0–7.2 ppm) of sulphate in Alpine waters by a known subtraction method.

Bibliography

Eysenbach, W., B. Suttkus, and G. Heller, 1975, 'Quantitative Bestimmung von Sulphat und Sulphid (nach Oxydation) mit Hilfe des Orion-Ionalizers', *Z. Anal. Chem.*, **277**, 183.

Goertzen, J. O., and J. D. Oster, 1972, 'Potentiometric titration of sulphate in water and soil extracts using a lead electrode', *Soil Sci. Soc. Amer. Proc.*, **36**, 691.

Heistand, R. N., and C. T. Blake, 1972, 'Titrimetric determination of trace sulphur in petroleum using a lead ion selective electrode', *Mikrochim. Acta*, 212.

Hicks, J. E., J. E. Fleenor, and H. R. Smith, 1974, 'The rapid determination of sulphur in coal', *Anal. Chim. Acta*, **68**, 480.

Kister, G., M. Planchon, A. M. Catterini, J. Chanal, and A. Puech, 1976, 'Potentiométrie a point d'équilibre et dosage de l'ion sulfate', *Trav. Soc. Pharm. Montpellier*, **36**, 129.

Mascini, M., 1973, 'Titration of sulphate in mineral waters and sea water by using the solid-state lead electrode', *Analyst*, **98**, 325.

Mohan, M. S., and G. A. Rechnitz, 1973, 'Preparation and properties of the sulphate ion selective membrane electrode', *Anal. Chem.*, **45**, 1323.

Orion, 1975, *Analytical Methods Guide*, 7th ed., Orion Research Inc., Cambridge, Mass.

Rechnitz, G. A., Z. F. Lin, and S. B. Zamochnick, 1967, 'Potentiometric measurements with SO_4^{-2} and PO_4^{-3} sensitive membrane electrodes', *Anal. Lett.*, **1**(1), 29.

Ross, J. W., and M. S. Frant, 1969, 'Potentiometric titrations of sulphate using an ion-selective lead electrode', *Anal. Chem.*, **41**, 967.

Selig, W., 1970, 'Micro and semimicro determination of sulphur in organic compounds by potentiometric titration with lead perchlorate', *Mikrochim. Acta*, 168.

Reynolds, R. C., 1971, 'Analysis of Alpine waters by ion electrode methods', *Water Resources Res.*, **7**, 1333.

Determination of Nitrate using a Liquid Ion-exchange Electrode

Electrodes of this type are the Orion 93-07, which replaces the earlier model 92-07, the Philips IS 561-NO_3, the Activion 003 15 013 and the Simac N/1C. The Orion 93-07 electrode has an interchangeable module containing the exchanger and the internal reference solution. In the others, the exchanger is encapsulated in a plastic membrane and the electrode has to be charged with reference solution.

APPARATUS

Nitrate electrode; reference electrode; pH meter; magnetic stirrer

It is recommended that a reference electrode containing chloride ions in the internal filling solution is not immersed directly in the sample since leakage of chloride from the liquid junction can interfere with the analysis. A mercury–mercurous sulphate reference electrode is suitable or, alternatively, a double-junction reference electrode with 1 mol l^{-1} potassium sulphate solution in the other chamber.

REAGENTS

Water

Distilled water which has been passed through a mixed-bed deionization unit such that its specific conductivity is $<0.2 \text{ }\mu\text{S cm}^{-1}$ at room temperature is suitable for the preparation of standard and reagent solutions.

Standard nitrate solution A (1000 ppm)

Dissolve exactly 1.630 g of analytical reagent grade potassium nitrate in water and transfer to a 1-litre calibrated flask. Make up to the mark with water.

$1 \text{ ml} \equiv 1 \text{ mg nitrate}$

Prepare other standards as required by dilution of solution A.

Standard nitrate-nitrogen solution A (1000 ppm)

Dissolve exactly 7.222 g of analytical reagent grade potassium nitrate in water and transfer to a 1-litre calibrated flask. Make up to the mark with water.

$1 \text{ ml} \equiv 1 \text{ mg nitrate-nitrogen}$

Prepare other standards as required by dilution of solution A.

Standard nitrate-nitrogen solution B (100 ppm)

Pipette 50 ml of nitrate-nitrogen solution A into a 500-ml calibrated flask and make up to the mark with water.

$1 \text{ ml} \equiv 0.1 \text{ mg nitrate-nitrogen}$

Prepare other standards as required by dilution of this solution.

Ionic strength adjustment solution

Dissolve 10.5 (±0.1) g of analytical reagent grade potassium sulphate and 3.11 (±0.01) g of analytical reagent grade silver sulphate in approximately 800 ml of water. Add 25 ml of 0.1N sulphuric acid and make up to 1 litre with water.

Preservative solution

Dissolve 0.1 g of phenylmercuric acetate in 20 ml of dioxane and make up to 100 ml with water. Store in a glass bottle fitted with a dropping pipette.

SAMPLE COLLECTION

Collect sufficient sample in glass or polyethylene bottles for at least duplicate analyses. To each 100 ml of samples which are known to contain organic matter, e.g. river or ground waters, add 2 drops of preservative solution immediately after collection to prevent biological degradation of the nitrate.

CONDITIONING AND STORAGE OF ELECTRODES

Assemble the electrode, if a sensing module is stored separately from the body of the electrode, and immerse its tip in an approximately 100 ppm nitrate solution for about 30 min before use. If the electrode has been stored dry, treat it in a similar manner. If the electrode is used on a day-to-day basis, store it in a small volume of deionized water between batches of analyses. For longer periods of inactivity, store the electrode dry at room temperatures or, in the case of the Orion electrode, return the sensing module to its storage capsule.

CONCENTRATION RANGE AND UNITS

All the commercial electrodes have a similar response range: linear down to nitrate concentrations of the order of 2–10 ppm and extending to concentrations one-tenth of these values in a non-linear but reproducible response curve. The exact onset of the non-linear range should be determined for each electrode by carrying out the procedure described under *Preparation and Use of a Calibration Graph* so that the appropriate method can be followed for samples whose concentrations are near the limit of the linear range.

$$10^{-3} \text{ mol l}^{-1} \equiv 62 \text{ ppm nitrate} \equiv 14.0 \text{ ppm nitrate nitrogen}$$
$$7.14 \times 10^{-5} \text{ mol l}^{-1} \equiv 4.43 \text{ ppm nitrate} \equiv 1 \text{ ppm nitrate nitrogen}$$
$$1.613 \times 10^{-5} \text{ mol l}^{-1} \equiv 1 \text{ ppm nitrate} \equiv 0.225 \text{ ppm nitrate nitrogen}$$

ANALYTICAL PROCEDURES

Method A: (ca. 600–4 ppm nitrate or 150–1 ppm nitrate nitrogen)

Step 1A Add by pipette 25 ml of a standard solution of concentration s_1 to a 100-ml beaker containing a magnetic stirrer bar. The standard solution should be representative of the lower end of the expected concentration range of the samples.

Step 2A Add by pipette 25 ml of the ionic strength adjustment solution.

Step 3A Remove the nitrate electrode from the solution in which it is immersed and place it in a holder together with the reference electrode. Rinse both electrodes with deionized water, remove surplus water with a tissue and immerse them in the test solution. Note the depth of immersion and keep this level constant throughout.

Step 4A Stir the solution at a moderate rate (keep this constant throughout the procedure) and, when the electrodes give a steady reading, note the e.m.f., E_1.

Step 5A Repeat Step 1A with a second standard whose concentration, s_2, is representative of the higher end of the expected concentration range of the samples and is at least equal to $2 s_1$.

Step 6A Repeat Steps 2A–4A, noting the steady reading E_2.

Step 7A *Either* calculate the calibration slope k, which should be about -56 mV for a tenfold increase in concentration, from

$$k = \frac{E_1 - E_2}{\log s_1 - \log s_2}$$

or

prepare a temporary calibration graph by plotting the points, ($\log s_2$, 0) and ($\log s_1, E_1 - E_2$).

Step 8A Add by pipette 25 ml of sample solution to a 100-ml beaker containing a stirrer bar. The temperature of the sample should be within 1 °C of the temperature of the standard solutions.

Step 9A Repeat Steps 2A–4A, noting the steady reading E_x.

Step 10A Calculate the e.m.f. difference $\Delta = (E_x - E_2)$ and *either* use this value to calculate the concentration c_x in the sample from

$$c_x = s_2 \text{ antilog } (\Delta/k)$$

or

read c_x directly off the calibration graph.

Step 11A Repeat Steps 8A–10A for all the samples to be analysed. The frequency with which the e.m.f. values E_1 and E_2 of the two standard solutions have to be measured depends on the precision required from the analysis but it is recommended that they are re-measured with every six to eight samples

Worked Example

The steady readings noted for two standard nitrate solutions of concentrations 8 and 40 ppm were 490 and 450 mV, respectively. The steady reading of the sample was 438.5 mV.

The slope k is first calculated from

$$k = \frac{(490 - 450)}{\log \frac{8}{40}} = \frac{40}{-0.699}$$

$$= -57.2 \text{ mV per decade}$$

$$c_x = 40 \times \text{antilog} \left(\frac{-11.5}{-57.2}\right)$$

$$= 63.5 \text{ ppm}$$

Method B: (ca. 4.0–0.4 ppm nitrate or 1.0–0.1 ppm nitrate nitrogen)

Step 1B Pipette 25 ml of 0.4 ppm standard nitrate solution (or 0.1 ppm nitrate nitrogen solution) into a 100-ml beaker containing a magnetic stirrer bar.

Step 2B Add by pipette 25 ml of ionic strength adjustment solution.

Step 3B As Step 3A.

Step 4B As Step 4A. Note the steady reading E_1.

Step 5B Repeat Step 1B using a 4.0 ppm standard nitrate solution (or a 1.0 ppm nitrate nitrogen solution).

Step 6B Repeat Steps 2B–4B, noting the steady reading E_2.

Step 7B Pipette 25 ml of sample solution into a 100-ml beaker containing a stirrer bar. The temperature of the sample should be within 1 °C of the temperature of the standard solutions.

Step 8B Repeat Steps 2B–4B, noting the steady reading E_x.

Step 9B Repeat Steps 1B–4B with two additional standard solutions whose concentrations are either representative of the samples or are suitably placed between the first two standards. For example, suitable standards are 0.8 and 2.0 ppm nitrate or 0.2 and 0.6 ppm nitrate nitrogen.

Step 10B Plot the e.m.f. reading for each standard solution against the logarithm of the corresponding concentration.

Step 11B Read the concentration of the sample directly from the calibration graph.

Preparation and Use of a Calibration Graph

Prepare a calibration graph by carrying out Steps 1A–4A of the analytical procedure then Steps 5A and 6A with at least four other standard solutions spanning the range 0.4–400 ppm nitrate or 0.1–150 ppm nitrate nitrogen. Plot the e.m.f. values against the logarithms of the concentrations as described on p. 117; although this graph is not used for the calculation of the results, it is advisable to determine its shape following each fresh assembly of an electrode or period of prolonged dry storage. The slope k obtained in Method A of the analytical procedure should be compared with that obtained from the corresponding portion of the calibration graph; a decrease of 5% in k at constant temperature is an indication that the membrane requires replacement (Mahendrappa, 1969).

SOURCES OF ERROR

Effect of other substances

Table 1 lists the anions whose interference effects have been investigated and expressed as selectivity coefficients. The values quoted for the Orion 92-07 electrode are expected to be valid for the Orion 93-07 electrode which also has a membrane containing a substituted o-phenanthroline complex. The coefficients refer to the equation

$$E = E^\circ + k \log (c_{NO_3} + K_{NO_3, X} c_X^{1/z_X})$$

where c_{NO_3} and c_X are the molar concentrations of nitrate and interferent, X, respectively, and z_X is the charge on the interferent.

From Table 1 it can be seen that interference would be expected from bicarbonate and chloride ions if the electrode was used directly in samples of natural waters. Langmuir and Jacobson (1970) found that an Orion 92-07 electrode at nitrate levels greater than 50 ppm did not suffer from any interference effects in natural waters containing up to 260 ppm bicarbonate ion but at lower nitrate levels an average positive bias of 1.8 ppm was observed. Bunton and Crosby (1969) considered that interference effects from up to 100 ppm chloride or 400 ppm calcium carbonate did not produce serious errors at nitrate-nitrogen concentrations above 5 ppm but the relative errors at 1 ppm were very high. For example, relative

Table 1 Selectivity coefficients for nitrate electrodes

Anion	Selectivity coefficients, $K_{NO_3,X}$	
	Orion 92-07[a,b]	Philips IS 561-NO_3[c]
ClO_4^-	10^3	6×10^2
I^-	20	10
Br^-	0.9	3×10^{-1}
S^{2-}	0.57	
NO_2^-	6×10^{-2}	7×10^{-2}
CN^-	2×10^{-2}	
HCO_3^-	2×10^{-2}	10
Cl^-	$6 \times 10^{-3} - 3 \times 10^{-2}$	10^{-2}
OAc^-	6×10^{-3}	
$S_2O_3^{2-}$	6×10^{-3}	
SO_3^{2-}	6×10^{-3}	
F^-	9×10^{-4}	5×10^{-3}
SO_4^{2-}	6×10^{-4}	
$H_2PO_4^-$	3×10^{-4}	
PO_4^{3-}	3×10^{-4}	
HPO_4^{2-}	8×10^{-5}	

[a] Ross (1969).
[b] Langmuir and Jacobson (1970).
[c] Philips (1975).

errors of 21% and 52% were observed using an Orion 92-07 electrode in solutions containing 400 ppm calcium carbonate and 100 ppm chloride, respectively, at a nitrate-nitrogen concentration of 1 ppm. Øien and Selmer-Olsen (1969), using an Orion 92-07 nitrate electrode for the analysis of samples to which had been added copper sulphate to the extent of 0.01 mol l^{-1}, found that up to 5000 ppm bicarbonate in solutions containing 44.3 ppm nitrate did not significantly alter the measured nitrate response. Myers and Paul (1968), using the same model of electrode, found that the presence of 305 ppm bicarbonate changed the response of the electrode in solutions containing 4.4 and 44.3 ppm nitrate to apparent levels of 18.6 and 62.0 ppm, respectively. The interference effects from bicarbonate and chloride are reduced to insignificant levels in the analytical procedure by the sulphuric acid and silver ions added with the buffer solution.

Mehendrappa (1969) investigated the interference from nitrate ion and concluded that the error of 6% which exists in solutions containing equal concentrations of nitrate and nitrite can be reduced to 0.4% by the addition of sulphamic acid. The cations NH_4^+, K^+, Na^+, Ag^+, Ca^{2+}, Mg^{2+}, Ba^{2+}, Zn^{2+}, and Al^{3+} cause no detectable effect (Milham et al., 1970).

A positive bias of up to 40% relative to other methods of analysis for nitrate was observed by Mack and Sanderson (1971) in soil extracts containing suspended clay or colloidal material. Those workers concluded that filtration was not necessary provided a polyacrylamide flocculating agent was added just prior to the immersion of the electrodes.

PRECISION

Potterton and Shults (1967) reported that the relative standard deviation varied from 3.2 to 0.9% over the range 0.62–6,200 ppm nitrate in solutions of sodium nitrate. The relative standard deviations for the determination of 2–100 ppm nitrate nitrogen in soil extracts were in the range 6–2% (Mehendrappa, 1969). Relative standard deviations of less than 2% were obtained in similar types of samples by Øien and Selmer-Olsen (1969) but this precision was much reduced in samples containing concentrations of nitrate nitrogen below 2 ppm. A reproducibility of about 1% was reported by Langmuir and Jacobson (1970) for the analyses of natural waters having nitrate concentrations in the range 1.5–123 ppm.

ACCURACY

Bunton and Crosby (1969) carried out recovery tests on drinking water and sewage effluent using the electrode, a standard method using brucine, a standard method using 2,4-xylenol and a reduction method with Devarda's alloy. The five drinking waters had nitrate concentrations in the range 0.25–6.4 ppm and recoveries in the range 100–106% were obtained when 2.5 ppm nitrate was added to each sample. The four sewage effluents had nitrate concentrations in the range 6–19 ppm and recoveries in the range 96–102% were obtained when the nitrate concentrations in the samples were approximately doubled. Keeney, Byrnes and Genson (1970) measured the recoveries obtained using an Orion 92-07 electrode at two levels, 10 and 50 ppm, of added nitrate for a number of well and surface waters whose initial nitrate concentrations varied from ∼0–9.5 ppm. The recoveries obtained with the well waters were in the ranges 94–105% and 88–105% for the 10 and 50 ppm additions, respectively, while those for the surface waters were generally lower, 89–95% and 83–98%, respectively, for the same two spiking levels.

The recovery of 25 ppm nitrate nitrogen spikes from soil extracts was greater than 97% (Øien and Selmer-Olsen, 1969). Extensive recovery tests were carried out by Milham et al. (1970) on leaf extracts in which the recovery of nitrate nitrogen was measured and compared with a value obtained by a Devarda's alloy reduction procedure. Recoveries of 2–500 ppm spikes were not significantly different from 100%.

COMPARISON WITH OTHER METHODS

Mahendrappa (1969) compared the performance of the electrode for the determination of nitrate nitrogen in soil extracts with both a distillation technique and a colorimetric method using phenoldisulphonic acid (PDS). The standard deviations obtained by the electrode were generally lower than those obtained by the other two techniques and the procedure was much simpler. In addition, the nitrate electrode is not so prone to interference from chloride as the PDS method, an important factor to consider in the analysis of many natural waters. The Kjeldahl steam distillation is time consuming and the nitrate concentration in the sample must be obtained as the difference of two ammonia determinations: as such, the accuracy of this method is dependent on the relative concentrations of ammonia and nitrate in the sample.

Langmuir and Jacobson (1970) considered that the electrode had the following advantages over the standard brucine method: measurements were unaffected by sample colour or turbidity; the analyses are rapid and one person can analyse 30–40 samples per hour; the analytical procedure is simple yet the accuracy and precision obtained is comparable to that of the brucine method.

In recovery tests of nitrate added to drinking water, Bunton and Crosby (1969), found that the results obtained using an Orion 92-07 electrode were better than those obtained by either a standard method using brucine, a standard method using 2,4-xylenol orange or a reduction method with Devarda's alloy. In similar tests using samples of sewage effluent, the same authors found that the recoveries by the electrode were consistently more reliable than those by the other three methods, where low values were invariably obtained.

OTHER APPLICATIONS

Analysis of Soil Extracts

Extensive use has been made of the electrode for the determination of nitrate nitrogen in soil extracts following an extraction procedure involving an intimate mixing of the soil sample with an aqueous extractant. The methods of extraction do not vary greatly but there are considerable differences in the solutions used (Table 2). Myers and Paul (1968) found that the analysis of distilled water, dilute copper sulphate–silver sulphate solution and saturated calcium sulphate solution extractions for nitrate nitrogen gave similar results and no extractant proved superior to distilled water.

Table 2 Extracting solutions for nitrate nitrogen in soils

Extracting solution	Reference
Potassium fluoride, 0.1 mol l^{-1}	Orion (1975)
Distilled water Saturated calcium sulphate Copper sulphate–silver sulphate	Myers and Paul (1968)
Distilled water Buffer solution	Milham *et al.* (1970)
Potassium sulphate, 0.25 mol l^{-1}, pH 2	Mack and Sanderson (1971)
Copper sulphate, 0.02 N	Øien and Selmer-Olsen (1969)

Bibliography

Bunton, N. G., and N. T. Crosby, 1969, 'The determination of nitrate in waters and effluents using a specific ion electrode', *Water Treatment Exam.*, **18**, 338.

Keeney, D. R., B. H. Byrnes, and J. J. Genson, 1970, 'Determination of nitrate in waters with the nitrate–selective ion electrode', *Analyst*, **95**, 383.

Langmuir, D., and R. L. Jacobson, 1970, 'Specific-ion electrode determination of nitrate in some freshwaters and sewage effluents', *Environ. Sci. Technol.*, **4**, 834.

Mack, A. R., and R. B. Sanderson, 1971, 'Sensitivity of the nitrate-ion membrane electrode in various soil extracts', *Can. J. Soil Sci.*, **51**, 95.

Mahendrappa, M. K., 1969, 'Determination of nitrate nitrogen in soil extracts using a specific ion activity electrode', *Soil Sci.*, **108**, 132.

Milham, P. J., A. S. Awad, R. E. Paull, and J. H. Bull, 1970, 'Analysis of plants, soils and waters for nitrate by using an ion-selective electrode', *Analyst*, **95**, 75.

Myers, R. J. K., and E. A. Paul, 1968, 'Nitrate ion electrode method for soil nitrate nitrogen determination', *Can. J. Soil. Sci.*, **48**, 369.

Øien, A., and A. R. Selmer-Olsen, 1969, 'Nitrate determination in soil extracts with the nitrate electrode', *Analyst*, **94**, 888.

Orion, 1975, *Methods Manual for Series 93 Electrodes*, Orion Research Inc., Cambridge, Mass.

Philips, 1975, 'Ion-selective plastic membrane electrodes.' *Technical Leaflet*, IS-561-series.

Potterton, S. S., and W. D. Shults, 1967, 'An evaluation of the performance of the nitrate-selective electrode', *Anal. Lett.*, **1**, 11.

Ross, J. W., 1969, 'Solid-state and liquid membrane ion-selective electrodes', in *Ion-selective Electrodes*, R. A. Durst (ed.), National Bureau of Standards Special Publication 314, US Department of Commerce, Washington, D.C., Ch. 2.

Sommerfeldt, T. G., R. A. Milne, and G. C. Kozub, 1971, 'Use of the nitrate-specific ion electrode for the determination of nitrate nitrogen in surface and ground water', *Comm. Soil Sci. Plant Anal.*, **2**, 415.

Determination of Boron as Fluoroborate using a Fluoroborate-selective Liquid Ion-exchange Electrode

Electrodes suitable for this analysis are the Orion Model 92-05 and 93-05.

APPARATUS

Fluoroborate electrode; double-junction reference electrode with plastic body; pH meter; magnetic stirrer; plastic beakers; plastic ion-exchange columns

Borosilicate glass containers should not be used. The outer compartment of the double-junction reference electrode should be filled with 0.1 mol l^{-1} potassium chloride solution. A suitable plastic-bodied reference electrode is the Orion 90-02.

The ion-exchange columns recommended for this analysis are non-standard items and a simple method for their construction (Carlson and Paul, 1969) is outlined below. The material of construction is principally standard polyethylene tubing. The curvature initially present in this tubing can be removed prior to construction by immersing the tubing briefly in boiling water, straightening it and then cooling it under cold tap water. Tapered outlets can be formed by gently heating a narrow section of tubing over a low Bunsen flame, drawing it out to form a constriction and then cutting the tubing above the constriction.

Boron ion-exchange column

Cut a 60-mm length of $\frac{1}{4}$-in i.d., $\frac{1}{16}$-in wall polyethylene tubing which is tapered at one end and seal the untapered end into a plastic reservoir which can be constructed from an inverted 60-ml capacity polyethylene bottle with the bottom removed.

Strong acid exchange column

Cut a 150-mm length of $\frac{1}{4}$-in i.d., $\frac{1}{16}$-in wall polyethylene tubing having a taper at one end. Fit a reservoir made from an approximately 100-mm length of $\frac{1}{2}$-in i.d.,

$\frac{1}{16}$-in wall polyethylene tubing onto the untapered end. It will be necessary to use a short length (~10 mm) of intermediate-diameter tubing to sleeve the column such that it seals onto the reservoir.

Weak acid exchange column

Cut a 70-mm length of $\frac{1}{2}$-in i.d., $\frac{1}{16}$-in wall polyethylene tubing tapered at one end. Fit the untapered end into a polyethylene reservoir similar to that described for the boron exchange column.

REAGENTS

Water

Deionized distilled water whose specific conductivity is less than $0.2\ \mu\text{S cm}^{-1}$ is suitable for the preparation of reagent and standard solutions.

Standard boron solution A (100 ppm)

Dissolve 0.572 g of analytical reagent grade boric acid in water and make up to 1 litre in a calibrated flask. Store in a plastic bottle.

$$1\ \text{ml} \equiv 100\ \mu\text{g boron}$$

Additional standards can be prepared by dilution of this stock solution.

Hydrofluoric acid, 10%

Dilute concentrated analytical reagent grade hydrofluoric acid 1:5 with water and store in a polyethylene bottle.

Hydrochloric acid, ~ 3 mol l^{-1}

Dilute concentrated analytical reagent grade acid 1:3 with water and store in a polyethylene bottle.

Ammonium hydroxide, ~ 3 mol l^{-1}

Dilute concentrated analytical reagent grade ammonia solution (sp. gr. 0.88) 1:5 with water and store in a polyethylene bottle.

Sodium hydroxide, ~ 0.3 mol l^{-1}

Dissolve 12 g of analytical reagent grade sodium hydroxide in water and make up to 1 litre. Store in a polyethylene bottle.

Calcium chloride, ~ 0.15 mol l^{-1}

Dissolve 33 g of analytical grade calcium chloride in water and make up to 1 litre. Store in a polyethylene bottle.

Ion-exchange resins

Amberlite XE-243 boron-specific resin, 40–80 mesh wet screen fraction. Carlson and Paul (1969) used BioRex 70, 50–100 mesh, and Dowex 50W-X8, 50–100 mesh, as the weak and strong acid resins, respectively, but equivalent resins may be equally suitable.

Packing and regeneration of columns

Packing and regeneration of boron exchange column

Place a small cotton-wool plug above the tapered outlet of the column and add sufficient Amberlite XE-243 resin to form a column 30-mm high. The resin should be added in the presence of water and the column tapped to aid settlement of the resin. Complete by adding a small cotton-wool plug to the top of the resin.

Regenerate this column after each determination by washing with the following sequence of solutions: 3 ml of 3 mol l^{-1} hydrochloric acid, 3 ml of water, 3 ml of 3 mol l^{-1} ammonium hydroxide solution and, lastly, 3 ml of water. The flow rate through all the columns should be approximately 1 ml per minute.

Packing and regeneration of strong acid exchange column

Place a small cotton-wool plug in the bottom of the column and add, with water, sufficient resin to form a column approximately 120 mm long. Allow the resin to settle under gravity and complete the column by the addition of a second cotton-wool plug.

Regenerate this column after each determination by washing with the following sequence of solutions: 10 ml of 3 mol l^{-1} hydrochloric acid, 2 ml of water, 10 ml of 0.15 mol l^{-1} calcium chloride and, finally, two 10-ml portions of water.

Packing and regeneration of weak acid exchange column

Pack the column in the same manner as described for the strong acid column, adding sufficient weak acid resin to form a column approximately 40 mm long.

Regenerate this column after each determination by washing with the following sequence of solutions: 3 ml of 3 mol l^{-1} ammonium hydroxide and two 5-ml portions of water.

CONCENTRATION RANGE AND UNITS

The method is suitable for samples containing 2–500 µg of boron and the volume of sample added to the column should be between 5 and 15 ml, i.e. the concentration range of samples is 0.15–100 ppm boron.

10^{-3} mol $l^{-1} \equiv 10.81$ ppm boron;   1 ppm boron $\equiv 9.25 \times 10^{-5}$ mol l^{-1}

ANALYTICAL PROCEDURE

Step 1 Add, accurately, a known volume of sample, V_i ml, directly onto the boron column if the sample is neutral or alkaline. If the sample is acidic, add it to the weak acid column and direct the effluent onto the top of the boron column. Discard the effluent from the boron column in both cases. (Note that the elution rate in all the columns should be approximately 1 ml min^{-1}.)

Step 2 Wash the weak acid–boron column arrangement with 4–5 ml of water, or, if only the boron column is used, wash it with approximately 2 ml of water. (Note that the water can be conveniently added from polyethylene wash bottles.)

Step 3 Add 2 ml of 3 mol l^{-1} ammonium hydroxide solution to the boron column and wash with 2–3 ml of water. Discard the effluent.

Step 4 Add 2 ml of 10% hydrofluoric acid to the boron column. Wait 10 min before washing the column with approximately 5 ml of water. Discard the effluents.

Step 5 Place the boron column above the strong acid column such that the former drains onto the latter. Add exactly 10 ml of 0.3 mol l^{-1} sodium hydroxide solution to the boron column and collect the eluate from the strong acid column in a 50-ml polyethylene beaker containing a magnetic stirrer bar.

Step 6 Add exactly a further 10 ml of sodium hydroxide solution to the boron column and collect the eluate in the same 50-ml beaker as in Step 5. Cover the top of the beaker with a watch glass or Petri dish and put aside until the samples in the batch have been collected. This is done to safeguard against errors from changes in electrode calibration which could occur if the analyses of sample and standard solutions were widely separated in time.

Step 7 Regenerate the columns as described previously.

Step 8 Repeat Steps 1–7 for all the samples and one standard boron solution whose concentration is the same as the highest standard used in the calibration procedure. Collect the eluates from Steps 5 and 6 in beakers of the same size containing identical magnetic stirrer bars.

Step 9 Remove the electrodes from the solution in which they are standing, rinse thoroughly with water and then remove surplus water with a paper tissue.

Step 10 Immerse the electrodes in the final eluate and stir at a moderate rate; keep the depth of immersion and the stirring rate constant throughout.

Step 11 Note the steady e.m.f., E_1, for the sample whose volume, added in Step 1, was V_1 ml. For samples in the lower end of the range it is convenient to detect when the e.m.f. is steady from a recorder trace.

Step 12 Repeat Steps 9–11 for the other eluates, noting the steady e.m.f. values $E_2, E_3, \ldots$ etc. for the samples and E_s for the standard solution.

Step 13 Calculate the e.m.f. difference Δ between the readings for each sample and that for the standard solution, i.e.,

$$\Delta_i = E_i - E_s$$

Step 14 From the calibration graph, read off the amount of boron in the sample, s_i, corresponding to a difference Δ from the standard solution. To express this amount (μg) as ppm boron in the sample

$$\text{ppm boron} = \frac{s_i}{V_i}$$

PREPARATION OF THE CALIBRATION GRAPH

The calibration graph is linear for samples containing approximately 20–500 μg of boron. In samples containing less than 20 μg, the loss of linearity is attributed to the presence in the eluate of fluoroborate ion from the liquid membrane of the electrode and at high concentrations it is due to overloading the capacity of the boron exchange column.

Use at least four standard boron solutions covering the expected concentration range of the samples. Carry out Steps 1–6 of the analytical procedure with 10 ml of a standard boric acid solution, adding it to the weak-acid column in Step 1. Repeat this procedure with the remaining standard solutions, regenerating the columns after each standard. It is advisable to start with the most dilute standard solution and follow the sequence of increasing concentrations. Measure the e.m.f. values observed with each standard solution as in Steps 9–11 of the analytical procedure. Calculate the difference Δ between the reading for each standard and the reading for the standard of highest concentration. Plot the values of Δ on the y-axis versus the logarithms of the corresponding boron contents (μg) on the x-axis. Note that the e.m.f. values become more negative with increasing concentrations.

Although the measured e.m.f. values of the standards can vary from day to day — most probably due to changes in E° of the liquid ion-exchange electrode — the shape of the calibration curve is relatively constant, and adequate precision can be obtained by including a reference solution in each batch of samples (Carlson and Paul, 1969). However, it is advisable to include two boron standards at least once a week to check the shape of the calibration graph. Because of the length of time (~50 min) required for each solution, the usual procedure of preparing a within-batch calibration curve for the non-linear region has not been followed.

SOURCES OF ERROR

Temperature

While e.m.f. values are being measured, the temperature of the fluoroborate solutions should be within 1 °C of the temperature at which the calibration graph was established. Accurate temperature control is not necessary for the preliminary ion-exchange stages but samples should be within 5 °C of the laboratory temperature.

Effect of other substances

The principal interfering ions were shown (Carlson and Paul, 1968) to be iodide and, to a lesser extent, nitrate, but these should not be present to any significant extent in the final eluate as the ion-exchange procedure has been designed to isolate fluoroborate from all known interfering ions. The exception is fluoride, which coexists with fluoroborate in the boron exchange column, but this is reduced to manageable concentrations by the inclusion of calcium ions in the final strong acid column where the majority of the fluoride is precipitated on the surface of the resin as calcium fluoride.

Precision

Carlson and Paul (1969) determined the standard deviation of an individual result from ten replicates of standard solutions containing 1—540 μg of boron. The standard deviations for solutions containing 1.08, 5.40, 108 and 540 μg were 1.58, 0.74, 0.49 and 0.54 mV, respectively. The same authors quoted relative standard deviations of less than 3% for well-water samples containing between 0.72 and 1.85 ppm boron.

ACCURACY

Boron concentrations of four well waters were determined (Carlson and Paul, 1969) by the electrode procedure and a colorimetric method using curcumin. The boron concentrations (0.72, 0.79, 1.17 and 1.85 ppm) determined by the two techniques differed by less than 2%.

TIME REQUIRED FOR ANALYSIS

The time required for analysis is about 50 min with column flow rates of 1 ml per minute. The total time required for a batch of analyses could be approximately halved if duplicate sets of exchange columns were used simultaneously, since the greatest time is taken in the ion-exchange procedure. Simple statistical tests of comparison (Student's t-test) carried out on the results obtained from portions of the same standard solution to test for the significance of column dependence.

OTHER USES OF THE ELECTRODE

Wilde (1973) determined boron as tetrafluoroborate ion in extracts of sodium carbonate—sodium fluoride fusions of aluminium oxide—boron carbide. The method was tested using an NBS borosilicate standard containing 3.96% boron, and a mean of 3.93% with a relative standard deviation of 0.67% in the percentage composition was found from eight replicate analyses.

Smith and Manahan (1969) used a fluoroborate electrode to follow the titration of tetrafluoroborate at 2 °C with tetraphenylarsonium chloride and considered that concentrations as low as 5×10^{-3} mol l^{-1} could be detected with a relative error of 1%.

COMPARISON WITH OTHER METHODS

Two colorimetric methods using curcumin and carmine are commonly used as standard procedures for the determination of boron. The curcumin method is applicable to boron in the range 0.10—1.0 ppm but suffers from interference from nitrate concentrations of more than 20 ppm and turbidity may cause difficulties in hard-water samples. The carmine method can be used over a wider range, 0.1—10 ppm, but this method is particularly lengthy, requiring waiting periods in excess of 90 min for sample cooling and colour development. The precision of the carmine method is comparable to that obtained by the electrode and both of these techniques appear to be more precise than the curcumin method.

Bibliography

Carlson, R. M., and J. L. Paul, 1968, 'Potentiometric determination of boron as tetrafluoroborate', *Anal. Chem.*, **40**, 1292.

Carlson, R. M., and J. L. Paul, 1969, 'Potentiometric determination of boron in agriculture samples', *Soil Sci.*, **108**, 266.

Smith, M. J., and S. E. Manahan, 1969, 'Low-temperature precipitation titration of perchlorate and tetrafluoroborate with tetraphenylarsonium chloride and ion-selective electrodes', *Anal. Chim. Acta*, **48**, 315.

Wilde, H. E., 1973 'Potentiometric determination of boron in aluminium oxide—boron carbide using an ion specific electrode', *Anal. Chem.*, **45**, 1526.

Determination of Perchlorate

The only commercial electrode sold for the determination of perchlorate is the Orion 93-81, which supersedes the Orion 92-81. There have been few reports of this electrode being used for regular analytical purposes. The analytical outline that is given here is for its use as an end-point detector in the precipitation titration of perchlorate with tetraphenylarsonium chloride.

APPARATUS

Perchlorate electrode; double-junction reference electrode; pH meter, 50-ml burette; magnetic stirrer

A double-junction reference electrode is preferred to a conventional calomel electrode because of the possibility of precipitation of potassium perchlorate in the liquid junction. The outer compartment of the double-junction reference should be filled with 1.0 mol l^{-1} ammonium nitrate solution.

REAGENTS

Water

Pass distilled water through a mixed-bed deionization column.

Standard tetraphenylarsonium chloride solution (0.05 mol l^{-1})

Dissolve 10.470 g of analytical reagent grade tetraphenylarsonium chloride in about 400 ml of water and adjust the solution to pH 5 ± 1 with dilute hydrochloric acid or sodium hydroxide. Make up to the mark in a 500-ml calibrated flask with water.

    1 ml ≡ 4.973 mg perchlorate

Where only reagent grade tetraphenylarsonium chloride hydrochloride is available, weigh out 11.4 g of this material, dissolve it in approximately 250 ml of water and filter the solution through a Whatman No. 42 filter paper (or equivalent).

Make up to about 500 ml with water and adjust to pH 5 ± 1 as before. Standardise this solution by titrating it against 75-ml portions of 0.025 mol l^{-1} standard perchlorate solution.

Standard perchlorate solution (0.025 mol l^{-1})

Weigh out exactly 2.937 g of analytical reagent grade ammonium perchlorate and dissolve in water. Make up to the mark in a 1-litre calibrated flask.

   1 ml ≡ 2.487 mg perchlorate

Alternatively, prepare standard perchlorate solution by neutralization of analytical reagent grade perchloric acid. Add by graduated pipette 2.7 (±0.1) ml of 60% analytical reagent grade perchloric acid to a beaker containing about 100 ml of deionized water. Pipette 20.0 ml of 1 mol l^{-1} sodium hydroxide solution into the beaker, stir and complete the neutralization of the acid by titrating with 0.1 mol l^{-1} sodium hydroxide solution to an end-point as indicated by methyl orange. Transfer the contents of the beaker to a 1-litre calibrated flask and make up to the mark with water. The concentration of perchlorate, c mol l^{-1}, is

$$c = 0.02 + 10^{-4} v \text{ mol l}^{-1}$$

where v is the volume of 0.1 mol l^{-1} sodium hydroxide solution required to complete the titration.

CONDITIONING AND STORAGE

The sensing module of the Orion 93-81 electrode should be detached from the electrode and kept in its storage capsule between periods of use. The electrode can be used immediately after the module has been screwed onto the body and the tip rinsed with water. Like most liquid ion-exchange electrodes, its response may be improved by a short conditioning period in dilute perchlorate solution.

CONCENTRATION RANGE AND UNITS

No precise limits to the concentration range have been reported. Sample solutions containing 2–3 g l^{-1} of perchlorate were suitable.

   10^{-3} mol l^{-1} ≡ 99.46 ppm

   1 ppm ≡ 1.005×10^{-5} mol l^{-1}

ANALYTICAL PROCEDURE

Pipette 75 ml of a solution containing about 200 mg of perchlorate into a 200-ml beaker containing a magnetic stirrer bar. Rinse the electrodes with deionized water, immerse them in the solution in the beaker and stir at a moderate rate. Carry out a potentiometric titration (see p. 86), adding the tetraphenylarsonium chloride solution from a 50-ml burette and plotting the e.m.f. against the volume added. Allow at least 30 sec before taking a reading of e.m.f. in the region of the end-point. An end-point break of about 150–200 mV should be observed.

Tetraphenylarsonium chloride was also used by Smith and Manahan (1969) for the titration of sodium perchlorate solutions at 2 °C in a jacketed cell with the Orion 92-81 electrode as the end-point indicator. By operating at 2 °C the useful limit of this titration was extended to solutions containing about 0.1 g l^{-1} of perchlorate.

SOURCES OF ERROR

Effect of other substances

The method is not suitable for the determination of perchlorate in the presence of anions which are precipitated with tetraphenylarsonium chloride, amongst which are iodide, periodate and permanganate. The presence of large concentrations of other anions produces an asymmetric titration curve and can introduce considerable error into the location of the equivalence point. It is recommended that the titrant be standardized against perchlorate solutions containing the appropriate salts in roughly equivalent amounts to those known to be present in the samples. Baczuk and Dubois (1968) reported errors of <5% for the determination of perchlorate in 1:1 mixtures containing 200 mg of ammonium perchlorate and 200 mg of another salt. The latter included potassium chloride, bromide, nitrate, chromate, sulphate and sodium fluoride. Potassium chlorate (200 mg) produced an error of 19.7% which was considered to be caused by non-quantitative precipitation of tetraphenylarsonium chlorate.

PRECISION AND ACCURACY

An overall relative standard deviation of 0.073% was reported by Baczuk and Dubois (1968) for the analysis of two American potash samples and a batch of reagent grade potassium perchlorate. The perchlorate content of the two potash samples as determined by the electrode, agreed to within 0.2% of the value obtained by a titanium(III) reduction method but there was a 1% difference in the analyses of the potassium perchlorate by the two methods.

COMPARISON WITH OTHER METHODS

Most volumetric methods for the determination of perchlorate are indirect, involving reduction to chloride which is subsequently titrated with silver or the change in concentration of the reductant is determined by a back titration. These methods are subject to interference from halides, cyanide or reducible species in the sample. Direct colorimetric methods have been reported having greater sensitivity than the electrode method but requiring an organic phase for the development of the absorbing species. A conductometric titration (Baczuk and Bolleter, 1967) using tetraphenylarsonium chloride as titrant, has the same reported accuracy and presision as the potentiometric titration and appears to suffer less from interference by other anions.

OTHER APPLICATIONS

Hseu and Rechnitz (1968) used the Orion 92-81 electrode for the determination of the solubilities of a number of sparingly soluble perchlorates by direct potentiometry. In these analyses, the ionic strength of both the standard solution and the samples was adjusted to 0.1 mol l^{-1} with sodium chloride.

Bibliography

Baczuk, R. J., and W. T. Bolleter, 1967, 'Conductometric titration of perchlorate with tetraphenylarsonium chloride', *Anal. Chem.*, **39**, 93.

Baczuk, R. J., and R. J. Dubois, 1968, 'Conductometric titration of perchlorate with tetraphenylarsonium chloride and a perchlorate ion specific electrode', *Anal. Chem.*, **40**, 685.

Hseu, T. M., and G. A. Rechnitz, 1968, 'Analytical Study of a perchlorate ion selective membrane electrode', *Anal. Lett.*, **1**(10), 629.

Smith, M. J., and S. E. Manahan, 1969, 'Low temperature precipitation titration of perchlorate and tetrafluoroborate with tetraphenylarsonium chloride and ion-selective electrodes', *Anal. Chim. Acta*, **48**, 315.

Appendix 1

Theoretical Values of the Nernstian Slope for a Univalent Cationic Determinand

°C	k	°C	k	°C	k	°C	k
10	56.18	18	57.77	26	59.36	34	60.95
11	56.38	19	57.97	27	59.56	35	61.15
12	56.58	20	58.17	28	59.76	36	61.34
13	56.78	21	58.37	29	59.95	37	61.54
14	56.98	22	58.57	30	60.15	38	61.74
15	57.18	23	58.76	31	60.35	39	61.94
16	57.38	24	58.96	32	60.55	40	62.14
17	57.57	25	59.16	33	60.75		

Values are calculated from $k = 10^3 \, RT \ln(10)/F$ mV per tenfold increase in activity, where the gas constant $R = 8.31432$ J K^{-1} mol^{-1}, T = the absolute temperature, $F = 9.64845 \times 10^4$ C mol^{-1} is Faraday's constant and $\ln(10) = 2.302585$.

The slope factors for polyvalent and anionic species are given by k/z, where k is the tabulated value and z is the charge, with sign, on the determinand.

Appendix 2

Debye-Hückel A and B Coefficients for Electrolytes in Water

$t\,°C$	10	15	18	20	25	30	35	38	40
A	0.499	0.503	0.505	0.507	0.512	0.516	0.521	0.524	0.526
B	0.326	0.327	0.328	0.328	0.329	0.330	0.331	0.332	0.332

Values are calculated from

$$A = \frac{1.8249 \times 10^6}{\sqrt{(\epsilon \cdot T)^3}} \quad \text{and} \quad B = \frac{50.293}{\sqrt{\epsilon \cdot T}},$$

where ϵ = the dielectric constant for water and $T = 273.16 + t$ is the absolute temperature.

Appendix 3

Equipment Manufacturers

No attempt has been made to provide a comprehensive survey of all the manufacturers of electrodes, pH meters and titrators, but list includes all the main international suppliers. The companies listed are those whose products have been the subject of some specific comment or description in the text and inclusion does not imply endorsement. More complete listing of manufacturers are available in the surveys published by *Analytical Chemistry* and *Chemistry and Industry*, as well as from many trade magazines. In each case the parent company is named, followed where appropriate by its British and US agents. The asterisks in the first five columns indicate the areas of activity concerned.

Key
P = pH and reference electrodes
E = ion-selective electrodes
L = laboratory instrumentation
I = industrial instrumentation
T = titration apparatus

MANUFACTURER AND ADDRESS

P	E	L	I	T	
*	*	*	–	–	Activion Glass Ltd., Halstead, Essex CO9 2EX.
*	–	*	–	–	Balsbaugh Laboratories Inc., 25 Industrial Park Rd., S. Hingham, Mass. 02043. Martron Associates Ltd., 81 Station Rd., Marlow, Bucks.
*	*	*	*	*	Beckman Instruments Inc., 2500 Harbor Blvd., Fullerton, Calif. 92634. Beckman–RIIC Ltd., Eastfield Industrial Estate, Glenrothes KY7 4NG.
*	*	*	–	–	Coleman Instruments, 2000 York Rd., Oakbrook, Ill. 60521. Perkin-Elmer Ltd., Post Office Lane, Beaconsfield, Bucks. HP9 1QA.
*	*	*	–	–	Corning Glass Works, Houghton Park, Corning, N.Y. 14830. Corning–EEL, Evans Electroselenium Ltd., Halstead, Essex, CO9 2DX.

```
P E L I T
```
\- * — — — EDT Supplies Ltd., 65 Ivy Crescent, London W4 5NG.
* * * * — Electrofact N.V., P.O. Box 163, Radium Weg 20, Amersfoort, Netherlands. Serck Controls, Eastern Avenue, Gloucester GL4 7BZ. Electrofact, 3407 Rose Avenue, Ocean, N.J. 07712.
* * * * * Electronic Instruments Ltd., Hanworth Lane, Chertsey, Surrey KT16 9LF. Kent Cambridge Scientific Inc., 8020 Austin Ave., Morton Grove, Ill. 60053.
* * — * — The Foxboro Company, Neponset Ave., Foxboro, Mass. 02035. Foxboro–Yoxall Ltd., Redhill, Surrey RH1 2HL.
* — — * — Dr. W. Ingold A.G., CH-8902 Urdorf-Zürich, Industiezone Nord, Switzerland. Ingold Electrodes, Lexington, Mass. 02173.
* * * — — Instrumentation Laboratory Inc., 113 Hartwell Ave., Lexington, Mass. 02173. Instrumentation Laboratory (UK) Ltd., Station House, Stamford New Rd., Altrincham, Cheshire.
* — — * * Ionics Inc., 65 Grove St., Watertown, Mass. 02172. Techmation Ltd., 58 Edgware Way, Edgware, Middlesex HA8 8JP.
* * * * — Leeds & Northrup Co., Sumneytown Pike, North Wales, Pa. 19454. Leeds & Northrup Ltd., Wharfdale Rd., Tyseley, Birmingham B11 2DJ.
* * * * * Metrohm A.G., CH-9100 Herisau, Switzerland. Roth Scientific Co. Ltd., Zurcourt House, 27 Osborne Rd., Farnborough, Hants. Brinkmann Instruments Inc., Cantiague Rd., Westbury, N.Y. 11590.
— — — — * Mettler Instruments A.G., CH-8606 Greifensee-Zürich, Switzerland. A. Gallenkamp and Co. Ltd., P.O. Box 290, Technico House, Christopher St., London EC2P 2ER. Mettler Instrument Corp., P.O. Box 100, Princeton, N.J. 08540.
* * * * — Orion Research Inc., 380 Putnam Ave., Cambridge, Mass. 02139. M.S.E. Scientific Instruments, Manor Royal, Crawley, Sussex RH10 2QQ.
* * * * — N.V. Philips' Gloeilampenfabrieken, Eindhoven, Netherlands. Pye–Unicam Ltd., York St., Cambridge, England. Pye–Ether Ltd., Caxton Way, Stevenage, Herts SG1 2DG.
* — — * — Polymetron Ltd., CH-8634 Hombrechticon, Switzerland. G. W. Thornton and Sons Ltd., 10 Eden Place, Cheadle, Cheshire SK8 1AU.
* — * * * Pye–Unicam Ltd., York St., Cambridge.
* * * * * Radiometer A/S, Emdruprej 72, DK-2400 Copenhagen NV, Denmark. V. A. Howe and Co., 88 Peterborough Rd., London SW6 3EP; The London Co., 811 Sharon Drive, Westlake, Ohio 44145.
* * — * — Serck Controls, Eastern Avenue, Gloucester GL4 7BZ.
* * * * — Simac Instrumentation Ltd., Lyon Rd., Hersham, Walton-on--Thames, Surrey KT12 3PU.
* * — — — [Schott] Jenaer Glaswerk Schott & Gen., Hattenbergstrasse 10,

P E L I T Postfach 24 80, Mainz, G.F.R. H. V. Skan Ltd., 425/433 Stratford Rd., Shirley, Solihull, Warwickshire.

* * * * * Tacussel électronique, Solea, 72-78 rue d'Alsace, 69100 Villeurbanne, France. Clandon Southern Instruments Ltd., Lysons Ave., Ash Vale, Aldershot GU12 5QF.

Appendix 4

Tables of the Function $\dfrac{1}{(\text{antilog}\ \dfrac{E_2 - E_1}{k}) - 1}$

$\|E_2 - E_1\|$ $\|z\|=1$	54	55	56	57	58	59	60	$\|E_2-E_1\|$ $\|z\|=2$
				$k, \|z\|=1$				
1.0	22.955	23.390	23.824	24.258	24.692	25.127	25.561	0.5
2.0	11.233	11.450	11.667	11.884	12.101	12.318	12.535	1.0
3.0	7.328	7.473	7.617	7.762	7.906	8.051	8.195	1.5
4.0	5.377	5.485	5.594	5.702	5.810	5.919	6.027	2.0
5.0	4.208	4.295	4.381	4.468	4.554	4.641	4.728	2.5
6.0	3.430	3.502	3.574	3.646	3.718	3.790	3.862	3.0
6.4	3.187	3.255	3.322	3.390	3.457	3.524	3.592	3.2
7.0	2.875	2.937	2.998	3.060	3.122	3.183	3.245	3.5
7.4	2.695	2.754	2.812	2.870	2.928	2.987	3.045	3.7
8.0	2.460	2.514	2.567	2.621	2.675	2.729	2.783	4.0
8.4	2.322	2.373	2.424	2.475	2.526	2.578	2.629	4.2
9.0	2.138	2.185	2.233	2.281	2.329	2.376	2.424	4.5
9.4	2.028	2.074	2.119	2.165	2.211	2.256	2.302	4.7
10.0	1.881	1.923	1.966	2.009	2.052	2.095	2.138	5.0
.2	1.835	1.877	1.919	1.961	2.003	2.045	2.087	.1
.4	1.792	1.833	1.874	1.915	1.956	1.998	2.039	.2
.6	1.750	1.790	1.831	1.871	1.911	1.952	1.992	.3
.8	1.710	1.749	1.789	1.828	1.868	1.908	1.947	.4
11.0	1.671	1.710	1.749	1.787	1.826	1.865	1.904	5.5
.2	1.634	1.672	1.710	1.748	1.786	1.824	1.862	.6
.4	1.598	1.635	1.672	1.710	1.747	1.785	1.822	.7
.6	1.563	1.599	1.636	1.673	1.710	1.747	1.783	.8
.8	1.529	1.565	1.601	1.637	1.674	1.710	1.746	.9
12.0	1.498	1.532	1.568	1.603	1.639	1.674	1.710	6.0
	27	27.5	28	28.5	29	29.5	30	
				$k, \|z\|=2$				

$\|E_2-E_1\|$			$k, \|z\|=1$					$\|E_2-E_1\|$
$\|z\|=1$	54	55	56	57	58	59	60	$\|z\|=2$
12.2	1.465	1.500	1.535	1.570	1.605	1.640	1.675	6.1
.4	1.435	1.469	1.504	1.538	1.572	1.607	1.641	.2
.6	1.406	1.439	1.473	1.507	1.541	1.574	1.608	.3
.8	1.377	1.411	1.444	1.477	1.510	1.543	1.577	.4
13.0	1.350	1.383	1.415	1.448	1.480	1.513	1.546	6.5
.2	1.323	1.355	1.387	1.420	1.452	1.484	1.516	.6
.4	1.298	1.329	1.361	1.392	1.424	1.456	1.487	.7
.6	1.272	1.304	1.335	1.365	1.397	1.428	1.459	.8
.8	1.248	1.279	1.309	1.340	1.371	1.401	1.432	.9
14.0	1.225	1.255	1.285	1.315	1.345	1.376	1.406	7.0
.2	1.202	1.231	1.261	1.291	1.321	1.350	1.380	.1
.4	1.179	1.209	1.238	1.267	1.297	1.326	1.355	.2
.6	1.158	1.187	1.216	1.244	1.273	1.302	1.331	.3
.8	1.137	1.165	1.194	1.222	1.251	1.279	1.308	.4
15.0	1.116	1.144	1.172	1.201	1.229	1.257	1.285	7.5
.2	1.097	1.124	1.152	1.179	1.207	1.235	1.263	.6
.4	1.077	1.104	1.132	1.159	1.186	1.214	1.241	.7
.6	1.058	1.085	1.112	1.139	1.166	1.193	1.220	.8
.8	1.040	1.067	1.093	1.120	1.146	1.173	1.199	.9
16.0	1.022	1.048	1.074	1.101	1.127	1.153	1.179	8.0
.2	1.005	1.031	1.055	1.082	1.109	1.134	1.160	.1
.4	0.988	1.013	1.039	1.064	1.090	1.115	1.141	.2
.6	0.971	0.996	1.022	1.047	1.072	1.097	1.122	.3
.8	0.955	0.980	1.005	1.030	1.055	1.079	1.104	.4
17.0	0.939	0.964	0.988	1.013	1.038	1.062	1.087	8.5
.2	0.924	0.948	0.972	0.997	1.021	1.045	1.070	.6
.4	0.909	0.933	0.957	0.981	1.005	1.029	1.053	.7
.6	0.895	0.918	0.942	0.965	0.989	1.013	1.036	.8
.8	0.880	0.904	0.927	0.950	0.974	0.997	1.020	.9
18.0	0.866	0.889	0.912	0.935	0.958	0.982	1.005	9.0
.2	0.853	0.875	0.898	0.921	0.944	0.967	0.990	.1
.4	0.839	0.862	0.884	0.907	0.929	0.952	0.975	.2
.6	0.826	0.848	0.871	0.893	0.915	0.938	0.960	.3
.8	0.814	0.835	0.857	0.879	0.901	0.924	0.946	.4
19.0	0.801	0.823	0.844	0.866	0.888	0.910	0.932	9.5
.2	0.789	0.810	0.832	0.853	0.875	0.896	0.918	.6
.4	0.777	0.798	0.819	0.841	0.862	0.883	0.905	.7
.6	0.765	0.786	0.807	0.829	0.849	0.870	0.892	.8
.8	0.754	0.775	0.795	0.816	0.837	0.858	0.879	.9
20.0	0.743	0.763	0.784	0.804	0.825	0.846	0.866	10.0
.2	0.732	0.752	0.772	0.793	0.813	0.834	0.854	.1
.4	0.721	0.741	0.761	0.781	0.802	0.822	0.842	.2
	27	27.5	28	28.5	29	29.5	30	

$$k, |z|=2$$

$\|E_2-E_1\|$			$k, \|z\|=1$					$\|E_2-E_1\|$
$\|z\|=1$	54	55	56	57	58	59	60	$\|z\|=2$
20.6	0.711	0.731	0.750	0.770	0.790	0.810	0.830	10.3
.8	0.700	0.720	0.740	0.759	0.779	0.799	0.819	.4
21.0	0.690	0.710	0.729	0.749	0.768	0.788	0.807	10.5
.2	0.681	0.700	0.719	0.738	0.757	0.777	0.796	.6
.4	0.671	0.690	0.709	0.728	0.747	0.766	0.785	.7
.6	0.661	0.680	0.699	0.718	0.737	0.756	0.775	.8
.8	0.652	0.671	0.689	0.708	0.727	0.745	0.764	.9
22.0	0.643	0.661	0.680	0.698	0.717	0.735	0.754	11.0
.2	0.634	0.652	0.671	0.689	0.707	0.726	0.744	.1
.4	0.625	0.643	0.661	0.680	0.698	0.716	0.734	.2
.6	0.617	0.635	0.652	0.670	0.688	0.706	0.724	.3
.8	0.608	0.626	0.644	0.661	0.679	0.697	0.715	.4
23.0	0.600	0.618	0.635	0.653	0.670	0.688	0.706	11.5
.2	0.592	0.609	0.627	0.644	0.661	0.679	0.696	.6
.4	0.584	0.601	0.618	0.636	0.653	0.670	0.687	.7
.6	0.576	0.593	0.610	0.627	0.644	0.661	0.679	.8
.8	0.569	0.585	0.602	0.619	0.636	0.653	0.670	.9
24.0	0.561	0.578	0.594	0.611	0.628	0.645	0.661	12.0
.2	0.551	0.570	0.587	0.603	0.620	0.636	0.653	.1
.4	0.546	0.563	0.579	0.595	0.612	0.628	0.645	.2
.6	0.539	0.555	0.572	0.588	0.604	0.620	0.637	.3
.8	0.532	0.548	0.564	0.580	0.596	0.613	0.629	.4
25.0	0.525	0.541	0.557	0.573	0.589	0.605	0.621	12.5
.2	0.518	0.534	0.550	0.566	0.582	0.597	0.613	.6
.4	0.512	0.527	0.543	0.559	0.574	0.590	0.606	.7
.6	0.505	0.521	0.536	0.552	0.567	0.583	0.598	.8
.8	0.499	0.514	0.529	0.545	0.560	0.576	0.591	.9
26.0	0.493	0.508	0.523	0.538	0.553	0.569	0.584	13.0
.2	0.486	0.501	0.516	0.531	0.547	0.562	0.577	.1
.4	0.480	0.495	0.510	0.525	0.540	0.555	0.570	.2
.6	0.474	0.489	0.504	0.518	0.533	0.548	0.563	.3
.8	0.468	0.483	0.498	0.512	0.527	0.542	0.557	.4
27.0	0.462	0.477	0.491	0.506	0.521	0.535	0.550	13.5
.2	0.457	0.471	0.485	0.500	0.514	0.529	0.543	.6
.4	0.451	0.465	0.480	0.494	0.508	0.523	0.537	.7
.6	0.446	0.460	0.474	0.488	0.502	0.516	0.531	.8
.8	0.440	0.454	0.468	0.482	0.496	0.510	0.525	.9
28.0	0.435	0.449	0.462	0.476	0.490	0.504	0.518	14.0
.2	0.429	0.443	0.457	0.471	0.485	0.499	0.513	.1
.4	0.424	0.438	0.452	0.465	0.479	0.493	0.507	.2
.6	0.419	0.433	0.446	0.460	0.473	0.487	0.501	.3
.8	0.414	0.428	0.441	0.454	0.468	0.481	0.495	.4
	27	27.5	28	28.5	29	29.5	30	
			$k, \|z\|=2$					

$|E_2-E_1|$			$k, |z|=1$					$|E_2-E_1|$
$|z|=1$	54	55	56	57	58	59	60	$|z|=2$
29.0	0.409	0.422	0.436	0.449	0.462	0.476	0.489	14.5
.2	0.404	0.417	0.431	0.444	0.457	0.470	0.484	.6
.4	0.400	0.413	0.426	0.439	0.452	0.465	0.478	.7
.6	0.395	0.408	0.421	0.434	0.447	0.460	0.473	.8
.8	0.390	0.403	0.416	0.429	0.442	0.455	0.468	.9
30.0	0.386	0.398	0.411	0.424	0.437	0.450	0.463	15.0
31.0	0.364	0.376	0.388	0.400	0.413	0.425	0.437	15.5
32.0	0.343	0.355	0.367	0.378	0.390	0.402	0.414	16.0
33.0	0.324	0.335	0.347	0.358	0.369	0.381	0.392	16.5
34.0	0.307	0.317	0.328	0.339	0.350	0.361	0.372	17.0
35.0	0.290	0.300	0.311	0.321	0.332	0.343	0.353	17.5
40.0	0.222	0.231	0.239	0.248	0.257	0.266	0.275	20.0
45.0	0.172	0.179	0.187	0.194	0.201	0.209	0.216	22.5
50.0	0.135	0.141	0.147	0.153	0.159	0.166	0.172	25.0
	27	27.5	28	28.5	29	29.5	30	
				$k, |z|=2$				

Index

Asterisked items may be consulted in each method in Part II. **Bold** numbers refer to main entries.

*Accuracy, 61, 71, 105, 107, 126
Acetate buffer, 84, 187, 228, 233, 236, 249, 320, 324, 329, 335, 363
Acetic acid, 85, 149
Acetone, in titrations, 83, 96, 370
Acetylacetone, masking, 125, 190
Acid—base titrations, 81, 87, 89, 93, 147
Acidity determination, **147**
Action limits, 77
Activity, 6, 9, 81
Activity coefficient, 6, 101, 117
  mean, 8
  single-ion, 8
Addition, known, *see* Known Addition
Addition titration, 83, 91, 222, 320
Adsorption, 95, 110, 213
Air, analysis of, 319
Air conditioners, 113, 283
Alkali error, 28, 144
Alkali metals determination, 124, *see also individual entries*
Alkalinity determination, **147**
Aluminium determination, 82, 224, 233, **248**
Aluminosilicate glass, 28, 155
Amalgams, 16, 18, 41
Amines, 23, 36, 161, 164, 168, 181, 284
Amino acids, 23
Ammonia
  buffer, 162, 229, 233
  determination, **279**
  electrode, 29, **34**, 37, 279, 290
Ammonium electrode, 28, 32, 286
Analate addition, 320
Antimony—antimony oxide electrode, 54, 58, 83, 95, 120, 135

Antimony titration, 85
Argentodicyanide indicator, 308, 362
Arsenate titration, 82
Arsenite determination, 82, 85, 350
Asbestos fibre junction, 43
Ascorbic acid
  determination, 350
  reagent, 110, 344, 356, 361
Association constant, 20, 94
Autoprotolysis constant, 20, 87

Back titration, 226, 231, 392
Barium
  electrode, 33, 82, 84
  titration, 82, 84, 235
Baseline, 75
Benzoquinone, 22
Between-batch standard deviation, 69
Biological degradation of samples, 110, 137, 375
Biological samples, analysis, 3, 184, 192, 319
Bismuth nitrate, reagent, 303, 310
Blanks, 75
Boiler water and feed-water, analysis, 167, 259, 285, 330, 339
Borax buffer, 138
Boric acid determination, 153
Boron determination, 153, **383**
Bromate determination, 85
Bromide
  determination, 82, **343**
  electrode, 24, 27, 343
Bromocresol green indicator, 367
Bromocresol purple indicator, 215, 240, 244
Buffers, pH, 45, 84, **136**, 162, 196, 233,

288, 317, *see also* Acetate buffer, Ammonia buffer, Phosphate buffer, Tris buffer

Cadmium
  determination, 84, 224, **239**
  electrode, 24, 84, 226, 239
  titrant, 226, 357
Cadmium sulphide, 24, 239
Caesium, 28, 82
Calcium
  determination, 84, 86, **186**, 224, 233, 242
  electrode, 32, 84, 186, 226
Calcium chloride, activity coefficients, 8
Calcium sulphate, extractant, 381
*Calibration, 5, 72, **117**
*Calibration graph, 77, **117**
Calomel electrode, *see* Reference electrodes
Capillary junction, 43
Carbon dioxide
  determination, **254**
  electrode, 29, 37, 254
  interference, 95, 108, 145
Carbonate buffer, 138
Carbonate determination, 254
CDTA, 84, 227, 235, 248, 314
Cell, electrochemical, 9
Ceramic frit junction, 43, 137, 205
Ceric sulphate, reagent, 85
Chart recorders, 51, 59, 130
Chelating agents, 83, **229**
Chel DP determination, 227
Chloranil, 23
Chloride
  determination, 75, **323**, **333**
  electrode, 24, 323, 333
  liquid ion-exchange electrode, 33
  titration, 82, 332
Chromate titration, 82, 246
Chromium, 233
Citrate buffer, 137, 317
Clinical analysis, 168, 184, 286
Cobalt titration, 84
Colloids, 12, 46
Combination electrodes, 54, 110, 127, 135, 138, 205, 214, 313, 323, 343, 355
Compleximetric titrations, 83, 90, 97, 192, 195, 222, 224, **226**, 242, 248
Complexing agents, 124, 226
Concentrated samples, 112, 166
*Concentration units, 112

*Conditioning of electrodes, 110
Conductivity, 5
Confidence limits, 63
Control chart, 76, 125
Copper
  determination, **214**
  electrode, 24, 84, 214, 226, 233
  titration, 84, 229
Copper selenide, 24, 214
Copper sulphate, extractant, 381
Copper sulphide, 24, 214
Co-precipitation, 95, 340
Correlation coefficient, 78
Criterion of detection, 68
Critical points, 68
Cryptates, 84
Cyanide
  antidote, 302
  determination, **298**, **306**, 350
  electrode, 34, **298**
  titration, 212, 304, 305
Cyclohexylamine buffer, 162, 233

Davies equation, 7
Debye–Hückel
  equation, 7
  A and B coefficients, 295
Deep-freeze, 110, 264, 292
Degrees of freedom, 64
Derivative titrations, 59, 89
Detergents, 38, 267
Determinand, definition, 5
Dichromate titration, 85, 277
Dielectric constant, 7, 96, 397
Diethylamine, 162, 181
Diisopropylamine, 181
Dimethylamine, 162
Dioxane, in titrations, 83, 369
Dip cells, 57
Diphenylthallium(I) sulphate, 82
Direct activity scale, 50, 53, 91, 114, 120
Dispensers, 107
Double-junction reference electrode, *see* Reference electrodes
Drift, 129

Earthing, 14, 130
EDTA
  complexes, 227
  determination, 229
  reagent, 84, 195, 222, 252, 279, 291, 299, 308
EGTA, 84, 224, 227

*Electrodes, 16, *see also* Gas-sensing electrodes, Glass electrodes, Hydrogen electrode, Liquid ion-exchange electrodes, Metal electrodes, quinhydrone electrode, Reference electrodes, Solid-state electrodes
Electrodes of the second kind, 18
Electrometers, 54
Emulsions, 43, 58
End-point, 5, 58, 86, 230
Enzyme electrodes, 34
Equitransferent junction, 12
Equivalence point, 5, **86**
Errors, statistical combination, 63
Ethanol, in titrations, 248, 320
Extractants, 381

Faraday, 394
Fats, fouling by, 58
Ferrocyanide, determination, 85, 246
Ferrous iron, 85, 153, 248
Fibre junction, 43
Fibres, fouling by, 58
Flammable atmospheres, 55
Flocculating agents, 380
Flow cells, 56, 106, **115**, 129, 145, 160, 180, 186, 254, 333
Fluoride electrode, 24, 37, 42, 82, 248, **313**, 350
Fluoride, in pH measurements, 22, 114, 135, 152
Fluoroborate electrode, 33, **383**
Fluorosilicate, 317
Foods, analysis of, 269, 277, 296, 304
Fouling, 20, 43, 57, 128, 135, 145
F-test, 67

Gas constant, 394
Gas-sensing electrodes, **34**, 56, 106, 110, 127, 254, 263, 271, 279, 290
Gas stripping, 125
Gaussian distribution, 61
Glass electrodes, 27, 34, 42, 58, 83, 115, 135, 147, 155, 160, 171, 180, 263, 271, 279
Glycerol, reagent, 273
Glycine buffer, 196
Gold redox electrode, 20, 22
Graphite electrode, 27
Ground, *see* Earthing
Ground-glass sleeve junction, 43, 128, 324
  in combination electrodes, 135

Gran titrations, 83, **89**, 96, 114, 125, 147, 151, 222, 229, 309, 350
Guntelberg equation, 7

Half-cell, 10, 39
Half-cell potential, 10, 40
Henderson equation, 12
Henry's Law, 276
Heterogeneous membranes, 24
Hexacyanoferrate(II), 85, 246
Hexamethylenetetramine buffer, 233, 248
Hydrazine, 350
Hydrofluoric acid, 22, 37, 267, 275, 317
Hydrogen electrode, 10, **21**, 54, 95
Hydrogen ions, in liquid junction, 12, *see also* pH control, pH electrode, pH measurement, pH meter
Hydrogen peroxide, 85
Hydrogen sulphide, 37, 205
Hydroxide
  electrode, 19
  in liquid junctions, 12
Hydroxylamine, determination, 350
Hydroxyquinoline, 242
Hypobromite, titration, 85
Hysteresis, 166

Indicator electrode, 5, 17
Indifferent electrolyte, 7
Industrial monitors, 54, 60, 108
Inflection, point of, 59, 86, 114
In-line, 42, 57
Input resistance, 14
In-stream, 42
*Interferences, 11, 18, 26, 34, 36, 45, 73, 81, 96, 125, 127
Iodate, 85, 272, 349, 350
Iodide
  electrode, 24, 27, 82, **343**
  titration, 82, 85
Iodine, determination, 349
Ion association, 8, 166
Ion-dipole interactions, 30
Ion-exchange, 24, 28, 383
Ionic strength, 7, 101, 124
Ionic transport, 23
Ion size parameter, 7
Iron, determination, 224, 233
Irreversible reactions, 98
Isopotential, 53, 142
Isopropanol, 370

J-tube junctions, 43

Karl Fischer titration, 54
Kjeldahl digestion, 290, 381
Known addition, 51, 72, 81, **98**, 122, 124, 199, 207, 212, 222, 296, 320
Known subtraction, 81, 98, **102**, 122, 373
Kryptofix, 84

Lanthanum, determination, 26, 224, 233
Lanthanum fluoride, 24, 313
Lanthanum nitrate, reagent, 82, 319, 367, 370
Lead
  electrode, 16, 24, 33, 82, **243**, 366
  titration, 224, 235, 246, 363
Lead nitrate, reagent, 303, 310, 356, 367
Lead perchlorate, titrant, 82, 356
Lead phosphate, co-precipitation, 340
Lead sulphide, 24, 243
Light, effects of, 26, 115, 131, 224, 328, 348
Limit of detection, 18, 26, 32, **68**, 97, 106, 116, 226, 237
Linear regression, 77
Linear titration plots, 83, **93**, 96, 114
Linearity, test for, 78
Liquid ion-exchange electrodes, **29**, 56, 74, 96, 101, 110, 115, 116, 171, 186, 194, 227, 374, 383, 390
Liquid junction, **11**, 38, 43, 114, 135
Liquid junction potential, **10**, 142, 320
Lithium
  determination, 82, 84, **155**, 320
  electrode, 28, 33, 155
Lithium chloride, 42
Lithium chloroacetate, 42

Macrocyclic ligands, 30
Magnesium
  determination, 84, 194, 233
  masking, 190
Manganese, titration, 84, 233
Masking, 96, 124, 190
Mean, definition, 61
Mean ionic activity coefficient, 8
Measuring cylinders, 107
Membrane electrodes, 23

Memory effects, 127
Mercuric nitrate, titrant, 352
Mercuric sulphide, 340
Mercurous chloride electrode, 340
Mercury, 16, 82, 224, 235
Mercury catalyst, 291
Mercury—mercuric oxide electrode, 20
Mercury—mercurous chloride electrode, see Reference electrodes, Calomel electrode
Mercury—mercurous sulphate electrode, see Reference electrodes
Metabisulphite, 271
Metal electrodes, 13, 16
Metal oxides, 19
Metals, analysis, 252, 289
Methanol, in titrations, 83, 96, 366
Methylamine, 36
Microcomputer, 51, 59
Minerals, analysis, 252, 319, 349
Mixing, 130
Mobility, ionic, 11
Molybdate, titration, 82
Monactin, 33

Nernst equation, 6
Nernstian response, 5, 18, 119, 394
Neutral carriers, **30**, 33, 158, 160, 171, 186, 243, 287
Nickel, 84, 233
Nitrate
  determination, 82, 286, **374**
  electrode, 33, 82, 269, 320, 374
Nitrate—nitrogen, 375
Nitric oxide, 263
Nitrite, 85, 125, **263**
Nitrogen, total, 290
Nitrogen dioxide, 37, 263
Nitrous acid, 263
NO_x electrode, 263
Noise, 43, 129
Nonactin, 33
Non-Nernstian response, 19, 114, 121, 217, 326, 336
Normal distribution, 61
NTA, determination, 224, 227, 242
Null-point potentiometry, 84, 212, 320

Off-scale readings, 127
Ohm's Law, 14, 52
Oils, fouling by, 58
On-line, 39, 42, 57, 160
On-stream, 44, 57

Osmotic effects, 38, 267, 275, 295
Oxalate
　determination, 85, 246
　reagent, 264
Oxidation, 16, 26, 74, 108
Oxides
　metal, 19
　passive, 20
Oxygen, effects of, 17, 21, 23, 41, 97, 275, 348

Palladium electrode, 22
Palladium annulus junction, 44
Perchlorate
　electrode, 33, 82, **390**
　titration, 82, 390
Permanganate, 85, 264
Petroleum, analysis, 276, 372
pH control, 124
pH electrode, 19, 22, 27, 135, 147
pH measurement, 42, 54, **135**
pH meter, 39, **49**, 114
Phenanthroline, 33, 378
Phenylmercuric acetate, 375
Phosphate
　•determination, 82, 246
　removal, 154, 370
Phosphate buffer, **138**, 288, 299, 308
Phthalate buffer, 138, 288
Pickling bath, analysis, 319
pIon meter, 39, **49**, 89, 91, 120
Piperidine buffer, 84
Pipework cells, 57
Platinum electrode, 20, 22, 83, 248
Plugs, 54, 127
Pneumatic signals, 55
Polishing of electrodes, 127, 128, 213, 216
Polypropylene, 38, 205
Potassium
　electrode, 28, 33, 41, 82, 84, **171**
　titration, 82, 84
Potassium dichromate, titrant, 85, 277
Potassium hydrogen phthalate
　buffer, 138, 288
　standard, 148
Potassium permanganate, titrant, 85, 264
Potential
　cell, 9, 13
　half-cell, 10, 40
　liquid-junction, **10**, 142, 320
　single electrode, 10
　standard, 10, 16, 40

Potentiometric selectivity coefficient, 31
Precipitates, fouling by, 58
Precipitation titrations, 83, 89, 90, 95, 212, 242, 243, 248, 320, 332, 362, 366, 388, 390
*Precision, 61, 71, 116, 118, 125
Preservatives, 110, 375
Pressure effects, 42, 57, 130
Printers, 51
PTFE, 38, 205
Pumps, see Flow cells
Pyrophosphate, titration, 246

Quinhydrone electrode, **22**, 83, 95, 135, 319

Rare earths, 28, 82, 320
Recorders, 51, 59, 130
Recovery tests, 71, 76, 126
Redox measurements, 20, 57
Reduction, 26
Reference electrodes, 10, 25, 38, 53, 111, 127, 350
　calomel, 18, **39**, 53, 111, 128, 136
　double-junction, **45**, 147, 172, 205, 243, 307, 343, 352, 355, 366, 374, 383, 390
　internal, 25, 27, 29, 115
　mercury—mercuric oxide, 20
　mercury—mercurous chloride, see Calomel electrode
　mercury—mercurous sulphate, 39, **41**, 111, 128, 205, 324, 333, 352, 374
　pressure effects, 42, 57, 130
　remote junction, 46, 53, 114
　sealed, **44**, 136
　silver—silver chloride, 27, 29, 35, **40**, 111, 115, 288
　Thalamid, 18, 27, 39, **41**, 136
Regression line, 77
Relative standard deviation, 63
Remote junction, 46, 53, 114
Resins, fouling by, 58
Resistance, 14, 29
*Response time, 126, 129
Reversible reactions, 13, 21
River water, analysis, 108, 286, 318, 375
Rocks, analysis, 319, 349
Rubidium, 28, 82

Samarium determination, 224

*Sampling, 108
SAOB, 356
Scratched membranes, 128
Sea water, analysis of, 259, 286, 365, 372
Selectivity, 31, see also Interferences
*Selectivity coefficients, 28, 31, 74
Selenium catalyst, 290
Semi-antilogarithmic graph paper, 92
Semi-logarithmic graph paper, 118
Servo-potentiometer, 56
Significance testing, 64
Silicone rubber, 24, 38, 116
Silver
   determination, **205**
   electrode, 17, 24, 82, 205
   titration, 82, 96, 362
Silver bromide, 24
Silver chloride, 24, 323
Silver iodide, 24, 34, 298
Silver–silver bromide electrode, 18, 343
Silver–silver carbonate electrode, 18
Silver–silver chloride electrode, 18, 25, 27, 54, 112, 323, 333, see also Reference electrodes
Silver–silver iodide electrode, 18, 299, 343
Silver–silver oxalate electrode, 18
Silver–silver oxide electrode, 20
Silver–silver sulphide electrode, 18, 355
Silver sulphate, reagent, 375, 381
Silver sulphide, 24, 115, 205, 214, 239, 243, 298, 306, 323, 343, 352, 355
Silver thiocyanate, 24, 352
Single electrode potential, 10
Single ionic activity coefficient, 8
Single-point titrations, 98
Sleeve-type junction, 43, 205
Slope control, 121
Slope factor, 5, 81, 99, 119
Soap, 43
Soda lime, 136, 148, 256
Sodium
   determination, 84, **160**
   electrodes, 28, 33, 84, 155, 160
Sodium salicylate, 344, 356
Soil, analysis, 46, 168, 191, 259, 268, 285, 296, 349, 372, 379, 381
Solid-state electrodes, **24**, 56, 73, 101, 110, 115
Solubility product, 18, 25, 87, 326, 336
Solvent
   in liquid ion-exchange electrodes, 30
   in titrations, 83, 96, 370

Specific conductivity, 5
Spiking, see Recovery tests
Stability constants, 94, 100, 226
Standard addition, see Known addition
Standard deviation, **61**, 125
Standard potentials, 10, 16, 21, 25, 40, 52, 73, 81
*Standard solutions, 76
Standard subtraction, 81, 98, **102**, 122
Stirring, 42, 96, 106, 114, 129, 222, 318, 371
*Storage of electrodes, 110
Strontium, 84, 235
Student's t-test, 65
Sulphamic acid, reagent, 277, 379
Sulphate, titration, 82, **366**
Sulphide
   determination, 26, 82, **355**
   electrode, 24, 37, 82, 355
   removal of, 303, 310
   solutions, 108
Sulphide antioxidant buffer, 356
Sulphite, 85, **271**, 350
Sulphur determination, 246, 372
Sulphur dioxide electrode, 29, 37, **271**
Sulphuric acid, 264, 272
Syringes, use of, 59, 107, 343, 368

Temperature
   compensation, 49, 52, 142
   control, 114, 116, 333
   *effects, 39, 45, 113, 129
TEPA, 84, 227, 236
Tetra(chlorophenyl)borate, 33, 171
Tetrafluoroborate electrode, 33, 383
Tetraphenylantimony sulphate, titrant, 82, 320
Tetraphenylarsonium chloride, titrant, 82, 388, 390
Tetraphenylborate, 82
Tetroxalate buffer, 138
Thalamid electrode, 18, 27, 39, **41**, 136
Thallium, 28
Thiocyanate
   electrode, 24, 352
   titration, 82
Thiols, titration, 82, 212
Thiosulphate, titration, 85
Thorium, titration, 82, 224, 236, 319
Thymoquinone, 23
Tin titration, 85
TISAB, 314
Titrations, **81**, 123, 277, see also Compleximetric titrations, Gran

titrations, Precipitation titrations
   addition, 83, 91, 222, 320
   competitive, 84, 86
   derivative, 59, 89
   to fixed pH, 60, 87, 151
   to fixed potential, 60, 87, 123, 248, 363
   linear plots, 83, **93**, 96, 114
   redox, 83, 91, 97
   single-point, 98
Titrators, automatic, 39, 58, 237, 371
Toluquinone, 23
*Tracing faults, 127
Transference number, 13
Trien, 227
Triethanolamine buffer, 41, 84, 228
Tris buffer, 159, 162, 229
t-test, 65
Tungstate, titration, 82, 246

Ultrasonic cleaning, 58

Uranium oxide, 28
Urine, analysis, 167, 184, 246

Valinomycin, 33, 171
Variance, 63
Voltmeters, 54
Volume errors, 107

Warning limits, 77
Water
   autoprotolysis, 20, 94
   CO_2-free, 136
   density, 108
   NO_x-free, 263
Water hardness, 33, 84, **194**, 233
Within-batch standard deviations, 69

Zinc, 16, 84, 224, 233
Zinc sulphide, 361
Zirconium, 224, 236

546.22 790786
M629p

Midgley, Derek
Potentiometric water analysis

MEMORIAL LIBRARY
MARS HILL COLLEGE
MARS HILL, NORTH CAROLINA 28754